Karl Popper, Science and Enlightenment

Karl Popper, Science and Enlightenment

Nicholas Maxwell

First published in 2017 by
UCL Press
University College London
Gower Street
London WC1E 6BT

Available to download free: www.ucl.ac.uk/ucl-press

Text © Nicholas Maxwell, 2017
Images © Nicholas Maxwell, 2017

A CIP catalogue record for this book is available
from The British Library.

This book is published under a Creative Commons Attribution Non-commercial Non-derivative 4.0 International license (CC BY-NC-ND 4.0). This license allows you to share, copy, distribute and transmit the work for personal and non-commercial use providing author and publisher attribution are clearly stated. Attribution should include the following information:

Nicholas Maxwell, *Karl Popper, Science and Enlightenment*. London, UCL Press, 2017. https://doi.org/10.14324/111. 9781787350397

Further details about Creative Commons licenses are available at
http://creativecommons.org/licenses/

ISBN: 978–1–787350–40–3 (Hbk.)
ISBN: 978–1–787350–41–0 (Pbk.)
ISBN: 978–1–787350–39–7 (PDF)
ISBN: 978–1–787350–38–0 (epub)
ISBN: 978–1–787350–37–3 (mobi)
ISBN: 978–1–787350–36–6 (html)
DOI: https://doi.org/10.14324/111.9781787350397

To my good friend Leemon McHenry

Acknowledgements

The Prologue, 'An Idea to Help Save the World', adapted from the Preface to the 2nd edition of *What's Wrong With Science?* (Pentire Press, 2009), was first published in *Sublime*, Issue 17, 2009, pp. 90-93. An earlier version of Chapter 1, 'Karl Raimund Popper', was first published in *British Philosophers, 1800-2000*, ed., P. Dematteis, P. Fosl and L. McHenry, Bruccoli Clark Layman, Columbia, 2002, pp. 176-194. Chapter 2, 'Popper, Kuhn, Lakatos and aim-oriented empiricism' is an updated version of an article first published in *Philosophia 32*, nos. 1–4, 2005, pp. 181-239 (Springer). An earlier version of Chapter 3, 'Einstein's Rational Discovery of Special and General Relativity' was first published in *The British Journal for the Philosophy of Science 44*, 1993, pp. 275-305 (Oxford University Press). Chapter 8, 'Does probabilism solve the great quantum mystery?', is an updated version of an article first published in *Theoria, vol.* 19/3, no. 51, 2004, pp. 321-336. Chapter 9, 'Science, reason, knowledge and wisdom: a critique of specialism', is a modified version of an article first published in *Inquiry 23*, 1980, pp. 19-81 (Taylor and Francis). Chapter 10, 'Karl Popper and the Enlightenment Programme' is an updated version of chapter 11 of *Karl Popper: A Centenary Assessment. Vol. 1: Life and Times, Values in a World of Facts*, ed. I. Jarview, K. Milford and D. Miller, Ashgate, London, pp. 177-190. The author is grateful to the editors and publishers of the above journals and books for permission to publish these updated essays. The Introduction and Chapters 4–7 have not been published before.

Contents

List of figures   viii
Prologue: An idea to help save the world   ix

Introduction   1

1. Karl Raimund Popper   8
2. Popper, Kuhn, Lakatos and aim-oriented empiricism   42
3. Einstein, aim-oriented empiricism, and the discovery of special and general relativity   90
4. Non-empirical requirements scientific theories must satisfy: simplicity, unity, explanation, beauty   125
5. Scientific metaphysics   143
6. Comprehensibility rather than beauty   170
7. A mug's game? Solving the problem of induction with metaphysical presuppositions   187
8. Does probabilism solve the great quantum mystery?   215
9. Science, reason, knowledge and wisdom: a critique of specialism   233
10. Karl Popper and the Enlightenment Programme   291

Notes   327
References   356
Index   369

List of figures

2.1	Aim-oriented empiricism	43
5.1	Another version of aim-oriented empiricism	160
9.1	Specio-universalism, integrating specialized and fundamental problem solving	236
9.2	Specio-universalist academic inquiry devoted to helping people realize what is of value in life	240
10.1	Aim-oriented rationality applied to the task of making progress towards a civilized world	314

Prologue: An idea to help save the world

Here is an idea that just might save the world. It is that science, properly understood, provides us with the methodological key to the salvation of humanity.

A version of this idea can be found buried in the works of Karl Popper. Famously, Popper argued that science cannot verify theories, but can only refute them. This sounds very negative, but actually it is not, for science succeeds in making such astonishing progress by subjecting its theories to sustained, ferocious attempted falsification. Every time a scientific theory is refuted by experiment or observation, scientists are forced to try to think up something better, and it is this, according to Popper, which drives science forward.

Popper went on to generalize this falsificationist conception of scientific method to form a notion of rationality, *critical rationalism*, applicable to all aspects of human life. Falsification becomes the more general idea of *criticism*. Just as scientists make progress by subjecting their theories to sustained attempted empirical falsification, so too all of us, whatever we may be doing, can best hope to achieve progress by subjecting relevant ideas to sustained, severe *criticism*. By subjecting our attempts at solving our problems to criticism, we give ourselves the best hope of discovering (when relevant) that our attempted solutions are inadequate or fail, and we are thus compelled to try to think up something better. By means of judicious use of criticism, in personal, social and political life, we may be able to achieve, in life, progressive success somewhat like the progressive success achieved by science. We can, in this way, in short, learn from scientific progress how to make personal and social progress in life. Science, as I have said, provides the methodological key to our salvation.

I discovered Karl Popper's work when I was a graduate student doing philosophy at the University of Manchester, in the early 1960s. As an undergraduate, I was appalled at the triviality, the sterility, of so-called "Oxford philosophy". This turned its back on all the immense and agonizing problems of the real world – the mysteries and grandeur of the universe, the wonder of our life on earth, the dreadful toll of human suffering – and instead busied itself with the trite activity of analysing the meaning of words. Then I discovered Popper, and breathed a sigh of relief. Here was a philosopher who, with exemplary intellectual integrity and passion, concerned himself with the profound problems of human existence, and had extraordinarily original and fruitful things to say about them. The problems that had tormented me had in essence, I felt, already been solved.

But then it dawned on me that Popper had failed to solve his fundamental problem: the problem of understanding how science makes progress. In one respect, Popper's conception of science is highly unorthodox: all scientific knowledge is conjectural; theories are falsified but cannot be verified. But in other respects, Popper's conception of science is highly orthodox. For Popper, as for most scientists and philosophers, the basic aim of science is knowledge of truth, the basic method being to assess theories with respect to evidence, *nothing being accepted as a part of scientific knowledge independently of evidence*. This orthodox view – which I came to call *standard empiricism* – is, I realized, false. Physicists only ever accept theories that are *unified* – theories that depict the same laws applying to the range of phenomena to which the theory applies. Endlessly many empirically more successful *disunified* rivals can always be concocted, but these are always ignored. This means, I realized, that science does make a big, permanent and highly problematic assumption about the nature of the universe independently of empirical considerations and even, in a sense, in violation of empirical considerations: namely, that the universe is such that all grossly disunified theories are false. Without some such presupposition as this, the whole empirical method of science breaks down.

It occurred to me that Popper, along with most scientists and philosophers, had misidentified the basic aim of science. This is not truth per se. It is rather truth presupposed to be unified, presupposed to be explanatory or comprehensible (unified theories being *explanatory*). Inherent in the aim of science there is the metaphysical – that is, untestable – assumption that there is some kind of underlying unity in nature, that the universe is, in some way, physically comprehensible.

But this assumption is profoundly problematic. We do not *know* that the universe is comprehensible. This is a conjecture. Even if it is comprehensible, almost certainly it is not comprehensible in the way science presupposes it is today. For good Popperian reasons, this metaphysical assumption must be made explicit within science and subjected to sustained criticism, as an integral part of science, in an attempt to improve it.

The outcome is a new conception of science, and a new kind of science, which I call *aim-oriented empiricism*. This subjects the aims, and associated methods, of science to sustained critical scrutiny, the aims and methods of science evolving with evolving knowledge. Philosophy of science (the study of the aims and methods of science) becomes an integral, vital part of science itself. And science becomes much more like natural philosophy in the time of Newton, a synthesis of science, methodology, epistemology, metaphysics and philosophy.

The aim of seeking *explanatory truth* is, however, a special case of a more general aim, that of seeking *valuable truth*. And this is sought in order that it be used by people to enrich their lives. In other words, in addition to metaphysical assumptions, inherent in the aims of science there are *value* assumptions, and *political* assumptions, assumptions about how science should be used in life. These are, if anything, even more problematic than metaphysical assumptions. Here, too, assumptions need to be made explicit and critically assessed, as an integral part of science, in an attempt to improve them.

Released from the crippling constraints of standard empiricism, science would burst out into a wonderful new life, realizing its full potential, responding fully both to our sense of wonder and to human suffering, becoming both more rigorous and of greater human value.

And then, in a flash of inspiration, I had my great idea. I could tread a path parallel to Popper's. Just as Popper had generalized falsificationism to form critical rationalism, so I could generalize my aim-oriented empiricist conception of scientific method to form an aim-oriented conception of rationality, potentially fruitfully applicable to all that we do, to all spheres of human life. But the great difference would be this. I would be starting out from a conception of science – of scientific method – that enormously improves on Popper's notion. In generalizing this, to form a general idea of progress-achieving rationality, I would be creating an idea of immense power and fruitfulness.

I knew already that the line of argument developed by Popper, from falsificationism to critical rationalism, was of profound importance for our whole culture and social order, and had far-reaching implications and application for science, art and art criticism, literature, music, academic inquiry quite generally, politics, law, morality, economics, psychoanalytic theory, evolution, education, history – for almost all aspects of human life and culture. The analogous line of argument I was developing, from aim-oriented empiricism to aim-oriented rationalism, would have even more fruitful implications and applications for all these fields, starting as it did from a much improved initial conception of the progress-achieving methods of science.

The key point is extremely simple. It is not just in science that aims are profoundly problematic; this is true in life as well. Above all, it is true of the aim of creating a good world – an aim inherently problematic for all sorts of more or less obvious reasons. It is not just in science that problematic aims are misconstrued or "repressed"; this happens all too often in life, too, both at the level of individuals, and at the institutional or social level. We urgently need to build into our scientific institutions and activities the aims-and-methods-improving methods of aim-oriented empiricism, so that scientific aims and methods improve as our scientific knowledge and understanding improve. Likewise, and even more urgently, we need to build into all our other institutions, into the fabric of our personal and social lives, the aims-and-methods-improving methods of aim-oriented rationality, so that we may improve our personal, social and global aims and methods as we live.

One outcome of the twentieth century is a widespread and deep-seated cynicism concerning the capacity of humanity to make real progress towards a genuinely civilized, good world. Utopian ideals and programmes, whether of the far left or right, that have promised heaven on earth, have led to horrors. Stalin's and Hitler's grandiose plans led to the murder of millions. Even saner, more modest, more humane and rational political programmes, based on democratic socialism, liberalism or free markets and capitalism, seem to have failed us. Thanks largely to modern science and technology, many of us today enjoy far richer, healthier and longer lives than our grandparents or great grandparents, or those who came before. Nevertheless, the modern world is confronted by grave global problems: the lethal character of modern war; the spread and threat of armaments, conventional, chemical, biological and nuclear; rapid population growth; severe poverty of millions in Africa, Asia and elsewhere; destruction of tropical rainforests and other natural habitats; rapid extinction of species; annihilation of languages and cultures. And over everything hangs the menace of climate change, threatening to intensify all the other problems (apart, perhaps, from population growth).

All these grave global problems are the almost inevitable outcome of the successful exploitation of science and technology plus the failure to build aim-oriented rationality into the fabric of our personal, social and institutional lives. Modern science and technology make modern industry and agriculture possible, which in turn make possible population growth, modern armaments and war, destruction of natural habitats and extinction of species, and global warming. Modern science and technology, in other words, make it possible for us to achieve the goals of more people, more industry and agriculture, more wealth, longer lives, more development, housing and roads, more travel, more cars and aeroplanes, more energy production and use, more and more lethal armaments (for defence only, of course!). These things seem inherently desirable and, in many ways, are highly desirable. But our successes in achieving these ends also bring about global warming, war, vast inequalities across the globe, destruction of habitats and extinction of species.

All our current global problems are the almost inevitable outcome of our long-term failure to put aim-oriented rationality into practice in life. Were we to do so, we would actively seek to discover problems associated with our long-term aims, actively explore ways in which problematic aims can be modified in less problematic directions, and at the same time develop the social, political, economic and industrial muscle able to change what we do, how we live, so that our aims become less problematic, less destructive in both the short and long term. Were we to do all this, we would at least be taking active steps to anticipate, and to avert the development of, grave global problems of the kind we face today. As it is, we have failed even to appreciate the fundamental need to improve aims and methods as the decades go by. Conventional ideas about rationality are all about means, not about ends, and are not designed to help us improve our ends as we proceed. Our current global problems are the outcome. Implementing aim-oriented rationality is essential if we are merely to survive in the long term. To repeat, the idea spelled out in this book, if taken seriously, just might save the world.

Einstein put his finger on what is wrong when he said, "Perfection of means and confusion of goals seems, to my opinion, to characterize our age." This outcome is inevitable if we restrict rationality to means, and fail to demand that rationality – the authentic article – must quite essentially include the sustained critical scrutiny of ends.

Scientists, and academics more generally, bear a heavy burden of responsibility for allowing our present impending state of crisis to develop. Putting aim-oriented rationality into practice in life can be painful, difficult and counterintuitive. It involves calling into question some of our most cherished aspirations and ideals. We have to learn how to live in aim-oriented rationalistic ways. And here, academic inquiry ought to have taken a lead. The primary task of our schools and universities, indeed, ought to have been, over the decades, to help us learn how to improve aims and methods as we live. Not only has academia failed miserably to take up this task, or even see it as necessary or desirable, even worse, perhaps, academia has failed itself to put aim-oriented rationality into practice. Science has met with such astonishing success because it has put something like aim-oriented empiricism into scientific practice, but this has been obscured and obstructed by the conviction of scientists that science ought to proceed in accordance with standard empiricism – with its fixed aim and fixed methods. Science has achieved success despite, and not because of, general allegiance of scientists to standard empiricism.

The pursuit of scientific knowledge dissociated from a more fundamental concern to help humanity improve aims and methods in life is, as we have seen, a recipe for disaster. This is the crisis behind all the others. We are in deep trouble. We can no longer afford to blunder blindly on our way. We must strive to peer into the future and steer a course less doomed to disaster. Humanity must learn to take intelligent and humane responsibility for the unfolding of history.

Introduction

Karl Popper is famous for having proposed that science advances by a process of conjecture and refutation. He is also famous for defending the open society against those he saw as its arch enemies – Plato and Marx.

Popper's contributions to thought are of profound importance, but they are not the last word on the subject. They need to be improved. My concern in this book is to spell out what is of greatest importance in Popper's work, what its failings are, how it needs to be improved to overcome these failings, and what implications emerge as a result.

The basic theme of the book has already been summarized in the Prologue. In what follows I spell out this theme in greater detail. The book consists of a collection of essays that dramatically develop Karl Popper's views about natural and social science, and how we should go about trying to solve social problems.

Criticism of Popper's falsificationist philosophy of natural science leads to a new philosophy of science, which I call aim-oriented empiricism.[1] This makes explicit metaphysical theses concerning the comprehensibility and knowability of the universe that are an implicit part of scientific knowledge – implicit in the way science excludes all theories that are not explanatory, even those that are more successful empirically than accepted theories. Aim-oriented empiricism has major implications, not just for the academic discipline of philosophy of science, but for science itself.

Popper generalized his philosophy of science of falsificationism to arrive at a new conception of rationality – critical rationalism – the key methodological idea of Popper's profound critical exploration of political and social issues in his *The Open Society and Its Enemies* (1966a) and *The Poverty of Historicism* (1961). This path of Popper, from scientific method to rationality and social and political issues, is followed here, but the starting point is aim-oriented empiricism rather than falsificationism. Aim-oriented empiricism is generalized to form a new conception of rationality – aim-oriented rationalism – which has far-reaching implications for political and social issues, for the nature of social inquiry and the humanities, and indeed for academic inquiry as a whole. The strategies for tackling social problems that arise from aim-oriented rationalism improve on Popper's recommended strategies of piecemeal social engineering and critical rationalism, associated with his conception of the open society. This book thus sets out to develop Popper's philosophy in new and fruitful directions.

The theme of the book, in short, is to discover what can be learned from scientific progress about how to achieve social progress towards a better world. That there is indeed much to be learned from scientific progress about how to achieve social progress was the big idea of the eighteenth-century Enlightenment. This was immensely influential. But the *philosophes* of the Enlightenment made mistakes, and these mistakes, inherited from the Enlightenment, are built into the institutional and intellectual structure of academic inquiry today. In his two great works, *The Logic of Scientific Discovery* (1959) and *The Open Society and Its Enemies* (1966a), Popper corrected some of the mistakes of the Enlightenment – mistakes about the nature of scientific method and rationality. But Popper left other mistakes undetected and uncorrected. The present book seeks to push the Popperian research programme further, and correct what Popper left uncorrected.

The fundamental idea that emerges is that there is an urgent need to bring about a revolution in academic inquiry so that it takes up its proper task of promoting *wisdom* and not just acquiring knowledge – wisdom being the capacity to realize what is of value in life for oneself and others, thus including knowledge and technological know-how, but much else besides. I have devoted much of my working life to trying to get this idea across. The essays that follow provide a record of this life work.

Most philosophers of science see their work as contributing to a *meta-discipline*. The object of study is science, and the task is to describe, explain and understand this object. It is no more the proper task of the philosopher of science to *criticize* science, or to make suggestion as to how science can be *improved*, than it is the task of the astronomer to criticize the moon. But this standard *meta-discipline* way of conceiving the subject entirely misconstrues what ought to be the proper relationship between science and the philosophy of science. A major implication of the view to be expounded and defended here, aim-oriented empiricism, is that the rationality of science requires that the philosophy of science – critical exploration of views concerning the aims and methods of science – is an integral, influential part of science itself, both being influenced by, and influencing, science. In other words, in order to be rigorous, science must include some imaginative and critical exploration of problematic aims and methods. The very act of setting up the philosophy of science as a meta-discipline, distinct from science itself, looking down on science from above, as it were, describing and seeking to explain and understand what goes on, but in no way interfering with, contributing to or criticizing science, serves to undermine the very thing the discipline seeks to understand, namely the rationality of science. The orthodox meta-discipline approach not only makes the subject sterile (in that it can have nothing to contribute to science itself), it makes it quite impossible to solve the fundamental problem of the discipline – the rationality of science. Indeed, the discipline, so conceived, actually becomes the source of a pervasive and damaging *irrationality* in science.[2]

I make no apology, therefore, for criticizing science, for attempting to contribute to and improve science, in what follows. Philosophy of science pursued within the framework of aim-oriented empiricism might be compared to the work Weierstrass, Dedekind and others did in the late nineteenth and early twentieth century in bringing greater rigour to mathematics: they made mathematics more rigorous, and contributed to mathematics itself, at one and the same time. Somewhat analogously, I seek to increase the rigour of science, and make a contribution to science at the same time. I might add that in criticizing science and suggesting how it might be improved (made more rigorous and of greater human value) I am again developing a tendency to be found scattered among Popper's works. Despite – or perhaps because of – his great admiration for science at its best, Popper does not shrink from criticizing what he sees as deplorable aspects of science: specialization, authoritarianism, submission to mere intellectual fashion, failure to grapple with the fundamental problems of cosmology. Again, Popper depicts graphically some of the bad consequences, for science itself, of attempting to put bad inductivist methods into scientific practice. But all this is paradoxically at odds with a major tenet of Popper's philosophy of science, namely his proposed solution to the problem of demarcation. Popper holds that an idea, in order to be scientific, must be empirically *falsifiable*. Philosophies of science, because they are not empirically falsifiable, are not a part of science. They are to be severely *demarcated* from science. Thus Popper, in actively criticizing aspects of science, violates the precepts of his own philosophy of science. All this changes dramatically once Popper's philosophy of science has been amended to become the doctrine espoused here, aim-oriented empiricism.[3]

The points just made concerning the proper relationship between science on the one hand and the philosophy of science on the other, will turn out to have a major bearing on further developments of the argument concerning social inquiry. I argue that social inquiry needs to be construed, not primarily as social *science*, but rather as social *philosophy* or social *methodology*. Social inquiry is not to be related to the social world as astronomy is to the moon, or geology is to the earth. Social inquiry is not, fundamentally, engaged in seeking to acquire knowledge about social phenomena. Rather, social inquiry needs to take the relationship between science and the philosophy of science (as specified by aim-oriented empiricism) to be the model, the ideal, of how social inquiry ought to be related to society. What the philosophy of science is to science, so social inquiry is to society. The proper task of social inquiry is to help worthwhile social endeavours improve their problematic aims and methods as they proceed, just as the proper task of the philosophy of science is to help science improve its problematic aims and methods as it proceeds. On this view, indeed, the philosophy of natural science is just that small, but crucial, bit of social inquiry that deals with the worthwhile social endeavour of *natural science*.

Let me now indicate, in a little more detail, the contents of the chapters of this book.

Chapter 1 gives an account of Karl Popper's life and work. I make it quite clear that, in my view, Popper is the greatest philosopher of the twentieth century. I am nevertheless critical of aspects of his work – it would be a betrayal of his "critical philosophy" not to be. In a preliminary way, I indicate what are, in my view, unsolved problems inherent in his views, and outline what needs to be done to overcome these difficulties. Subsequent chapters seek to improve Popper's philosophy in some key respects in order to overcome these defects.

In Chapter 2 I argue that aim-oriented empiricism is a kind of synthesis of the views of Popper, Thomas Kuhn and Imre Lakatos, but also a dramatic improvement over all three views. Aim-oriented empiricism stands in sharp contrast to standard empiricism, versions of which are defended by Popper, Kuhn and Lakatos, and are taken for granted by most scientists and philosophers of science. According to standard empiricism, the basic intellectual aim of science is truth,[4] and the basic method is to assess claims to knowledge of truth impartially with respect to evidence. Considerations of simplicity, unity or explanatory character may legitimately influence preference for a theory for a time, but not in such a way that the universe itself is permanently presumed to be simple, unified or comprehensible. Choice of theory may be biased in the direction of a paradigm or metaphysical view for a time, but in the end empirical considerations must decide what theories are accepted and rejected. The key tenet of all versions of standard empiricism is that no assumption about the universe can be accepted permanently as a part of scientific knowledge independent of evidence, let alone in conflict with evidence. But physics, in accepting unified fundamental physical theories only, and persistently rejecting empirically more successful disunified rivals, does thereby make a persistent metaphysical assumption about the world: some kind of unified pattern of physical law runs through all phenomena. Standard empiricism is thus untenable. Aim-oriented empiricism, by contrast, acknowledges that persistent scientific acceptance of unified theories means that science implicitly accepts that the universe itself possesses some kind of underlying unity. Rigour demands that this highly problematic, implicit metaphysical conjecture be made explicit, so that it can be critically assessed, and so that alternatives can be developed and assessed, in the hope of improving the specific assumption that physics makes at any time. Aim-oriented empiricism holds that we need to represent this highly problematic assumption in the form of a hierarchy of assumptions, these becoming less and less substantial, and so less and less problematic, and more nearly such that their truth is required for science to be possible at all, as one goes up the hierarchy. In this way, we form a framework of relatively stable and unproblematic assumptions, high up in the hierarchy, within which much more substantial and problematic assumptions, low down in the hierarchy, may be critically assessed, and improved, in the light of which best help promote empirical knowledge, and comply with assumptions higher up in the hierarchy. This is the view that provides a triumphant synthesis of, and improvement over, the views of Popper, Kuhn and Lakatos – more Popperian than Popper, more Kuhnian than Kuhn and more Lakatosian than Lakatos.

Chapter 3 argues that Einstein was the first scientist to put something like aim-oriented empiricism explicitly into scientific practice in discovering special and general relativity. The method of discovery of aim-oriented empiricism played a crucial role in Einstein's discovery of these theories. And not only did Einstein implement aim-oriented empiricism in scientific practice: after his discovery of general relativity, Einstein came to *advocate* a view that came closer and closer to aim-oriented empiricism as the years passed.

Chapter 4 solves the problem of what it means to say of a scientific theory that it is simple, unified or explanatory. This problem was recognized by Popper (1963, p. 241), but Popper did not know how to solve it. Einstein recognized the problem too, but did not know how to solve it, either. It is one of the great successes of aim-oriented empiricism that it provides the means for the problem to be solved.

Chapter 5 gives a careful and more detailed argument in support of aim-oriented empiricism, attention being given to some of the difficulties that arise in connection with the view, and the argument in support of the view.

Chapter 6 compares and contrasts views about simplicity, unity, explanatory power or "beauty" associated with aim-oriented empiricism on the one hand, and a view put forward by James McAllister on the other hand (see his influential *Beauty and Revolution in Science* [McAllister, 1996]).

Chapter 7 argues that aim-oriented empiricism succeeds in doing what Popper's falsificationism fails to do, namely solve the problem of induction.

Chapter 8 takes up a theme close to Popper's heart: the problems of interpreting quantum theory in a realist way – so that the theory can be understood to be about electrons, nuclei, atoms and other denizens of the quantum world, and is not doomed to be a mere *instrument* for the prediction of experimental results. I argue that *probabilism* is the key to solving the fundamental quantum mystery – the apparent capacity of quantum entities (electrons, atoms and so on) to behave *both* as particles and waves. Probabilism, here, is the doctrine that nature herself is probabilistic. What exists at one moment may only determine what exists next *probabilistically*, and not deterministically. This develops basic ideas of Popper about quantum theory, but in ways of which he strongly disapproved.

Chapter 9 criticizes specialism, the doctrine that academic inquiry quite properly consists of a great number of specialized disciplines, only specialized intellectual standards being worthwhile. This is opposed by the view that academic inquiry must engage in sustained exploration of fundamental problems that cut across disciplinary boundaries, this exploration being undertaken in a way that influences, and is influenced by, specialized research. This is an improvement of anti-specialist remarks scattered throughout Popper's works.

Chapter 10 argues that Popper's *The Logic of Scientific Discovery* and *The Open Society and Its Enemies*, taken together, constitute a major development of the Enlightenment programme of learning from scientific progress how to achieve social progress towards a more enlightened world. But what Popper has to say is not the last word on the subject. Popper's version of the Enlightenment programme needs further improvement, partly because Popper's conception of scientific method needs to be improved, but mainly because, in order to implement the programme, we need to apply scientific method, not to social science, but to *the social world itself*. How and why Popper's version of the Enlightenment programme needs to be improved is outlined in this chapter.

1
Karl Raimund Popper

Karl Popper is the greatest philosopher of the twentieth century. No other philosopher of the period has produced a body of work that is as significant. What is best in Popper's output is contained in his first four published books. These tackle fundamental problems with ferocious, exemplary integrity, clarity, simplicity and originality. They have widespread, fruitful implications, for science, for philosophy, for the social sciences, for education, for art, for politics and political philosophy.

In his first published book *The Logic of Scientific Discovery* (1959, first published in German in 1934), Popper argues that, although scientific theories cannot be verified, or even rendered probable, by evidence, they can be falsified. Science makes progress by putting forward falsifiable conjectures – theories which say as much as possible about the world, and which thus expose themselves as much as possible to the risk of empirical refutation; they are then subjected to a ruthless onslaught of attempted observational and experimental refutation. When finally a scientific theory is falsified empirically, the task then becomes to think up an even better theory, which says even more about the world. The new theory must predict all the success of the old theory, predict successfully the phenomena that falsified the old theory, and predict new phenomena as well. In his next book, *The Open Society and Its Enemies* (1966), written during the Second World War and first published in 1945, Popper tackles problems that arise in connection with creating an "open" society, one which tolerates diversity of views and ways of life. Popper argues that some of the greatest thinkers have been opposed to the "open" society, most notably Plato and Marx. In *The Poverty of Historicism* (1961), first published in 1957, Popper is concerned to demolish the view that social science should, or can, predict the way societies evolve. Popper spells out his view of how social science should be developed, closely modelled on the account of natural science given in *The Logic of Scientific Discovery*. The next book, *Conjectures and Refutations* (1963), is a collection of essays which restate in a more accessible way Popper's falsificationist view of science, and draw out implications for a range of philosophical problems. Further books include *Objective Knowledge* (1972), a collection of essays which draw on the analogy between Darwinian evolution and scientific progress, and which expound Popper's view that there exists, in addition to the material world and the psychological world, a third world of theories, problems and arguments; *The Self and Its Brain* (1977), written with the neurologist John Eccles, which applies Popper's "third world" view to the mind-body problem; and the three-volume *The Postscript* (1982a, 1982b, 1983), which amounts to a massive restatement and development of Popper's falsificationist conception of science. A volume of *The Library of Living Philosophers* (Schilpp, 1974) is devoted to Popper's work; this includes Popper's intellectual autobiography, published subsequently as a separate book with the title *Unended Quest* (1976a). This gives a fascinating and gripping account of the development of Popper's thought. *The Two Fundamental Problems of the Theory of Knowledge* was a precursor to *The Logic of Scientific Discovery*. Its two problems are the problem of induction (the problem of how theories can be verified by evidence), and the problem of demarcation (the problem of how science is to be demarcated from non-science). The book was not published until 1979 in German, and in 2009 in English.

Fundamental to Popper's philosophy is the idea that criticism lies at the heart of rationality. It would be a betrayal of Popper's philosophy to give an entirely uncritical exposition of his work; some criticism of key tenets of his philosophy will therefore be included in what follows.

1.1 Life

Karl Raimund Popper was born in Vienna on 28 July 1902. His parents were Jewish but converted to Protestantism before their children were born. Popper's father, Simon Carl Siegmund (1856–1932), was a doctor of law of the University of Vienna. He had a successful legal practice in Vienna, at which he apparently worked hard, but his real interests lay in the direction of scholarship and literature. Popper's mother, Jenny Schiff (1864–1938), came from a musical family, and was herself musical. Popper tells us that she played the piano beautifully; music had an important place in Popper's life.

During Popper's early childhood, his parents were prosperous. They lived in a large apartment in an eighteenth-century house in the centre of Vienna, where Popper's father conducted his legal practice. Popper's father had an enormous library, which included many works of philosophy; books were everywhere, Popper tells us, except in the dining room, where stood a concert grand piano.

As a young boy, Popper was much concerned with the poverty he saw all around him in Vienna. In his autobiography, Popper recounts an early brush with philosophy. His father had suggested he read some volumes of August Strindberg's autobiography. Finding that Strindberg gave much too much importance to words and their meanings, Popper tried to point this out to his father, and was surprised to discover that he did not agree. Popper later saw this as his first brush with a lifelong battle to combat the influential view that philosophy must concern itself with analysis of meaning.

Popper left school at 16 because of the tedium of the classes, and enrolled at the University of Vienna, initially as a non-matriculated student. Four years later, at the second attempt, he passed the exam to become a matriculated student. Any student could take any lecture course, and, initially, Popper sampled lectures in a wide range of subjects – history, literature, psychology, philosophy – before concentrating on physics and mathematics. In these fields Popper had excellent, if remote and autocratic, teachers: Hans Thirring, Wilhelm Wirtinger, Philipp Furtwängler and Hans Hahn. Later, Popper devoted himself to the study of the psychology of thinking, influenced by Karl Bühler and the writings of Otto Selz.

The First World War and its aftermath brought dramatic changes to conditions of life in Vienna. Popper's father lost much of his savings. Popper left home and moved into part of a disused military hospital converted by students into a primitive students' home, and joined socialist groups seeking political change. For a time, Popper thought of himself as a communist. But then, on 15 June 1919, an event occurred which Popper was later to describe as one of the most important in his life; it caused him to become critical of communism and Marxism, and, years later, led to the writing of his *The Open Society and Its Enemies*. The communists organized a demonstration with the intention of freeing communists held in a police station in Vienna. The police opened fire, and some of the demonstrators were killed. Popper was deeply shocked, and even felt some personal responsibility for the tragedy, in that he had endorsed a doctrine, Marxism, which required that there should be just such incidents, so that the struggle to overcome capitalism might be intensified.

Popper nevertheless continued to think of himself as a socialist, and to associate with socialist groups. In his autobiography, Popper celebrates these groups of working people for their dedication, their eagerness to become educated. Even though the times were troubled, the economic and political outlook bleak, Popper says that he and his friends were often exhilarated at the intellectual and political challenges that lay before them. For a time Popper worked as a labourer, but found the work too hard; he then tried his hand at cabinet making, but was distracted by the intellectual problems that he was working on. Popper also worked for the psychologist Adler, and as a social worker concerned with neglected children.

Shortly before submitting his dissertation for his PhD, the focus of Popper's interest switched from the psychology, to the methodology, of thought and problem solving, and in particular to the methodology of science. This came about partly as a result of long discussions with two friends, the philosophers Julius Kraft and Heinrich Gomperz. The dissertation was hastily written. Popper's examiners were Bühler and Schlick; Popper thought he had failed, but in fact he passed with distinction.

At this time a Pedagogic Institute was created in Vienna to train teachers in new methods of education. Popper decided to become a teacher, joined the course, held informal seminars for fellow students, and duly became qualified to teach physics and mathematics in secondary schools. He met, and later married, a fellow student, Josephine Henninger (Hennie), who also became a teacher.

While employed full-time as a teacher, Popper continued to work hard at epistemological and methodological problems of science, writing down his thoughts as an aid to research, rather than with the idea that the work might eventually be published. During this time, Popper got to know a number of people associated with the Vienna Circle, famous for promoting logical positivism. The Vienna Circle was essentially a seminar which one attended when invited by its convenor, Moritz Schlick. Rudolf Carnap, Otto Neurath, Herbert Feigl, Kurt Gödel, Friedrich Waismann, Victor Kraft, Karl Menger, Hans Hahn, Philipp Frank, Richard von Mises, Hans Reichenbach and Carl Hempel were among the members; Ludwig Wittgenstein, much admired by Schlick, was the intellectual godfather (together, perhaps, with Ernst Mach and Bertrand Russell). Visitors from abroad included A. J. Ayer and Frank Ramsey from England, Ernest Nagel and W. V. Quine from the USA, Arne Næss from Norway, and Alfred Tarski from Poland. But Popper was never invited to join the Circle (possibly because Schlick was aware of Popper's low opinion of Wittgenstein).[1] Nevertheless, Popper did attend, and give papers at, a number of fringe seminars, and his work was strongly influenced by, but also critical of, the doctrines of the Circle.

1.2 Early work

Two issues were of central concern to Popper. The first was the problem of how to distinguish science from pseudoscience. Popper was impressed by the difference between the theories of Marx, Freud and Adler on the one hand, and Einstein's general theory of relativity, on the other. The former theories seemed able to explain phenomena whatever happened; nothing, it seemed, could tell against these theories. Einstein's theory, by contrast, issued in a definite prediction: light travelling near the sun would pursue a curved path due to the gravitational field of the sun. If this did not happen, Einstein's theory would be refuted. Popper decided, around 1921 (he tells us) that this constituted the key difference between pseudo and genuine scientific theories: whereas the former were unrefutable, the latter were open to empirical refutation (see Popper, 1963, pp. 34–9; 1976a, p. 38; see also Hacohen, 2000, pp. 91–6).

The other problem that preoccupied Popper was that of the logic, or methodology, of scientific discovery: How does science acquire new knowledge? This was the problem that confronted Popper when his earlier interest in the psychology of thinking transmuted into interest in the logic of thinking, the logic of discovery.

Suddenly, Popper tells us, he put two and two together (Popper, 1976a, p. 79). His earlier solution to the first problem also solves the second problem. There is no such thing as the verification of theories in science; there is only refutation. Scientists put forward theories as empirically falsifiable conjectures or guesses, and these are then subjected to sustained attempted empirical refutation. Science advances through a process of trial and error, of conjecture and refutation.

Encouraged by Feigl, Popper wrote the first volume of what was intended to be a two-volume work, entitled *Die beiden Grundprobleme der Erkenntnistheorie* (*The Two Fundamental Problems of the Theory of Knowledge*). The first volume was accepted for publication by Schlick and Frank, the editors of a series of publications written mostly by members of the Vienna Circle. The publisher, Springer, insisted the book must be shortened. But in the meantime Popper had finished the second volume. He offered a new work consisting of extracts from both volumes, but this was still judged by Springer to be too long. Popper's uncle, Walter Schiff, cut the manuscript by about a half, and this was finally published late in 1934 as *Logik der Forschung*. Thus emerged into the public domain, in the shadow of Hitler and impending war, what is, perhaps, the most important book on scientific method to be published in the last century. The book was only published in English translation, as *The Logic of Scientific Discovery* (with many additional appendices and footnotes), in 1959. (*Die beiden Grundprobleme der Erkenntnistheorie* was not published until much later, in 1979.)

1.3 *The Logic of Scientific Discovery*

The Logic of Scientific Discovery (*L.Sc.D.*) begins by spelling out what are, for Popper, the two fundamental problems concerning the nature of scientific inquiry. (1) The problem of induction: how can scientific theories be verified by evidence, in view of Hume's arguments which seem to show that this is impossible? (2) The problem of demarcation: How is science to be demarcated from non-science (pseudoscience and metaphysics)? As we saw above, Popper's solution to the second problem is that, in order to be scientific, a theory must be empirically *falsifiable*. This, for Popper, solves the first problem as well. Scientific laws and theories cannot be verified by evidence at all; they can only be falsified. However much evidence may be amassed in support of a theory, its probability remains zero. But despite this negative conclusion, science can still make progress. This comes about as a result of theories being proposed as conjectures, in response to problems; these conjectures are then subjected to a ruthless barrage of attempted empirical refutation. The purpose of observation and experimentation is not to verify, but to refute. When a theory is refuted empirically, this creates the problem of discovering a new conjecture, a new theory, which is even more successful than its predecessor in that it meets with all the success of its predecessor, successfully predicts the phenomena that refuted its predecessor, and predicts new phenomena as well. When such a theory is formulated, the task then becomes to try to refute this new theory in turn. Thus science advances, from one falsifiable conjecture to another, each successfully predicting more than its predecessor, but none ever having probability greater than zero. All theoretical knowledge in science is irredeemably conjectural in character. But science makes progress precisely because, in science, it is possible to discover that theories are false, and thus need to be replaced by something better.

Popper has been much criticized for not appreciating that even empirical refutations are not decisive: it is always a conjecture that a theory has been falsified, since it is always a conjecture that a given observation or experiment has yielded a falsifying result. But Popper has at least two replies to such criticisms.

First, there is a decisive logical asymmetry between verification and falsification. Any theory has infinitely many empirical consequences, for infinitely many times and places. We, however, can only ever verify finitely many of these consequences, and thus must forever be infinitely far away from verifying the theory. But we only need to discover one false empirical consequence of a theory in order to show decisively that the theory is false.

Second, Popper emphasizes that a theory is only falsifiable with respect to the adoption of a methodology. Given that a theory is empirically falsified, it is always possible to rescue the theory from falsification by adopting what Popper calls "conventionalist stratagems". These include explaining the experimental result away in some way, or modifying the theory, in an *ad hoc* way, so that it no longer clashes with the empirical result. Popper proposes that science should adopt methodological rules governing the way theories are to be accepted and rejected in science in the light of evidence: these rules need to be designed to expose theories to the maximum risk of empirical refutation. Conventionalist stratagems, in particular, are to be banned. Faced by a refutation, a theory may be modified so as to overcome the refutation, but only if the modification increases the empirical content, the degree of falsifiability, of the theory – the modified theory predicting more, excluding more potentially falsifying observational statements, than before. Scientists should always strive to put forward theories that say as much as possible about the empirical world, that expose themselves to the greatest risk of refutation, that have the highest possible degree of falsifiability. The supreme methodological principle of science, for Popper, "says that the other rules of scientific procedure must be designed in such a way that they do not protect any statement in science against falsification" (1959, p. 54).

Even though theories cannot be verified, they can be "corroborated". For Popper, corroboration is a measure of how well a theory has stood up to attempts to refute it. If a highly falsifiable theory has survived an onslaught of severe testing, then it has proved its worth. It deserves to be taken more seriously than an untested theory, or an unfalsifiable speculation.

According to Popper, then, science makes progress by means of wild imagining, bold guesswork, on the one hand, controlled by ferocious attempted empirical refutation on the other hand.

L.Sc.D. was influenced by the thought of the Vienna Circle, but also differs from, and is highly critical of, some of the main tenets of the Circle. Logical positivism sought to demarcate the meaningful from the meaningless, with only those propositions capable of being verified being meaningful, the hope being that all meaningful factual propositions would be scientific. Popper stressed that any such criterion would condemn scientific theories to being meaningless, since they could not be verified. Popper's demarcation problem differed from that of the positivists. For Popper, as we have seen, the problem was to demarcate science from non-science (pseudoscience and metaphysics); falsifiability, not verifiability, is the key requirement; but non-scientific, metaphysical theories, though neither verifiable nor falsifiable, may nevertheless be entirely meaningful, and may even have a fruitful role to play in the development of science. Metaphysical theses, such as atomism, may suggest, and may (as a result of acquiring precision) be transformed into, falsifiable scientific theories.

1.4 Criticism

Does L.Sc.D. succeed in solving its two basic problems? Three great merits of the book are its originality, its clarity and its tight structure: everything devolves from the key idea of falsifiability. This makes the book especially open to criticism, and to improvement. In the end, the book fails to solve its basic problems, due to its treatment of simplicity.

Popper claims that the more falsifiable a theory is, so the greater its degree of simplicity. (There is a second method for assessing degrees of simplicity, in terms of number of observation statements required to falsify the theories in question, but Popper stresses that if the two methods clash, it is the first that takes precedence.) It is easy to see that Popper's proposal fails. Given a reasonably simple scientific theory, T, one can readily increase the falsifiability of T by adding on an independently testable hypotheses, h_1, to form the new theory, $T + h_1$. This new theory will be more falsifiable than T but, in general, will be drastically less simple. And one can make the situation even worse, by adding on as many independently testable hypotheses as one pleases, h_2, h_3 and so on, to form new theories, $T + h_1 + h_2 + h_3 + ...$, as highly empirically falsifiable and as drastically lacking in simplicity as one pleases. Thus simplicity cannot be equated with falsifiability.

And there is a further, even more devastating point. Popper's methodological rules favour $T + h_1 + h_2 + h_3$ over T, especially if h_1, h_2 and h_3 have been severely tested and corroborated. But in scientific practice, $T + h_1 + h_2 + h_3$ would never even be considered, however highly corroborated it might be if considered, because of its extreme lack of simplicity or unity, its grossly *ad hoc* character. There is here a fundamental flaw in the central doctrine of *L.Sc.D.*

Later, in *Conjectures and Refutations* (1963), Popper put forward a new methodological principle which, when added to those of *L.Sc.D.*, succeeds in excluding theories such as $T + h_1 + h_2 + h_3$ from scientific consideration. According to Popper, a new theory, in order to be acceptable, "should proceed from some *simple, new, and powerful, unifying idea* about some connection or relation (such as gravitational attraction) between hitherto unconnected things (such as planets and apples) or facts (such as inertial and gravitational mass) or new 'theoretical entities' (such as field and particles)" (p. 241). $T + h_1 + h_2 + h_3$ does not "proceed from some *simple, new and powerful, unifying idea*" and is to be rejected on that account, even if more highly corroborated than T.

But the adoption of this "requirement of simplicity" (as Popper calls it) as a basic methodological principle of science, has the effect of permanently excluding from science all *ad hoc* theories (such as $T + h_1 + h_2 + h_3$) that fail to satisfy the principle, however empirically successful such theories might be if considered. This amounts to assuming permanently that the universe is such that no *ad hoc* theory, that fails to satisfy Popper's principle of simplicity, is true. It amounts to accepting, as a permanent item of scientific knowledge, the substantial metaphysical thesis that the universe is non-*ad hoc*, in the sense that no theory that fails to satisfy Popper's principle of simplicity is true. But this clashes with Popper's criterion of demarcation: that no unfalsifiable, metaphysical thesis is to be accepted as a part of scientific knowledge.

It is, in fact, important that Popper's criterion of demarcation is rejected, and the metaphysical thesis of non-*ad hocness* is explicitly acknowledged to be a part of scientific knowledge. The thesis, in the form in which it is implicitly adopted at any given stage in the development of science, may well be false. Scientific progress may require that it be modified. The thesis needs to be made explicit, in other words, for good Popperian reasons, namely so that it can be critically assessed, and perhaps improved. As long as Popper's demarcation criterion is upheld, the metaphysical thesis must remain implicit, and hence immune to criticism.

Popper's falsificationism can be modified, however, so that substantial metaphysical theses, implicit in methods that exclude *ad hoc* theories, are made explicit within science, and are thus rendered available to critical scrutiny and revision (as we shall see in subsequent chapters; see also Maxwell, 1974, 1998, 2017a, 2017b).

On publication, *L.Sc.D.* achieved a certain impact; it was quite widely reviewed and discussed, and led to Popper being invited to give lectures in England, Denmark and elsewhere. Popper himself later claimed that his criticisms of logical positivism led eventually to the downfall of that doctrine.

1.5 The Open Society

Having dealt with the methodology of the natural sciences, Popper turned his attention again to what had long been of concern to him: the intellectual defects of Marxism, and the philosophy of the social sciences. But before he could get very far with that work, Popper was offered a lectureship at the University of Canterbury, New Zealand. He accepted, and Popper and his wife left Vienna for New Zealand early in 1937.

For some years Popper had been privately highly critical of policies of socialists in Germany and Austria for playing into the hands of the Fascists and Hitler. This was due, in Popper's view, to the harmful influence of Marxism. But he had kept these criticisms to himself, as he felt any public criticism could only weaken the forces opposing Hitler. Then, in March 1938, Hitler occupied Austria, and Popper felt all grounds for restraint had disappeared. He decided to put his criticisms of Marxism, and his views on the social sciences, into a publishable form.

He began work on what was to become *The Poverty of Historicism*. But then, unexpectedly, sections on essentialism, and on totalitarian tendencies in Plato, grew and grew (driven by the desperation of the times), and Popper found he had a new work on his hands: it became what is perhaps his best known, most influential and greatest work, *The Open Society and Its Enemies* (1966a, first published 1945). Without referring anywhere to Hitler or Stalin, the book is, nevertheless, an urgent and passionate investigation into the problem and threat of totalitarianism, whether of the right or left. It seeks to understand what the appeal of totalitarianism can be, and why it should have come to be such a threat to civilization. Popper regarded the writing of the book as his contribution to the war effort.

In *The Open Society and Its Enemies* (*O.S.E.*), Popper argues that a fundamental problem confronting humanity is that of moving from a closed, tribal way of life to an open society. The closed society is a society that has just one view of the world, one set of values, one basic way of life. It is a world dominated by dogma, fixed taboos and magic, devoid of doubt and uncertainty. The open society, by contrast, tolerates diversity of views, values and ways of life. In the open society learning through criticism is possible just because diverse views and values are tolerated. For Popper, the open society is the civilized society, in which individual freedom and responsibility, justice, democracy, humane values, reason and science can flourish.

But moving from the closed to the open society imposes a great psychological burden on the individuals involved, "the strain of civilization". Instead of the security of the tribe, organic, dogmatic and devoid of doubt, there is all the uncertainty and insecurity of the open society, the painful necessity of taking personal responsibility for one's life in a state of ignorance, the lack of intimacy associated with the "abstract society", in which individuals constantly rub shoulders with strangers. This transition from the closed to the open society is, for Popper, "one of the deepest revolutions through which mankind has passed" (1966a, vol. 1, p. 175). Many cannot bear the burden of freedom and doubt, and long for the false security and certainties of the closed society. In particular, some of the greatest thinkers of Western civilization have given into this temptation and have, in one way or another, urged upon long-suffering humanity a return to something like a closed society under the guise of Utopia. This is true of Plato and Aristotle; and it is true, in more recent times, of Hegel and Marx. The lure of totalitarianism is built deep in our history and traditions.

The revolutionary transition from closed to open society first occurred, according to Popper, with the "Great Generation" of ancient Athens in the fifth century BC. Those to be associated with the birth and affirmation of the open society include Pericles, Herodotus, Protagoras, Democritus, Alcidamas, Lycophron, Antisthenes and, above all, Socrates.

It is from Plato, especially, that we learn of Socrates' passionate scepticism, his searching criticism of current beliefs and ideals, his conviction that first one must acknowledge one's own ignorance before one could hope to acquire knowledge and wisdom. But Socrates, Popper argues, was ultimately betrayed by Plato. The greatest advocate of the open society became, in Plato's *Republic*, the spokesman for a return to a closed society.

Popper's devastating account of Plato's "propaganda" for the closed society, in bald outline, amounts to this. Deeply disturbed by the democracy, and the beginnings of the open society, in contemporary Athens, Plato came to fear all social change as embodying decay and corruption. Synthesizing elements taken from Parmenides, from the Pythagoreans, and from Socrates, Plato turned these fears into an entire cosmology and social theory. Every kind of material object has its perfect copy, its ideal representation, as a Form in a kind of Platonic Heaven (Plato's famous theory of Forms). These Forms initiated the material universe by printing themselves on space, thus producing initial material copies. But, as time passes, copies of copies gradually become more and more corrupt, further and further removed from their ideal progenitors. And this is just as true in the social and moral sphere as the material. The primary task for the rulers of society is to arrest all social change, and try to keep society resembling, as far as possible, the ideal Forms of order, justice and the Good. Most people know only of imperfect material things; but a very few philosophers, as a result of studying mathematics (which enables us to acquire knowledge of abstract, perfect objects and not just their imperfect material copies), are able to come to see, intellectually, the Forms, and eventually the supreme Form of the Good (represented as the sun in Plato's famous myth of the cave in *The Republic*). Enlightened philosophers alone have seen the Form of the Good; they alone know what ideal form society should take, and how it can be protected from the corrupting effects of change. Philosophers, then, must rule, aided by guardians, a class of soldiers or police, who ensure that the rest of the population obeys the strictures of the ruling philosophers. Plato's republic is a nightmarish totalitarian, closed society, rigidly ordered, individual liberty, freedom of expression and discussion, art, democracy and justice ruthlessly suppressed. But Plato presents all this with great subtlety, with a kind of twisted logic, so that ostensibly he is arguing for a just, wise and harmonious society, one of legal and moral perfection. Popper even suggests that Plato wrote *The Republic* as a kind of manifesto, to aid his adoption as philosopher-ruler.

Popper's two big enemies of the open society are Plato (volume 1 of *O.S.E.*), and Marx (volume 2). Both uphold versions of historicism – the doctrine that history unfolds according to some fixed pattern, to some rigid set of laws of historical evolution. Plato, as we have seen, was a pessimistic historicist: historical change involves decay and degeneration, and all that enlightened philosopher-rulers can do is arrest change somewhat. Marx, by contrast, is an optimistic historicist: historical development will eventually result in socialism and freedom.

Popper traces a direct link from Plato to Marx, via Aristotle and Hegel. Prompted in part by his biological interests, Aristotle modified Plato's doctrine of the Forms so that it could give an account of biological growth and development. Aristotle inserts a Platonic Form into each individual object so that it becomes the essence of that object, an inherent potentiality which the object, through movement, change or growth, strives to realize. Thus the oak tree is inherent as a potentiality in the acorn. Germination and growth are to be understood as the acorn striving to realize its potentiality, thus becoming an oak tree.

In short, Aristotle modifies Plato's doctrine of the Forms so that the Form ceases to be the perfect copy of an object from which the object can only decay, and becomes instead an inherent potentiality which the object strives to realize. This modification potentially transforms Plato's pessimistic historicism of inevitable decay into an optimistic historicism of social growth, development and progress. But not until Hegel, did anyone fully exploit Aristotelianism in this way.

Popper depicts Hegel as a complete intellectual fraud. He agrees with Schopenhauer's verdict: "Hegel, installed from above, by the powers that be, as the certified Great Philosopher, was a flat-headed, insipid, nauseating, illiterate charlatan, who reached the pinnacle of audacity in scribbling together and dishing up the craziest mystifying nonsense" (quoted in Popper, 1966a, vol. 2, pp. 32–3). Hegel's great idea was to depict history as the process of Spirit, the Aristotelian essence and potentiality of the State and the Nation, striving to realize itself through war and world domination. Taking over and corrupting the antinomies of Kant's *Critique of Pure Reason* (1961), Hegel depicted history as a kind of pseudo-rational or logical dialectical process, *thesis* giving way to *antithesis*, which then results in *synthesis*. What matters is not individual liberty or democracy, but rather the triumph of the strongest State on the stage of history, its inner essence interpreted and directed by the Great Leader by means of dictatorial power.

Despite (or because of) his intellectual fraudulence, Hegel exercised – Popper argues – a powerful influence over the development of subsequent nationalist, historicist and totalitarian thought, of both the extreme right and the extreme left. Both Hitler and Stalin stumble onto the world stage out of Hegel, Popper implies (although neither is mentioned by name in *O.S.E.*, as indicted above). In particular, Hegel exercised a powerful and corrupting influence on Karl Marx.

For Popper, Marx is in a quite different category from Hegel. Popper pays tribute to Marx's sincerity, his humanitarianism, his intellectual honesty, his hatred of moralizing verbiage and hypocrisy, his sense of facts and his sincere quest for the truth, his important contributions to historical studies and social science, his burning desire to help the oppressed. Nevertheless, Marx is one of the most dangerous enemies of the open society, his thought disastrously corrupted by its Hegelian inheritance.

In a well-known passage in *Capital*, Marx declared that Hegel "stands dialectics on its head; one must turn it the right way up again" (quoted in Popper, 1966a., vol. 2, p. 102). And in another passage, Marx declared: "It is not the consciousness of man that determines his existence – rather, it is his social existence that determines his consciousness" (quoted ibid., p. 89). Whereas, for Hegel, an idealist, history is the dialectical development of ideas, for Marx history is determined by the dialectical development of material processes, in particular those associated with the means of production. Distinct historical phases – pre-feudal, feudal, capitalistic, post-revolutionary socialist – owe their existence to distinct phases in the means of production, and the social arrangements these phases generate. Each phase leads, as a result of inevitable dialectical processes, to its own destruction and the creation of the next phase. Thus capitalism concentrates wealth and ownership of the means of the production into fewer and fewer hands until, eventually, the workers unite, overthrow the capitalists and establish socialism. The historical processes of dialectical materialism work themselves out through class struggle; classes and the conflicts between them being determined by the means of production. It is the laws determining the evolution of the economic base that decide the path of history; ideas, democratic and legal institutions form an ideological superstructure, which reflects the economic base and the interests of the dominant class, but is powerless to influence the path of history. Marx condemned as "Utopian" those socialists who sought to bring about the revolution by means of political policies and plans. He held that the proper "scientific" approach to bringing about socialism is, first, to discover the dialectical laws governing the evolution of the economic base of society, and then to help this evolution along, in so far as this is possible, thus speeding up the coming of the final, inevitable socialist revolution.

Popper argues that a number of elements of Marxist thought are of value, if not taken too far. There is the idea that the social cannot be reduced to the psychological, sociology not being reducible to psychology. There is the thesis that much of history has been influenced by class struggle, and the idea that the means of production, economic circumstances, play an important role in influencing the development of other aspects of social and cultural life, even something as apparently remote from economic conditions as mathematics. Above all, there is the recognition and depiction of the appalling conditions of life of the poor in the unrestrained capitalist conditions of Marx's time, and the recognition, too, of the hypocrisy of much of the morality, the legal system and the politics of those times. Having described Marx's account of the working conditions of children as young as six years, Popper writes: "Such were the conditions of the working class even in 1863, when Marx was writing *Capital*; his burning protest against these crimes, which were then tolerated, and sometimes even defended, not only by professional economists but also by churchmen, will secure him forever a place among the liberators of mankind" (1966a, vol. 2, p. 122).

But these good points are, for Popper, more than counterbalanced by the dreadful defects, most of which stem from Marx's historicism, inherited from Hegel. For the central tenet of Marxism is the idea that the laws of dialectical materialism determine the evolution of the means of production, and this in turn determines the evolution of everything else, from class struggle to culture, religion, the law and politics. But this is manifestly false. For one thing, there is a two-way interaction between economic conditions and ideas; eliminate scientific and technological ideas, and the economy would collapse. For another, ideas can themselves influence the course of history, Marxism itself being an example. Historical predictions made by Marx, on the basis of his economic historicism, have been falsified by subsequent historical events. The Russian Revolution is, for example, entirely at odds with Marx's theory, as is the way in which the unrestricted capitalism of Marx's time has subsequently become both more economically successful and more just and humane as a result of diverse political interventions. Marx's economic historicism is not just false; it is pseudoscientific. Only for exceptionally simple systems, such as the solar system, is long-term prediction, based on scientific theory, possible. In the case of social systems, incredibly complex and open to the influence of a multitude of unpredictable factors, the idea that science should be able to deliver long-term predictions is hopelessly unwarranted. Marx's historicism leads him to turn good points into bad ones by exaggeration. "The history of all hitherto existing society is a history of class struggle" (quoted in Popper, 1966a, vol. 2, p. 111) is a good point if "all" is not taken too seriously, but as it stands is an oversimplification and exaggeration; it ignores, for example, power struggles within the ruling class. Again, Marx was surely right to see legal and political institutions of his time as being biased in the direction of the interests of the ruling classes; but he was wrong to condemn all legal and political institutions as inevitably having this function, as his economic historicism compelled him to do.

For Popper, the most damaging feature of Marx's historicism has to do, perhaps, with the severe limitations that it places on the power of politics, on the capacity of people to solve social problems. Marx is famous for his eleventh thesis on Feuerbach: "The philosophers have only interpreted the world in various ways; the point however is to change it" (quoted in Popper, 1966a, vol. 2, p. 84). But Marx's economic historicism leads immediately to a severely restricted view as to what political intervention can achieve. In *Capital* he declares: "When a society has discovered the natural law that determines its own movement, ... even then, it can neither overleap the natural phases of its evolution, nor shuffle them out of the world by a stroke of the pen. But this much it can do; it can shorten and lessen its birth-pangs" (quoted ibid., p. 86). Just those actions which were to improve the unrestrained capitalism of Marx's time beyond all recognition, namely political intervention and the actions of trade unions, are discounted at the outset by Marx's economic determinism as necessarily impotent. Political planning and policymaking for socialism is condemned by Marx, in line with his central doctrine, as inherently inefficacious and Utopian. One disastrous consequence of this was that when Marxists gained power in Russia, they found their literature contained no guidelines as to how to proceed. Another disastrous consequence was that Marxism, blind to the potency of political power, failed to anticipate the dangers inherent in handing over power to political leaders after the revolution, dangers which, after the Russian Revolution, became all too manifest.

The full force of Popper's criticism is devoted, however, to the central argument of *Capital* – an argument which seeks to establish the inevitable downfall of capitalism and the triumph of socialism. Popper presents Marx's arguments as having three steps, only the first of which is elaborated in *Capital*. The first step argues that an inevitable increase in the productivity of work leads to the accumulation of more and more wealth in the ruling class, and the greater and greater poverty and misery of the working class. The second step then argues that all classes will disappear except for a small, wealthy ruling class and a large impoverished working class, this situation inevitably leading to a revolution. The third step argues that the revolution will result in the victory of the working class, which in turn will result in the withering away of the state and the creation of socialism.

Popper demonstrates that none of these steps is inevitable by showing that alternative developments are entirely possible and, in many cases, have actually happened since Marx wrote *Capital*. Even if there is a tendency under capitalism for the means of production and wealth to be concentrated in fewer and fewer hands (as the first step assumes), the state can intervene to counteract this tendency by such means as taxation and death duties. And as far as the increasing poverty of workers is concerned, this can be counteracted by the formation of trade unions, by collective bargaining backed up by strikes. The brutal, unrestricted capitalism of Marx's time has since been transformed out of all recognition by just such interventionist methods. And Popper makes analogous, decisive points to demolish the second and third steps of Marx's argument. Even if the ruling class did become increasingly wealthy and the working class increasingly poor (as the second step assumes), this does not mean that all classes but these two would necessarily disappear, since landowners, rural workers, and a new middle class may well exist, given Marx's assumptions. And even if violence breaks out, this does not mean it would necessarily constitute the social revolution, as envisaged by Marx. And finally, even if it is granted that the workers unite and overthrow the ruling class (as the third step assumes), this does not mean that a classless society and socialism would necessarily result. It is all too easy to suppose that the new political leaders would seize and hold on to power, justifying this by exploiting and twisting the revolutionary ideology, and by invoking the threat of counter-revolutionary forces. And many other possible outcomes can be envisaged. It is in fact implausible to suppose that the victory of the working class would mean the creation of a classless society, and hence the withering away of the state. (This bald summary does not begin to do justice to the cumulative force of Popper's argument.)

Marx, as we have seen, condemned planning for socialism as Utopian; and in a sense, Popper agrees. Popper distinguishes two kinds of social planning or intervention, which he calls Utopian and piecemeal social engineering. Utopian social engineering seeks to attain an ideal social order, such as socialism, by bringing about holistic changes in society; such an approach is, Popper argues, doomed to failure. Piecemeal social engineering, by contrast, searches for and fights against "the greatest and most urgent evils of society": this is the approach that Popper advocates (1966a, vol. 1, ch. 9). Subsequently, during the course of criticizing Marx, Popper points out that piecemeal social engineering can take the form either of state intervention, or of the creation of legal, institutional checks on freedom of action. The latter is to be preferred, Popper argues, as the former carries with it the danger of increasing the power of the state (1966a, vol. 2, pp. 129–33).

There is very much more to Popper's *O.S.E.* than the above indicates. Central to the book is the idea that reason is a vital component of the open society; reason being understood as "critical rationalism", arrived at by generalizing Popper's falsificationist conception of scientific method. For Popper, both scientific method and rationality need to be understood in social terms. Popper criticizes Karl Mannheim's sociology of knowledge for overlooking the "social aspect of scientific method" (Popper, 1966a, vol. 2, ch. 23). Popper criticizes moral historicism, oracular philosophy and the revolt against reason, and the idea that history might have a meaning (ibid., chs. 22, 24 and 25, respectively). Both volumes have extensive footnotes containing fascinating discussion of a great variety of issues tangentially related to the main argument, such as the development of ancient Greek mathematics, the problem of putting an end to war, or the proper aims of a liberal education.

Popper's fiercely polemical book has provoked much controversy. His critical onslaughts against Plato, Aristotle, Hegel and Marx have been angrily repudiated by many scholars in these fields, or, much worse, just blandly ignored.

1.6 *The Poverty of Historicism*

The general doctrine of historicism is expounded and criticized by Popper in his *The Poverty of Historicism* (*P.H.*), first published in three parts in *Economica* in 1944 and 1945 (somewhat before *O.S.E.*) and only published in book form in 1957. Popper divides historicist views and arguments into two classes: those that hold that the methods of the social and natural sciences are quite different (the "anti-naturalist doctrines") and those that hold they are the same or similar (the "pro-naturalist doctrines").

Anti-naturalist doctrines can be summarized like this. Generalizations, experiments, predictions and understanding have roles in social science that are radically different from those they have in physics. Social phenomena exhibit novelty, complexity and a holistic aspect that is lacking in physical phenomena. These differences ensure that historicist social science, predicting in more or less rough outline the evolution of society, must employ methods that differ from those of natural science.

Pro-naturalist doctrines of historicism make much of the success of long-term predictions in astronomy. Just as states of the solar system can be predicted by natural science far into the future, so too historicist social science ought to be able to predict states of society far into the future. Such predictions will, however, employ social laws of succession, laws which specify how one characteristic phase of social development gives way to a subsequent phase.

Popper effectively criticizes the anti-naturalist doctrines of historicism. It is, however, Popper's criticism of the pro-naturalist standpoint that is the really important nub of the book. Historicist laws of succession are not laws at all, as these are understood in physics. They are *trends*. And *"trends are not laws"* (Popper, 1961, p. 115). A law provides a causal explanation of an event when the law *plus initial conditions* imply that the event occurs. Whenever a succession of causally connected events occurs in our environment, such as the wind shaking a tree and causing an apple to fall to the ground, laws (usually a number of quite different laws) *plus the specification of a sequence of initial conditions* are required to predict the sequence of events. Trends can, then, be explained by means of laws, but it is always laws *plus relevant initial conditions* which provide such explanations. And the crucial point is that, given some trend, in particular a social trend, initial conditions that must continue to exist if the trend is to continue are likely to be very many indeed, most of which will be easy to overlook. This ensures that trends, such as the growth of a population, which have persisted for centuries, may quite suddenly cease if some condition, necessary for the persistence of the trend, ceases to exist. "The poverty of historicism", Popper declares "is a poverty of imagination" (ibid., p. 130) – the poverty of being unable to imagine that conditions, necessary for the persistence of some trend, might suddenly themselves change. And this is highly relevant to the whole idea of piecemeal social engineering, for the piecemeal engineer may seek to change just such conditions, required for the persistence of some undesirable trend.

1.7 At the LSE

In 1945 Popper was appointed to a readership in logic and scientific method at the London School of Economics (LSE); he took up the appointment in 1946, and was promoted to a personal professorship in 1949. Initially the only philosopher at the LSE, Popper was subsequently joined by J. O. Wisdom in 1948, Joseph Agassi in 1957 (who left in 1960), John Watkins in 1958, W. W. Bartley III and Imre Lakatos in 1960, and Alan Musgrave in 1964. The Department at the LSE was famous for Popper's weekly seminar. Notoriously, visiting speakers rarely succeeded in concluding the announcement of the title of their talk before being interrupted by Popper. He subjected each speaker to a devastating critical attack, almost sentence by sentence; quite often, the subject of the seminar would be continued a week later. The seminars were always dramatic, sometimes farcical, but nevertheless created an overwhelming impression of Popper's passionate determination to get at the truth, even if conventions of politeness and good manners had to be sacrificed.

1.8 *Conjectures and Refutations*

In 1963 Popper published *Conjectures and Refutations* (*C.R.*), a collection of essays restating, extending and applying his views on scientific method, philosophy and rationality. This is perhaps the best introduction to Popper's work. Here is a quick survey of some of the items in the book.

In the Introduction Popper makes a number of important epistemological points. He notes the widespread tendency to believe in the false doctrine that truth is manifest – readily available and easy to come by. When truth turns out not to be so easy to obtain, epistemological optimists become pessimists, and deny that knowledge is possible at all, or resort to conspiracy theories to account for the inaccessibility of the truth. Both Descartes and Bacon are famous for their anti-authoritarian stance in epistemological matters; and yet, Popper points out, there is an unnoticed implicit authoritarianism in their views. For Bacon, and for the empiricists who followed him, the senses are authoritative sources of knowledge; for Descartes, and for the rationalists who followed him, reason is the authoritative source of knowledge. Popper, of course, argues against the idea that conjectural knowledge has any authoritative source.

In chapter two, Popper argues that philosophical problems have their roots in science and mathematics; and he argues against the Wittgensteinian view that philosophical problems are pseudo-problems that arise when ordinary language is misused.

In chapter three Popper distinguishes three views concerning human knowledge: essentialism, which holds that science can grasp the ultimate essence of things; instrumentalism, which holds that scientific theories are merely instruments for the prediction of observable phenomena; and realism, which holds that science puts forward falsifiable conjectures about aspects of reality that often go beyond what is observable. Popper criticizes the first two views, and defends the third view.

In chapter five Popper gives a magnificent account of the Presocratic philosophers – Thales, Anaximander, Anaximenes, Heraclitus, Parmenides – as proposing and critically assessing successive theories about the origins and ultimate constituents of the universe, and about the problem of how to understand change. The Presocratics, Popper argues, almost unintentionally created critical rationality, the tradition of proposing bold conjectures which are then subjected to criticism – a tradition that led eventually to modern science.

In chapter eight Popper tackles the problem of how philosophical or metaphysical doctrines can be rationally assessed given that they cannot be empirically falsified, like scientific theories. His solution is that philosophical doctrines can be assessed from the standpoint of the problems that they are intended to solve; even though irrefutable, they can nevertheless be criticized from the standpoint of the problems they seek to solve.

In chapter ten Popper restates and, as we have seen above, develops somewhat his falsificationist conception of scientific method. In this chapter Popper formulates and tries to solve what has subsequently come to be known as *the problem of verisimilitude*: What can we mean by scientific progress if science advances from one false theory to another? Popper's solution is that, given two theories, T_1 and T_2, even though both are false, nevertheless T_2, say, may be closer to the truth than T_1. Suppose, for example, that T_2 implies everything true that T_1 implies and more besides, but T_2 does not imply anything false that T_1 does not imply. Granted this, there is a perfectly good sense in which T_2 can be said to be "closer to the truth" than T_1, and thus an advance over T_1. Unfortunately it was subsequently shown by Tichy (1974) and Miller (1974) that this proposed solution to the problem does not work. If T_2 has more true implications than T_1 does, then T_2, necessarily, has some false implications which T_1 does not have. Popper's requirements for T_2 to be closer to the truth than T_1, when both are false, cannot be satisfied.

Chapter fifteen provides an exposition and decisive criticism of dialectic reasoning: it is thus an adjunct to the criticisms of Hegel and Marx to be found in *O.S.E.*

One of the themes running through *C.R.*, and through much of Popper's subsequent writings, is that the proper task of philosophy is to tackle, in an imaginative and critical way, real, fundamental problems having their roots outside philosophy in science, politics, art, life. This Popperian conception of philosophy stands in sharp contrast both to the pomposities and obscurities of much so-called "continental" philosophy, and to the poverty and aridity of philosophy in the so-called "analytic" tradition, which is restricted to ordinary language analysis, the analysis of meaning. Popper has fought against both rival conceptions of philosophy, and has sought to put into practice his own critical rationalist, problem-solving conception of philosophy. His first four books are exemplary in this respect, and have undoubtedly exercised an enormous, healthy influence on much subsequent philosophy, even though this influence has often not been acknowledged. A basic impulse behind these works might almost be summed up in a stray remark tossed out in *O.S.E.*: "We have to learn the lesson that intellectual honesty is fundamental for everything we cherish" (Popper, 1966a, vol. 2, p. 59).

1.9 The basic argument running through Popper's early work

It is important to appreciate the existence of a central backbone of argument running through these four books. In *L.Sc.D.*, as we have seen, Popper argues that all scientific knowledge is irredeemably conjectural in character, it being impossible to verify theories empirically. Science makes progress by proposing bold conjectures in response to problems, which are then subjected to sustained attempted empirical refutation. This falsificationist conception of scientific method is then generalized to form Popper's conception of (critical) rationality, a general methodology for solving problems or making progress. As Popper puts it in *L.Sc.D.*: "inter-subjective *testing* is merely a very important aspect of the more general idea of inter-subjective *criticism*, or in other words, of the idea of mutual rational control by critical discussion" (1959, p. 44, note 1*). But in order to make sense of the idea of *severe* testing in science, we need to see the experimentalist as having at least the germ of an idea for a rival theory up his sleeve (otherwise testing might degenerate into performing essentially the same experiment again and again). This means experiments are always *crucial experiments*, attempts at trying to decide between two competing theories. Theoretical pluralism is necessary for science to be genuinely empirical. And, more generally, in order to *criticize* an idea, one needs to have a rival idea in mind. Rationality, as construed by Popper, requires plurality of ideas, values, ways of life. Thus, for Popper, the rational society *is* the open society.

Given pre-Popperian conceptions of reason, with their emphasis on proof rather than criticism (and associated plurality of ideas), the idea that the rational society is the open society is almost a contradiction in terms. There is thus a very close link between *L.Sc.D.*, on the one hand, and *O.S.E.*, *P.H.* and *C.R.* on the other. And the direction of argument does not go in just one direction, from *L.Sc.D.* to *O.S.E.*: it goes in the other direction as well. For in *O.S.E.* (1966a, vol. 1, ch. 10), Popper argues that rationality, and scientific rationality as well, need to be conceived of in social and institutional terms (and the argument is echoed in *P.H.*, in connection with a discussion about the conditions required for scientific progress to be possible). *O.S.E.*, *P.H.* and *C.R.* illuminate and enrich the doctrines of *L.Sc.D.*

Above, in connection with the discussion of *L.Sc.D.*, it was argued that Popper's falsificationism ultimately fails, because of its failure to exclude highly falsifiable but grossly *ad hoc* theories from science. The scientific enterprise is obliged to conjecture that the universe is more or less comprehensible, having some kind of unified dynamic structure, only those theories being tentatively accepted which satisfy (a) empirical considerations, and (b) considerations having to do with simplicity, unity, comprehensibility. As science proceeds, we improve our (conjectural) knowledge of the kind of comprehensible unity which may exist in nature; the *aim* of science improves, and with it the *methods* of science. There is, in other words, a kind of positive feedback between improving knowledge and improving aims and methods, improving knowledge about how to improve knowledge. Science adapts its nature to what it finds out about the nature of the universe (which helps to account for the almost explosive growth of scientific knowledge).

This "evolving-aims-and-methods" view of science modifies quite considerably Popper's falsificationism. When generalized, it leads to an "evolving-aims-and-methods" view of rationality which in turn modifies quite considerably Popper's critical rationalism. These modifications, if adopted, have far-reaching implications for central doctrines of Popper's *L.Sc.D.*, *O.S.E.*, *P.H.* and *C.R.* (see Chapters 2–7 and 10, and Maxwell, 1984a, 1998, 2004a, 2017a, 2017b).

1.10 Popper's later work

Work published by Popper after *C.R.*, though containing much of great value, is not, perhaps, in quite the same league as that of his first four books. Much of this work restates, extends and further applies earlier ideas. Where Popper's subsequent work launches forth in new directions, these are not always well chosen. Battles against subjectivity, anti-realism and physical determinism lure Popper into defending opposing views that are exaggerated, sometimes, almost to the point of absurdity. A subtle shift of perspective, of allegiance, can be discerned as we move from Popper's earlier to his later work. In his early work, Popper speaks up on behalf of humanity, on behalf of any concerned person of good will, and against those traditional "great thinkers" and "experts" who threaten to beguile us and lead us to disaster. In his later work, the allegiances have shifted; now Popper speaks up on behalf of great science and great scientists, and against fraudulent academics, mostly philosophers and social scientists.

In 1970 there appeared *Criticism and the Growth of Knowledge*, edited by Imre Lakatos and Alan Musgrave, the fourth volume of the proceedings of a conference on philosophy of science held in London in 1968. This volume is devoted to a comparison of the views of Thomas Kuhn and Popper on the philosophy of science, and contains contributions from Kuhn, Popper, Watkins, Toulmin, Lakatos, Feyerabend and others. In his contribution Popper praises Kuhn for having discovered normal science, science which takes some "paradigm" for granted and devotes itself to puzzle solving. Popper points out that he had himself made the same discovery over thirty years earlier, as recorded in the preface to *L.Sc.D.* But the normal scientist "has been badly taught. He has been taught in a dogmatic spirit: he is the victim of indoctrination". Normal science is "a danger to science and, indeed, to our civilization" (Popper, 1970, p. 53).

In 1972 Popper published a second collection of his essays entitled *Objective Knowledge*. One of the essays makes the good point that common sense tends to combine two incompatible theses: common sense realism, and the epistemological view that knowledge comes flooding into our minds via the senses, rather like water being poured into a bucket – a view which Popper dubs "the bucket theory of the mind". Popper argues that these two theses clash, and that philosophers, registering this clash, have all too often held on to the bucket theory and rejected realism. But this, Popper argues, is exactly the wrong thing to do; one should hold on to realism, and reject the bucket theory.

Much of the rest of the book is devoted to developing and defending Popper's three-world view. There are, according to this view, three worlds: the physical world (world 1), the psychological or mental world (world 2), and the world of objective theories, propositions, arguments and problems (world 3). World 3 interacts with world 1 via world 2.

Popper argues that this interaction is demonstrated by the fact that scientific theories lead to new technology, world 1 phenomena, which would not exist were it not for the prior development of world 3 theories. Popper puts world 3 into a biological and evolutionary context: like the webs, nests and dams created by spiders, birds and beavers, so too world 3 is our creation but, once created, it acquires an objective existence independent of us.

This theme is continued in Popper's contribution to *The Self and Its Brain* (*S.B.*) (1977) a book written with the neurologist John Eccles. In this work, Popper develops a sustained argument in support of interactionism and his three-world view, and criticizes materialism, physicalism and the thesis that the physical world is (causally) closed. There is also an interesting chapter on the history of the mind-body problem, in which Popper argues for the questionable thesis that the problem was recognized independently of, and before the arrival of, anything like the modern scientific view of the world.

What is one to make of this three-world view? Popper is surely right to hold that the contents of theories need to be distinguished from their linguistic forms (and from the causal effects these linguistic forms can have on appropriately educated brains). Popper is also right, surely, to stress that in order to make human sense of human action, we need to attend to the contents of theories. But it is quite another matter to argue, as Popper does, that world 3 entities, such as contents of theories, exist as full-blooded, almost Platonic entities, poltergeistic intellectual objects capable of influencing material phenomena via their influence on conscious minds. Popper overlooks or ignores the possibility that the material world may be *causally* closed but not *explanatorily* closed. He overlooks, that is, the possibility that physical phenomena, such as those associated with human actions and human technology, can be explained and understood in two distinct (but perhaps interdependent) ways: (1) physically and causally, in terms of physical theory, and (2) "personalistically", in terms of the intentions, plans and ideas of people. Such a view would hold that personalistic explanation is compatible with, but not reducible to, physical explanation. This view would give to the contents of theories a vital role in the (personalistic) explanation of human actions and the development of technology, without in any way undermining the existence, in principle, of a purely physical, causal explanation of physical phenomena associated with human action and technology (see Maxwell, 2001a, ch. 5; 2016a, pp. 195–200).

Popper insists that his world 3 entities differ from Plato's Forms in that they are man-made, they consist of theories, including false theories, and problems, rather than reified concepts or essences, and there is no suggestion that world 3 objects can be known with certainty (Popper and Eccles, 1977, pp. 43–4). But even if Popper's world 3 entities do not have implausible epistemological Platonic features, they most certainly have highly implausible ontological and causal Platonic features, in that they have causal effects on the material world (via their influence on conscious minds). That the elderly Popper should espouse such an implausible Platonic doctrine almost seems like Plato's revenge for the youthful Popper's onslaught against him.

Chapter three of *Objective Knowledge* is called "Epistemology Without a Knowing Subject". Despite the title, Popper does not altogether neglect the personal dimension of the search for knowledge. What he does argue is that subjective knowledge is irrelevant to the study of scientific knowledge, only knowledge construed in objective, impersonal, world 3 terms being important. But this downplays the point that all of objective knowledge, stored in books and libraries, is of value only in so far as it is understood and used by people. Albert Einstein once remarked: "Knowledge exists in two forms – lifeless, stored in books, and alive in the consciousness of men. The second form of existence is after all the essential one; the first, indispensable as it may be, occupies only an inferior position" (Einstein, 1973, p. 80). Einstein's priorities seem saner than later Popper's. And altogether saner, more humane and down-to-earth than elderly Popper's spooky world 3 objects, is the viewpoint of the more youthful Popper of *O.S.E.*, which sees science and reason in personal, social and institutional terms, without any appeal being made to ghostly, quasi-Platonic Forms.

In 1974 Popper became the fourteenth subject of *The Library of Living Philosophers*, edited by P. A. Schilpp. This two-volume work opens with Popper's "Intellectual Autobiography" – subsequently published independently as *Unended Quest* (1976a) – continues with descriptive and critical papers on diverse aspects of Popper's work, by Quine, Putnam, Lakatos, Medawar, Watkins, Ayer, Margenau, Grünbaum, Kuhn and others, and concludes with Popper's replies. *Unended Quest* is a fascinating book, and gives a gripping account of Popper's lifelong, passionate engagement with his fundamental problems and concerns. It includes a marvellous discussion of the development of polyphonic music, and provides an account of Popper's battles with subjectivism in physics in connection with quantum theory, and with thermodynamics and the arrow of time. Popper also declares that it was he who killed logical positivism. This book, together with *C.R.*, provides the best introduction to Popper's philosophy.

In 1982–3 there appeared *The Postscript to The Logic of Scientific Discovery*, a three-volume work which extends and elaborates doctrines and arguments of *L.Sc.D.*, and much of which was written in the years 1951–6. The work reached the stage of proofs in 1956–7, but was abandoned because Popper suffered from detached retinas and had operations on both eyes, his sight for a time in question. It was only much later, under the editorship of W. W. Bartley III, and after some additions and rewriting, that the work finally appeared.

Volume one, entitled *Realism and the Aim of Science* (1983), restates and elaborates Popper's earlier views and arguments concerning induction, falsification, corroboration, demarcation, realism, metaphysics and probability. At one point Popper illuminatingly contrasts how a scientific paper might be written in the style of inductivism, and in the critical, problem-solving approach of falsificationism and critical rationalism (see pp. 47–51).

Volume two, entitled *The Open Universe: An Argument for Indeterminism* (1982a), sets out to refute determinism. Popper distinguishes between "scientific" and metaphysical determinism. "Scientific" determinism asserts that future states of physical systems can be predicted with any degree of precision by means of theories and initial conditions specified with sufficient precision (see p. 36). Metaphysical determinism asserts merely that "all events in this world are fixed, or unalterable, or predetermined" (p. 8). Popper spells out an argument which, he claims, refutes scientific determinism. Even given a universe in which all events occur in accordance with a deterministic physical theory, T, nevertheless a predictor, put within an isolated system, could not predict all future states of the system with unlimited precision. Even if such a predictor had unprecedented powers to acquire knowledge of initial conditions, and make predictions using T, nevertheless it could not acquire up-to-date information about its own state, because the attempt to do so would continually alter its state. This means it would not be possible for the predictor to predict future states of the system of which it forms a part. Popper goes on to argue against metaphysical determinism.

Although full of interesting points, there are two oddities about this discussion. First, as Popper admits, his refutation of "scientific" determinism does not refute a second version of "scientific" determinism which asserts that past states of physical systems can be predicted, employing prior initial conditions and physical theory. Second, Popper ignores a rather different third version of "scientific" determinism, which asserts that the universe is such that there is a discoverable, true, physical "theory of everything", T, which is deterministic. This version of determinism deserves to be called "scientific", because T is asserted by it to be scientifically discoverable; furthermore, once discovered, T will be falsifiable, and hence, by Popper's own standards, scientific. It is curious that Popper, who is elsewhere (as we have seen) opposed to instrumentalism and in favour of realism, should here discuss at length a version of "scientific" determinism which is thoroughly instrumentalistic in character, in that it makes assertions about *predictability*, and should ignore a version of "scientific" determinism which is much more in keeping with scientific realism, in that it makes an assertion about the nature of the universe. This oversight seriously weakens Popper's argument for indeterminism.

1.11 Quantum Theory

Volume three of *The Postscript* is called *Quantum Theory and the Schism in Physics* (*Q.T.S.P.*) (1982b). It is concerned with quantum theory and probability, interconnected issues which preoccupied Popper, on and off, throughout his working life.

Thus in *Logik der Forschung* (1934), Popper tackled two problems concerning probability: How are probabilistic statements or theories to be interpreted? And how can probabilistic theories be falsifiable given that they are in principle "impervious to falsification"? In response to the first problem, Popper defended a version of von Mises' objective, frequency interpretation of probability. In response to the second, he insisted that probabilistic statements become falsifiable as a result of a methodological decision to treat them as falsifiable.

Logik der Forschung also devoted a chapter to problems of quantum theory. The main task is to criticize Bohr's and Heisenberg's orthodox interpretation of quantum theory (which gives equal weight to the two "complementary" pictures of particle and wave), and to provide an alternative which interprets the theory as an objective, realistic statistical theory about *particles*. Popper criticizes Heisenberg's interpretation of his uncertainty relations, which interprets these relations as placing restrictions on (simultaneous) measurement. Popper argues that these relations need to be interpreted as "scatter relations", restricting what can be predicted, and not what can be measured. Indeed, Popper argues, not only *can* we make measurements – for example, simultaneous measurements of position and momentum – that are more precise than allowed by the uncertainty relations as interpreted by Heisenberg, we *need* to do this in order to test these relations.

These issues are restated and further developed in *The Postscript*, taking into account relevant developments in quantum physics itself, such as John Bell's proof that local hidden variable versions of quantum theory cannot reproduce all the predictions of orthodox quantum theory, and experiments, such as those of Aspect, which seem to have refuted these local hidden variable theories. The main change in Popper's views is his development of his "propensity" interpretation of probability, and his application of this to quantum theory.

Popper's propensity idea is perhaps best understood in terms of an example. Consider tossing a die on a table. There is a certain probability of obtaining a six, which may or may not equal 1/6. This is determined by such things as properties of the die (e.g. whether or not it is made of a homogeneous material), the procedure for tossing and the properties of the table. It is this combination of properties that is, for Popper, the propensity: it is a property, determining a probability associated with some repeatable event (such as tossing the die), "of the *whole repeatable experimental arrangement*" (Popper, 1982b, p. 71). In particular, then, the probabilistic statements of quantum theory can be interpreted as attributing propensities, not to individual electrons or photons as such, but rather to electrons or photons in the context of some specific, repeatable measurement.

Popper's views on quantum theory have been criticized by Paul Feyerabend (1968-9), on the grounds that Popper fiercely criticizes Bohr but ends up defending a view very close to Bohr's. Because propensities are properties defined in terms of experimental arrangements, this means that Popper's propensity interpretation of quantum theory, just like Bohr's interpretation, brings in measurement in an essential way. Popper's reply is that propensities relate to "physical situations" which may, but need not be, experimental arrangements (Popper, 1982b, p. 71). But this reply fails in two ways. First, the probabilistic predictions of standard quantum theory *are* restricted to measurements. If these predictions are to include "physical situations" that are not measurements, then they need to be specified, and need to have specified quantum observables associated with them, so that definite probabilistic predictions may be forthcoming: Popper provides nothing of this. Second, even if Popper did extend the interpretation of quantum theory in the way just indicated, the result would be a version of quantum theory which would reproduce most of the serious defects of the theory given Bohr's interpretation. These defects include being vague, ambiguous, *ad hoc* and non-explanatory, all resulting from the fact that the theory is made up of two incoherent parts, a quantum mechanical part, and a classical part specifying measurement or specific "physical situations".

It may, however, be possible to overcome these defects by modifying Popper's propensity version of quantum theory, so that quantum propensities determine probabilistically how quantum entities, such as electrons and photons, interact with each other (rather than with classically described, macroscopic, measuring instruments or "physical situations"). But this leads to a fully micro-realistic propensity version of quantum theory, very different from Popper's version (see Chapter 8, and Maxwell, 1982, 1988, 1994a, 2011a). Quantum theory emerges as a theory that is about, not particles, but a new kind of probabilistic entity, the "propensiton" (as it may be called), which is neither a particle nor a wave, even though it has some features of both. Furthermore, according to such a version of quantum theory, probabilistic transitions involve something like "wave-packet collapse" as a real physical process: for Popper, any such idea is just another part of "the great quantum muddle". But what this indicates is that here, as elsewhere in his work, Popper's ideas, even when wrong or inadequate, are nevertheless rich in fruitful suggestions and implications for further development.

1.12 Final years and reputation

After *The Postscript*, a number of collections of essays have appeared, restating and elaborating themes already indicated: *A World of Propensities* (1990), *In Search of a Better World* (1992), *The Myth of the Framework* (1994), *Knowledge and the Body-Mind Problem* (1994), *Lesson of this Century* (1997), *The World of Parmenides* (1998), *All Life is Problem Solving* (1999), and *After the Open Society* (2008).

Popper was knighted in 1965, and became a Companion of Honour in 1982. He retired from his position at the LSE in 1969. He became a Fellow of the Royal Society in 1976. Popper's wife, Hennie, died in 1985, after a long struggle with cancer. In his later years, Popper was showered with academic honours of various kinds: membership of many academic societies, honorary degrees, conferences dedicated to his philosophy, and honours, medals and prizes from various sources (see Miller, 1997, pp. 403–6). Popper died a week after a serious operation, on 17 September 1994.

Popper's reputation, after his death, has suffered a curious fate. Most philosophers and philosophers of science recognize the significance of Popper's work for twentieth-century philosophy of science, but seem to hold that it has little to contribute to the field in the second decade of the twenty-first century. Popper has become somewhat passé. Many scientists, on the other hand, hold Popper in high esteem, and even call upon his work during the course of debates about scientific matters – a point made recently by Godfrey-Smith (2016, p. 104).

As far as Popper's reputation among philosophers is concerned, a part of the problem is that too few philosophers have responded to a central feature and claim of Popper's work: that the basic task of philosophy is to tackle, and try to help solve, urgent, fundamental problems that have their roots outside the discipline, in science, politics, the arts, the environment, education – problems which, if solved, may well have widespread fruitful implications for such diverse fields. And as far as the philosophy of science is concerned, the discipline has developed in the twenty-first century in ways that would have horrified Popper. Philosophy of science has increasingly succumbed to what was, for him, the ultimate intellectual sin: specialization. It has fragmented into philosophy of physics, biology, chemistry, psychology, neuroscience and so on – mirroring specialization in science, and doing nothing to counteract it. The fundamental problems about the nature of the cosmos, our place in the cosmos, and our knowledge and understanding of it, that so gripped Popper's imagination, are increasingly ignored.

In the rest of this book, I set out to subject Popper's ideas to ferocious criticism in an attempt to improve them. It will emerge that Popper's ideas about science, reason, quantum theory, academic inquiry, how to make progress towards a more civilized world, all need radical revision. In order to do justice to what is best in Popper's philosophy, some of its tenets must be rejected – including what is, perhaps, its most famous tenet, the principle of demarcation. My hope is that Popper's whole approach to doing philosophy will come to be seen as highly relevant to the tasks that face us today: to provoke the thought of humanity into tackling, imaginatively, critically, and fruitfully, the gigantic problems that confront the modern world. Freed of some deadwood, Popper's work becomes highly relevant to today's global problems, both practical and intellectual.

Select bibliography of works by Popper

Books

Logik der Forschung: Zur Erkenntnistheorie der modernen Naturwissenschaft (Vienna: Julius Springer, 1934).
The Open Society and Its Enemies, 2 volumes: *Volume I: The Spell of Plato; Volume II: The High Tide of Prophecy: Hegel, Marx, and the Aftermath* (London: George Routledge, 1945; revised edition, Princeton: Princeton University Press, 1950).

The Poverty of Historicism (London: Routledge and Kegan Paul, 1957; Boston, MA: Beacon Press, 1957).
The Logic of Scientific Discovery (translation from the German of *Logik der Forschung* with additional footnotes and appendices, London: Hutchinson, 1959; New York: Basic Books, 1959).
Conjectures and Refutations: The Growth of Scientific Knowledge (New York: Basic Books, 1962; London: Routledge and Kegan Paul, 1963).
Objective Knowledge: An Evolutionary Approach (Oxford and New York: Clarendon Press, 1972).
Unended Quest: An Intellectual Autobiography (London: Fontana/Collins, 1976).
The Self and Its Brain: An Argument for Interactionism, with J. C. Eccles (Berlin, New York and London: Springer-Verlag, 1977).
Die beiden Grundprobleme der Erkenntnistheorie [1930–3], edited by T. E. Henson (Tübingen: Mohr, 1979).
Postscript to the Logic of Scientific Discovery, edited by W. W. Bartley, III: *Volume I: Realism and the Aim of Science* (Totowa, NJ: Rowman and Littlefield, 1982; London: Hutchinson, 1983).
Volume II: The Open Universe: An Argument for Indeterminism (Totowa, NJ: Rowman and Littlefield, 1982; London: Hutchinson, 1982).
Volume III: Quantum Theory and the Schism in Physics, (Boston, MA: Unwin Hyman, 1982; London: Hutchinson, 1983).
A World of Propensities (Bristol: Thoemmes, 1990).
In Search of a Better World: Lectures and Essays from Thirty Years, translated by L. J. Bennett (London and New York: Routledge, 1992).
Knowledge and the Body-Mind Problem: In Defence of Interaction, edited by M. A. Notturno (London and New York: Routledge, 1994).
The Myth of the Framework: In Defence of Science and Rationality, edited by M. A. Notturno (London and New York: Routledge, 1994).
The Lesson of this Century: With Two Talks on Freedom and the Democratic State, with an introduction by G. Bosetti, translated by P. Camiller (London and New York: Routledge, 1997).
The World of Parmenides: Essays on the Presocratic Enlightenment, edited by A. F. Petersen and J. Mejer (London and New York: Routledge, 1998).
All Life is Problem Solving, translated by P. Camiller (London and New York: Routledge, 1999).
After the Open Society: Selected Social and Political Writings, edited by J. Shearmur and P. N. Turner (London: Routledge, 2008).
The Two Fundamental Problems of the Theory of Knowledge, edited by T. E. Hansen and translated by A. Pickel (London: Routledge, 2009; first published in German in 1979).

Selected periodical publications (uncollected) and book chapters

"What can Logic do for Philosophy?" *Aristotelian Society, Supplementary Volume XXII: Logical Positivism and Ethics* (1948): 141–54.
"A Note on Natural Laws and So-called Contrary-to-fact Conditionals", *Mind* 58 (1949): 62–6.
"Indeterminism in Quantum Physics and in Classical Physics", *British Journal for the Philosophy of Science* 1 (1950): 117–33 and 173–95.
"The Principle of Individuation", *Aristotelian Society, Supplementary Volume XXVII: Berkeley and Modern Problems* (1953): 97–120.
Papers on "The Arrow of Time", "Irreversibility" and "Entropy", *Nature* 177 (1956): 538; 178 (1956): 382; 179 (1957): 1297; 181 (1958): 402–3; 207 (1965): 233–4; 213 (1967): 320; 214 (1967): 322.
"Probability Magic or Knowledge out of Ignorance", *Dialectica* 11 (1957): 354–72.
"The Propensity Interpretation of the Calculus of Probability, and the Quantum Theory", *Observation and Interpretation*, edited by S. Körner (London: Butterworth Scientific Publications, 1957): 65–79, 88–9.
"Irreversibility; or Entropy since 1905", *British Journal for the Philosophy of Science* 8 (1957): 151–5.
"The Propensity Interpretation of Probability", *British Journal for the Philosophy of Science* 10 (1959): 25–42.
"Philosophy and Physics", *Proceedings of the XIIth International Congress of Philosophy* (1961): 367–74.

"A Theorem on Truth-Content", *Mind, Matter and Method: Essays in Philosophy and Science in Honor of Herbert Feigl*, edited by P. Feyerabend and G. Maxwell (Minneapolis, MN: University of Minnesota Press, 1966): 343–53.

"Normal Science and Its Dangers", *Criticism and the Growth of Knowledge: Proceedings of the International Colloquium in the Philosophy of Science, London, 1965, Volume 4*, edited by I. Lakatos and A. Musgrave (London and New York: Cambridge University Press, 1970): 51–8.

"Replies to My Critics", *The Philosophy of Karl Popper, Volume 2*, edited by P. A. Schilpp (LaSalle, IL: Open Court, 1974): 962–1197.

"A Note on Verisimilitude", *British Journal for the Philosophy of Science* 27 (1976): 147–59.

"Natural Selection and the Emergence of Mind", *Dialectica* 32 (1978): 339–55.

Bibliographies

T. E. Hansen, "Bibliography of the writings of Karl Popper" [1925–1974], in *The Philosophy of Karl Popper*, Book II, edited by P. A. Schilpp (LaSalle, IL: Open Court, 1974): 1201–87.

M. Lube, *Karl R. Popper, Bibliographie 1925–2004* (Frankfurt: Peter Lang, 2005).

Biographies

M. H. Hacohen, *Karl Popper – The Formative Years, 1902–1945: Politics and Philosophy in Interwar Vienna* (New York: Cambridge University Press, 2000; Cambridge: Cambridge University Press, 2001).

D. Miller, "Sir Karl Raimund Popper, C. H., F. B. A.", *Biographical Memoirs of Fellows of the Royal Society* 43 (1997): 367–409.

J. Watkins, "Karl Popper: A Memoir", *The American Scholar* 66 (1977): 205–19.

J. Watkins, "Karl Raimund Popper 1902–1994", *Proceedings of the British Academy* 94 (1997): 645–84.

Other references to Popper's work

T. W. Adorno, H. Albert, R. Dahrendorf, J. Habermas, H. Pilot and K. R. Popper, *The Positivist Dispute in German Sociology*, translated by G. Adey and D. Frisby (Heinemann: London, 1976).

J. Agassi, *A Philosopher's Apprentice: In Karl Popper's Workshop* (Amsterdam and Atlanta, GA: Rodopi, 1993).

R. Bambrough, ed., *Plato, Popper and Politics* (New York: Barnes and Noble, 1967).

W. Berkson and J. Wettersten, *Learning from Error: Karl Popper's Psychology of Learning* (LaSalle, IL: Open Court, 1984).

M. Bunge, ed., *The Critical Approach to Science and Philosophy: In Honor of Karl R. Popper* (New York: Free Press of Glencoe, 1964).

P. Catton and G. Macdonald, eds., *Karl Popper: Critical Appraisals* (London: Routledge, 2004).

G. Curie and A. Musgrave, eds., *Popper and the Human Sciences* (The Hague: Nijoff, 1985).

S. Gattei, *Karl Popper's Philosophy of Science: Rationality without Foundations* (London: Routledge, 2009).

P. Godfrey-Smith, "Popper's Philosophy of Science: Looking Ahead", in *The Cambridge Companion to Popper*, edited by J. Shearmur and G. Stokes (Cambridge: Cambridge University Press, 2016): 104–24.

I. Jarvie, *The Republic of Science: The Emergence of Popper's Social View of Science 1935–1945* (Amsterdam: Rodopi, 2001).

I. Jarvie, K. Milford and D. Miller, eds., *Karl Popper: A Centenary Assesssment*, vols. I, II, and III (Aldershot: Ashgate, 2006).

I. Jarvie and S. Pralong, eds., *Popper's Open Society after Fifty Years* (London and New York: Routledge, 1999).

H. Keuth, *The Philosophy of Karl Popper* (Cambridge: Cambridge University Press, 2005).

I. Lakatos and A. Musgrave, eds., *Criticism and the Growth of Knowledge: Proceedings of the International Colloquium in the Philosophy of Science, London, 1965, Volume 4* (London and New York: Cambridge University Press, 1970).
P. Levinson, ed., *In Pursuit of Truth: Essays on the Philosophy of Karl Popper on the Occasion of His 80th Birthday* (Atlantic Highlands, NJ: Humanities, 1982).
B. Magee, *Popper* (London: Fontana/Collins, 1973).
D. Miller, "Popper's Qualitative Theory of Verisimilitude"; *British Journal for the Philosophy of Science* 25 (1974): 166–77.
D. Miller, *Critical Rationalism* (Chicago, IL: Open Court, 1994).
A. O'Hear, ed., *Karl Popper: Philosophy and Problems* (Cambridge: Cambridge University Press, 1995).
A. O'Hear, ed., *Karl Popper: Critical Assessments of Leading Philosophers*, vols. I, II and III (London: Routledge, 2004).
D. P. Rowbottom, *Popper's Critical Rationalism* (London: Routledge, 2011).
P. A. Schilpp, ed., *The Philosophy of Karl Popper*, 2 volumes (LaSalle, IL: Open Court, 1974).
J. Shearmur, *The Political Thought of Karl Popper* (London: Routledge, 1996).
J. Shearmur and G. Stokes, eds., *The Cambridge Companion to Popper* (Cambridge: Cambridge University Press, 2016).
C. Simkin, *Popper's Views on Natural and Social Science* (Leiden: Brill, 1993).
P. Tichy, "On Popper's Definition of Verisimilitude"; *British Journal for the Philosophy of Science* 25 (1974): 155–60.

Karl Popper's papers

The Hoover Institution, Stanford University, is the main repository for Popper's papers. These include manuscripts of Popper's works from 1927 to 1980, correspondence from 1932 to 1987, teaching materials from 1937 to 1970, conference proceedings, letters of reference and additional biographical material. The register can be accessed via the Karl Popper Web (www.tkpw.net). Microfilmed copies of the Hoover originals are held by The British Library of Political and Economic Science at the London School of Economics, the Karl Popper Institute (Vienna), the *Karl-Popper-Sammlung* (University of Klagenfurt, Austria), and the Popper Project (Central European University, Budapest).

2
Popper, Kuhn, Lakatos and aim-oriented empiricism

2.1 Introduction

In this chapter I argue that aim-oriented empiricism (AOE), a conception of natural science that I have spelled out and defended at some length elsewhere,[1] is a kind of synthesis of the views of Popper, Kuhn and Lakatos, but is also an improvement over the views of all three. Whereas Popper's falsificationism protects metaphysical assumptions implicitly made by science from criticism, AOE exposes all such assumptions to sustained criticism, and furthermore focuses criticism on those assumptions most likely to need revision if science is to make progress. Even though AOE is, in this way, more Popperian than Popper, it is also, in some respects, more like the views of Kuhn and Lakatos than falsificationism is. AOE is able, however, to solve problems which Popper's, Kuhn's and Lakatos's views cannot solve.

AOE stems from the observation that theoretical physics persistently accepts unified theories, even though endlessly many empirically more successful, but seriously disunified, *ad hoc* rivals can always be concocted. This persistent preference for and acceptance of unified theories, even against empirical considerations, means that physics makes a persistent untestable (metaphysical) assumption about the universe: the universe is such that no seriously disunified, *ad hoc* theory is true. Intellectual rigour demands that this substantial, influential, highly problematic and implicit assumption be made explicit, as a part of theoretical scientific knowledge, so that it can be critically assessed, so that alternative versions can be considered, in the hope that this will lead to an improved version of the assumption being developed and accepted. Physics is more rigorous when this implicit assumption is made explicit even though there is no justification for holding the assumption to be true. Indeed, it is above all because there is no such justification, and the assumption is substantial, influential, highly problematic and all too likely to be false, that it becomes especially important to implement the above requirement for rigour, and make the implicit (and probably false) assumption explicit.

Once it is conceded that physics does persistently assume that the universe is such that all seriously disunified theories are false, two fundamental problems immediately arise. What precisely ought this assumption to be interpreted to be asserting about the universe? Granted that the assumption is a pure conjecture, substantial and influential but bereft of any kind of justification, and thus all too likely in its current form to be false, how can rival versions of the assumption be rationally assessed, so that what is accepted by physics is improved?

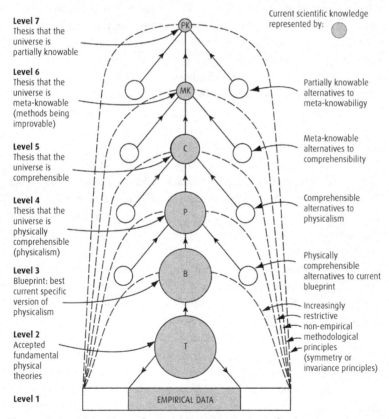

Figure 2.1 Aim-oriented empiricism (Source: author)

AOE is designed to solve, or help solve, these two problems. The basic idea is that we need to see physics (and science more generally) as making not one, but a hierarchy of assumptions concerning the unity, comprehensibility and knowability of the universe, the assumptions becoming less and less substantial as one goes up the hierarchy, and thus becoming more and more likely to be true (see Figure 2.1). The idea is that in this way we separate out what is most likely to be true, and not in need of revision, at and near the top of the hierarchy, from what is most likely to be false, and most in need of criticism and revision, near the bottom of the hierarchy. Evidence, at level 1, and assumptions high up in the hierarchy, are rather firmly accepted, as being most likely to be true (although still open to revision); this is then used to criticize, and to try to improve, theses at levels 2 and 3 (and perhaps 4), where falsity is most likely to be located.

At the top there is the relatively insubstantial assumption that the universe is such that we can acquire some knowledge of our local circumstances, sufficient to make life possible. If this assumption is false, we will not be able to acquire knowledge whatever we assume. We are justified in accepting this assumption permanently as a part of our knowledge, even though we have no grounds for holding it to be true. As we descend the hierarchy, the assumptions become increasingly substantial and thus increasingly likely to be false. At level 6 there is the assumption that the universe is such that we can discover how to improve methods for the improvement of knowledge. At level 5 there is the rather more substantial assumption that the universe is comprehensible in some way or other, the universe being such that there is just one kind of explanation for all phenomena. At level 4 there is the more specific, and thus more substantial assumption that the universe is *physically* comprehensible, it being such that there is some yet-to-be-discovered, true, unified "theory of everything". At level 3 there is the even more specific, and thus even more substantial, assumption that the universe is physically comprehensible in a more or less specific way, suggested by current accepted fundamental physical theories. Examples of assumptions made at this level, taken from the history of physics, include the following. The universe is made up of rigid corpuscles that interact by contact; it is made up of point-atoms that interact at a distance by means of rigid, spherically symmetrical forces; it is made up of a unified field; it is made up of a unified quantum field; it is made up of quantum strings. Given the historical record of dramatically changing ideas at this level, and given the relatively highly specific and substantial character of successive assumptions made at this level, we can be reasonably confident that the best assumption available at any stage in the development of physics at this level will be false, and will need future revision. At level 2 there are the accepted fundamental theories of physics, currently Einstein's theory of general relativity and the standard model.[2] Here, if anything, we can be even more confident that current theories are false, despite their immense empirical success. This confidence comes partly from the vast empirical content of these theories, and partly from the historical record. The greater the content of a proposition the more likely it is to be false; the fundamental theories of physics, general relativity and the standard model have such vast empirical content that this in itself almost guarantees falsity. And the historical record backs this up: Kepler's laws of planetary motion and Galileo's laws of terrestrial motion are corrected by Newtonian theory, which is in turn corrected by special and general relativity; classical physics is corrected by quantum theory, in turn corrected by relativistic quantum theory, quantum field theory and the standard model. Each new theory in physics reveals that predecessors are false. Indeed, if the level 4 assumption of AOE is correct, then all current physical theories are false, since this assumption asserts that the true physical theory of everything is unified, and the totality of current fundamental physical theory, general relativity plus the standard model, is notoriously disunified. Finally, at level 1 there are accepted empirical data, low-level, corroborated, empirical laws.

In order to be acceptable, an assumption at any level from 6 to 3 must (as far as possible) be compatible with, and a special case of, the assumption above in the hierarchy; at the same time it must be (or promise to be) empirically fruitful in the sense that successive accepted physical theories increasingly successfully accord with (or exemplify) the assumption. At level 2, those physical theories are accepted which are sufficiently (a) empirically successful and (b) in accord with the best available assumption at level 3 (or level 4). Corresponding to each assumption, at any level from 7 to 3, there is a methodological principle, represented by sloping dotted lines in the figure, requiring that theses lower down in the hierarchy are compatible with the given assumption.

When theoretical physics has completed its central task, and the true theory of everything, T, has been discovered, then T will (in principle) successfully predict all empirical phenomena at level 1, and will entail the assumption at level 3, which will in turn entail the assumption at level 4, and so on up the hierarchy. As it is, physics has not completed its task, T has not (yet) been discovered, and we are ignorant of the nature of the universe. This ignorance is reflected in clashes between theses at different levels of AOE. There are clashes between levels 1 and 2, 2 and 3, and 3 and 4. The attempt to resolve these clashes drives physics forward.

In seeking to resolve these clashes between levels, influences can go in both directions. Thus, given a clash between levels 1 and 2, this may lead to the modification or replacement of the relevant theory at level 2; but, on the other hand, it may lead to the discovery that the relevant experimental result is not correct for any of a number of possible reasons, and needs to be modified. In general, however, such a clash leads to the rejection of the level 2 theory rather than the level 1 experimental result; the latter are held onto more firmly than the former, in part because experimental results have vastly less empirical content than theories, in part because of our confidence in the results of observation and direct experimental manipulation (especially after expert critical examination). Again, given a clash between levels 2 and 3, this may lead to the rejection of the relevant level 2 theory (because it is disunified, *ad hoc*, at odds with the current metaphysics of physics); but, on the other hand, it may lead to the rejection of the level 3 assumption and the adoption, instead, of a new assumption (as has happened a number of times in the history of physics, as we have seen). The rejection of the current level 3 assumption is likely to take place if the level 2 theory, which clashes with it, is highly successful empirically, and furthermore has the effect of increasing unity in the totality of fundamental physical theory overall, so that clashes between levels 2 and 4 are decreased. In general, however, clashes between levels 2 and 3 are resolved by the rejection or modification of theories at level 2 rather than the assumption at level 3, in part because of the vastly greater empirical content of level 2 theories, in part because of the empirical fruitfulness of the level 3 assumption (in the sense indicated above).

It is conceivable that the clash between level 2 theories and the level 4 assumption might lead to the revision of the latter rather than the former. This happened when Galileo rejected the then current level 4 assumption of Aristotelianism, and replaced it with the idea that "the book of nature is written in the language of mathematics" (an early precursor of our current level 4 assumption). The whole idea of AOE is, however, that as we go up the hierarchy of assumptions we are increasingly unlikely to encounter error, and the need for revision. The higher up we go, the more firmly assumptions are upheld, the more resistance there is to modification.

AOE is put forward as a framework which makes explicit metaphysical assumptions implicit in the manner in which physical theories are accepted and rejected, and which, at the same time, facilitates the critical assessment and improvement of these assumptions with the improvement of knowledge, criticism being concentrated where it is most needed, low down in the hierarchy. Within a framework of relatively insubstantial, unproblematic and permanent assumptions and methods (high up in the hierarchy), much more substantial, problematic assumptions and associated methods (low down in the hierarchy) can be revised and improved with improving theoretical knowledge. There is something like positive feedback between improving knowledge and improving (low-level) assumptions and methods – that is, knowledge-about-how-to-improve-knowledge. Science adapts its nature, its assumptions and methods, to what it discovers about the nature of the universe. This, I suggest, is the nub of scientific rationality, and the methodological key to the great success of modern science.

The above is intended to be an introductory account of AOE. Further clarifications and details will emerge below when I come to expound AOE again during the course of arguing that the position can be construed to be a kind of synthesis of, and improvement over, the views of Popper, Kuhn and Lakatos.

In what follows I begin with Karl Popper and argue that AOE can be seen to emerge as a result of modifying Popper's falsificationism[3] to remove defects inherent in that position. AOE does not, however, break with the spirit of Popper's work; far from committing the Popperian sin of "justificationism", AOE is even more Popperian than Popper, in that it is a conception of science which exposes more to effective criticism than falsificationism does. Falsificationism, in comparison, shields substantial, influential and problematic scientific assumptions from criticism within science. Whereas falsificationism fails to solve what may be called the "methodological" problem of induction, AOE successfully solves the problem. And, associated with that success, AOE also solves the problem of what it means to assert of a physical theory that it is "simple", "explanatory" or "unified", a problem which falsificationism fails to solve.

The conception of science expounded by Thomas Kuhn in his *The Structure of Scientific Revolutions* (1970a) shares important elements with Popper's falsificationism. The big difference is that whereas Kuhn holds that "normal science" is an important, healthy and entirely rational (indeed, the most rational) part of science, Popper regards normal science as "dogmatic", the result of bad education and "indoctrination", something that is "a danger to science and, indeed, to our civilization" (Popper, 1970, p. 53). It is the apparent persistent dogmatism of normal science – the persistent retention of the current paradigm in the teeth of ostensible empirical refutations – that is so irrational, so unscientific, when viewed from a falsificationist perspective. AOE, however, though subjecting scientific assumptions to even greater critical scrutiny than

Popper's falsificationism, turns out to have features which are, in some respects, closer to Kuhn than to Popper. For, according to AOE, substantial and influential metaphysical assumptions are persistently accepted as a part of scientific knowledge in a way which seems much closer to the way paradigms are accepted, according to Kuhn, during normal science, than to the way falsifiable theories are to be treated in science, according to Popper. AOE depicts science as, quite properly, proceeding in a way that is reminiscent, in important respects, of Kuhn's normal science, something that is anathema to Popper's falsificationism. At the same time, AOE is free of some of the serious defects inherent in Kuhn's conception of science. Even though AOE science mimics some aspects of Kuhnian normal science, it nevertheless entirely lacks the harmful dogmatism of this kind of science, and avoids problems that arise from Kuhn's insistence that successive paradigms are "incommensurable".

Imre Lakatos's "methodology of scientific research programmes"[4] was invented, specifically, to do justice both to Popper's insistence on the fundamental importance of subjecting scientific theories to persistent, ruthless attempted empirical refutation, and to Kuhn's insistence on the importance of preserving accepted paradigms from refutation, scientists, not paradigms, being under test when ostensible refutations arise. It is, like AOE, a kind synthesis of the ideas of Popper and Kuhn. Just as AOE incorporates elements of Popper and Kuhn, so too it incorporates elements of Lakatos's research programme methodology. At the same time, AOE is an improvement over Lakatos's view; it solves problems which Lakatos's view is unable to solve. Whereas Lakatos's view provides no means for the assessment of "hard cores" (Lakatos's "paradigms") other than by means of the empirical success and failure of the research programmes to which they give rise, AOE specifies a way in which "hard cores" (or their equivalent) can be rationally, but fallibly assessed, independent of the kind of empirical considerations to which Lakatos is restricted. This has important implications for the question of whether or not there is a rational method of discovery. It also has important implications for the *strength* of scientific method. For Lakatos, notoriously, scientific method could only decide which of two competing research programmes was the better long after the event, when one had proved to be vastly superior, empirically, to the other. "Minerva's owl flies at night," as Lakatos put it, echoing Hegel. AOE provides a much more decisive methodology than Lakatos's, one which is able to deliver verdicts when they are needed, and not long after the event.

It may be thought that yet another critique of Popper, Kuhn and Lakatos is unnecessary, given the flood of literature that has appeared on the subject in the last 40 years or so.[5] My reply to this objection comes in two parts.

First, nowhere in this large body of critical literature can one find the particular line of criticism developed in the present chapter.[6] This line of criticism is, furthermore, especially fundamental and insightful in that it reveals, as other criticisms do not, what needs to be done radically to *improve* the views of Popper, Kuhn and Lakatos. Second, the improved view, namely AOE, that emerges from the criticism to be expounded here, has been entirely overlooked by the body of literature discussing and criticizing Popper, Kuhn and Lakatos. This is the decisive point. It is not enough merely to show that the views of Popper, Kuhn and Lakatos are defective. What really matters is to develop a view that overcomes these defects. That is what I set out to do here.

It is also true that, during the last 30 years, a substantial body of work has emerged on scientific method quite generally.[7] In none of these works does one find the criticism of Popper, Kuhn and Lakatos, expressed below, or the synthesis, namely AOE, which emerges from this criticism.[8] Furthermore, the methodological views developed in the works just cited all fall to the line of criticism deployed against Popper, Kuhn and Lakatos in the present chapter. There is no space to develop this last point here: it is, however, spelled out in Maxwell (1998, ch. 2) as far as works up to that date are concerned. One implication, then, of the present chapter is that philosophy of science took a wrong turning around 1974 when it failed to take up the line of argument spelled out here, an early version of which is to be found in Maxwell (1974).

2.2 Karl Popper

As I mentioned in the last chapter, Popper held that science proceeds by putting forward empirically falsifiable conjectures which are then subjected to severe attempts at falsification by means of observation and experiment. Scientific theories cannot be verified by experience, but they can be falsified. Once a theory is falsified, scientists have the task of developing a potentially better theory, even more falsifiable than its predecessor, at least as ostensibly empirically successful as its predecessor, and such that it is corroborated where its predecessor was falsified. In order to be accepted (tentatively) as a part of conjectural scientific knowledge, a theory must (at least) be empirically falsifiable. Non-falsifiable, metaphysical theses are meaningful, and may influence the direction of scientific research. There can even be what Popper has called "metaphysical research programmes" – programmes of research "indispensable for science, although their character is that of metaphysical or speculative physics rather than of scientific physics ... more in the nature of myths, or of dreams, than of science" (Popper, 1982a, p. 165). For Popper, metaphysical (that is, unfalsifiable) theses cannot be a part of (conjectural) scientific knowledge; such theses cannot help determine what is accepted and rejected as (conjectural) scientific knowledge, but they can influence ideas, choice of research aims and problems, in the context of scientific discovery. (For further details see Popper 1959, 1963, 1983).

Popper defended two distinct versions of falsificationism which, echoing terminology of Maxwell (1998), I shall call *bare* and *dressed* falsificationism. According to bare falsificationism, defended in Popper (1959), only empirical considerations, and such things as the falsifiability of theories and degrees of falsifiability, decide what is to be accepted and rejected in science. According to dressed falsificationism, a new theory, in order to be acceptable, "should proceed from some *simple, new, and powerful, unifying idea* about some connection or relation (such as gravitational attraction) between hitherto unconnected things (such as planets and apples) or facts (such as inertial and gravitational mass) or new 'theoretical entities' (such as field and particles)" (Popper, 1963, p. 241). This "requirement of simplicity" (as Popper calls it) is in addition to anything specified in *The Logic of Scientific Discovery* (*L.Sc.D.*). In *L.Sc.D.* Popper does, it is true, demand of a theory that it should be as simple as possible, but Popper there identifies degree of simplicity of a theory with degree of falsifiability. (There is a second, related notion, but Popper makes clear that if the two clash it is the falsifiability notion, just indicated, which takes priority [see Popper, 1959, p. 130]). Thus, in *L.Sc.D.*, in requiring of an acceptable theory that it should be as simple as possible, Popper is demanding no more than that it should be as falsifiable as possible. But Popper's "requirement of simplicity" of his *Conjectures and Refutations* (*C.R.*) (1963) is wholly in addition to falsifiability. A theory of high falsifiability may not "proceed from some simple, new, and powerful unifying idea", and vice versa. We thus have two versions of falsificationism before us: bare falsificationism of Popper's *L.Sc.D.* (1959), and dressed falsificationism of *C.R.* (1963, ch. 10), with the new "requirement of simplicity" added on to the 1959 doctrine.

I now give my argument for holding that neither doctrine is tenable. (I here elaborate the criticism of Popper sketched in the previous chapter.) My argument is *not* that Popper fails to show how theories can be verified, or rendered probable; nor is my argument that Popper fails to show how scientific theories can be falsified, in that falsification requires the *verification* of a low-level falsifying hypothesis (which, according to Popper, is not possible).[9] There is nothing "justificationist", in other words, about my criticism.[10] It amounts simply to this. Bare falsificationism fails dramatically to do justice to the way theories are selected in science (entirely independently of any question of verification, justification or falsification). Dressed falsificationism does better justice to scientific practice, but commits science to making substantial, influential and problematic assumptions that remain implicit, and cannot adequately be made explicit within science. Science pursued in accordance with dressed falsificationism is irrational, in other words, because it fails to implement the elementary, and quasi-Popperian, requirement for rationality that "assumptions that are substantial, influential, problematic and implicit need to be made explicit, so that they can be critically assessed and so that alternatives may be put forward and considered, in the hope that such assumptions can be improved" (Maxwell, 1998, p. 21). Dressed falsificationism fails, in other words, for good Popperian reasons: it fails to expose substantial, influential, problematic assumptions to *criticism* within science.

2.3 Refutation of bare falsificationism

Here, then, in a little more detail, is my refutation of bare falsificationism. Given any accepted physical theory, at any stage in the development of physics, however empirically successful (however highly corroborated) – Newtonian theory, say, or classical electrodynamics, quantum theory, general relativity, quantum electrodynamics, chromodynamics or the standard model – there will always be endlessly many rival falsifiable theories that can easily be formulated which will fit the available data just as well as the accepted theory. Taking Newtonian theory (NT) as an example of an accepted theory, here are two examples of rival theories. NT*: "Everything occurs as NT asserts, until the first second of the year 2100, when an inverse cube law of gravitation will abruptly hold." NT**: "Everything occurs as NT asserts, except for systems consisting of gold spheres, each having a mass of 1,000 tons, interacting with each other gravitationally in outer space, in a vacuum, within a spherical region of 10 miles: for these systems, Newton's law of gravitation is repulsive, not attractive." (For further examples and discussion, see Maxwell, 1998, pp. 47–54). It is easy to see that there are infinitely many such rivals to NT, just as empirically successful (at the moment) as

NT. The predictions of NT may be represented as points in a multidimensional space, each point corresponding to some specific kind of system (there being infinitely many points). NT has only been verified (corroborated) for a minute region of this space. In order to concoct a (grossly *ad hoc*) rival to NT, which is just as empirically successful as NT, all we need do is identify some region in this space that includes no prediction of NT that has been verified, and then modify the laws of NT arbitrarily, for just that identified region.

The crucial question now is this: On what basis does bare falsificationism reject all these falsifiable but unfalsified rival theories? According to bare falsificationism, T_2 is to be accepted in preference to T_1 if T_1 has been falsified, T_2 has greater empirical content (is more falsifiable) than T_1, T_2 successfully predicts all that T_1 successfully predicts, T_2 successfully predicts the phenomena that falsified T_1, and T_2 successfully predicts new phenomena not predicted by T_1 (see Popper, 1959, pp. 81–4 and elsewhere). Given NT, it is a simple matter to concoct rival theories, of the above type, that satisfy all the above bare falsificationist requirements for being more acceptable than NT. Most accepted physical theories yield empirical predictions that clash with experiments, and thus are ostensibly falsified. We can always concoct new theories, in the way just indicated, doctored to yield the "correct" predictions. We can add on independently testable auxiliary postulates, thus ensuring that the new theory has greater empirical content than the old one. And no doubt this excess content will be corroborated. For details of how this can be done, see Maxwell (1998, pp. 52–4). Such theories are, of course, grossly *ad hoc*, grossly "aberrant" as I have called them, but they satisfy Popper's (1959) requirements for being better theories than accepted physical theories.

It is worth noting that such "better" theories need not be quite as wildly *ad hoc* as the ones indicated above; sometimes such theories are actually put forward in the scientific literature, and yet are not taken seriously, even by their authors, let alone by the rest of the scientific community. An example is an *ad hoc* version of NT put forward by Maurice Levy in 1890, which combined in an *ad hoc* way two distinct modifications of Newton's law of gravitation, one based on the way Weber had proposed Coulomb's law should be modified, the other based on the way Riemann had proposed Coulomb's law should be modified (for details see North, 1965). By 1890, NT had been refuted by observation of the precession of the perihelion of the orbit of Mercury; attempts to salvage NT by postulating an additional planet, Vulcan, had failed. Levy's theory successfully predicted all the success of NT, and in addition successfully predicted the observed orbit of Mercury, just that which refuted NT; in addition, of course, it made predictions different from NT for further Sun–Mercury type systems not yet observed. Despite this, Levy's theory was not taken seriously for a moment, not even by Levy himself. How can bare falsificationism recommend rejection of such *ad hoc* versions of NT when they satisfy all the requirements of bare falsificationism for being more acceptable theories? No adequate answer is forthcoming, and it is this which spells the downfall of bare falsificationism – as Popper may himself have realized when he put forward dressed falsificationism in his *C.R.* (1963, ch. 10).

Note, again, that this criticism of Popper has nothing justificational about it whatsoever: it simply points to the drastic failure of bare falsificationism to do justice to what actually goes on in physics.

It may be objected that *ad hoc* rivals to NT of the kind just considered are so silly, so crackpot, that they do not *deserve* to be taken seriously within physics.[11] This is of course correct. The crucial point, however, is that bare falsificationism ought to be able to deliver this verdict, and this it singularly fails to do. Bare falsificationism actually declares of appropriately concocted *ad hoc* rivals to NT that these are *better, more acceptable* than NT.

But can a criticism of Popper that appeals to such silly, crackpot theories be taken seriously? I have two replies to this question. First, not all the *ad hoc* or aberrant variants are entirely silly. Levy's theory is perhaps an example. There are degrees of *ad hocness*, from the utterly crackpot and absurd, to a degree of *ad hocness* so slight, so questionable, in comparison, that the issue of whether the theory really is *ad hoc* or not may be hotly disputed by physicists themselves. (Such disputes arise especially during scientific revolutions.) This is an important point which will have a bearing on the argument of the next section. Second, it is, I submit, the very silliness of these crackpot theories that makes the above criticism of Popper so serious. If bare falsificationism favoured T_1 over T_2, while most scientists favoured T_2 over T_1, even though admitting that T_1 is nevertheless a good theory, almost as acceptable as T_2, bare falsificationism would not be in such trouble. What is lethal for bare falsificationism is that it declares T_1 to be better than T_2 in circumstances where scientists themselves (and all of us) can see that T_2 is vastly superior to T_1, T_1 being grossly *ad hoc*, aberrant, wholly crackpot and silly. Bare falsificationism favours theories that receive, and deserve, instant rejection: there could scarcely be a more decisive falsification of falsificationism than that.

2.4 Refutation of dressed falsificationism

Having argued that Popper's (1959) bare falsificationism is untenable, I turn my attention now to Popper's (1963, chapter 10) doctrine of dressed falsificationism. As I have mentioned, this adds on to the (1959) doctrine Popper's new "requirement of simplicity" (Popper, 1963, p. 241) (see section 2.2 above).

As long as there is no serious ambiguity as to what proceeding "from some simple, new, and powerful, unifying idea" means, it is at once clear that the new doctrine is able to exclude from science all the empirically successful but *ad hoc*, aberrant, crackpot, silly theories, of the kind discussed above. They do not proceed "from some simple ... unifying idea", and are to be rejected on that account, whatever their empirical success may be, even if this empirical success is greater than accepted scientific theories.

However, adopting Popper's new "principle of simplicity" as a basic methodological principle of science has the effect of permanently excluding from science all *ad hoc* theories that fail to satisfy the principle, however empirically successful such theories might be if considered. This amounts to assuming permanently that the universe is such that all *ad hoc* theories that fail to satisfy Popper's principle of simplicity are false – granted that a basic aim of science is truth. It amounts to accepting, as a permanent item of scientific knowledge, the substantial metaphysical thesis that the universe is non-*ad hoc*, in the sense that all theories that fail to satisfy Popper's principle of simplicity are false, however empirically successful they might turn out to be if considered. But this, of course, clashes with Popper's criterion of demarcation: that no unfalsifiable, metaphysical thesis is to be accepted as a part of scientific knowledge.

If the demarcation principle is upheld, then the metaphysical thesis just indicated, asserting that the universe is non-*ad hoc*, remains implicit in the permanent adoption of Popper's principle of simplicity as a basic methodological principle of science. (And this is the way Popper himself seems to have conceived the matter: he says of metaphysical research programmes that they are "often held unconsciously", and "are implicit in the theories and in the attitudes and judgements of the scientists" [Popper, 1982b, p. 161].) But in leaving the metaphysical thesis of non-*ad hocness* implicit in the methodological principle of simplicity, science violates an elementary requirement for rationality, already mentioned, according to which "assumptions that are substantial, influential, problematic and implicit need to be made explicit, so that they can be critically assessed and so that alternatives may be put forward and considered, in the hope that such assumptions can be improved" (Maxwell, 1998, p. 21). The non-*ad hoc* metaphysical assumption may, after all, be false. We may need to adopt a modified version of the assumption. It may be essential for the progress of science that this assumption is modified. Just this turns out to be the case, given certain formulations of the assumption, as we shall see below. In leaving the non-*ad hoc* metaphysical assumption implicit in the adoption of the methodological principle of simplicity, dressed falsificationism protects this substantial, influential and highly problematic assumption from *criticism*, from the active consideration of alternatives.[12]

Dressed falsificationism fails, in other words, for good Popperian reasons: it is inconsistent, in that the untestable, metaphysical thesis that the universe is non-*ad hoc* is accepted implicitly as a part of conjectural scientific knowledge, in conflict with the principle of demarcation; and it lacks rigour, in that it protects this implicit, metaphysical assumption from explicit criticism within the intellectual domain of science.

Here again, it should be noted, there is nothing justificationist about this criticism of Popper's dressed falsificationism. On the contrary, what the argument shows is that dressed falsificationism protects a substantial, influential, problematic but implicit assumption from criticism within science: Popper's doctrine fails for the good Popperian reason of restricting criticism.

It may be objected that adopting Popper's methodological principle of simplicity does *not* commit science to making a substantial metaphysical assumption about the universe – namely that it is such that no falsifiable theory, however empirically successful, which fails to satisfy the principle, is true. But I do not see how such an objection can be valid. Suppose, instead of adopting Popper's principle, science adopted the principle that in order to be acceptable, a new physical theory must postulate that the universe is made up of atoms. This methodological principle is upheld in such a way that even though theories are available which postulate fields rather than atoms, and which are much more empirically successful than any atomic theory, nevertheless these rival field theories are all excluded from science. Would it not be clear that science, in adopting and implementing the methodological principle of atomicity in this way, is making the assumption that the universe is made up of atoms, whether this is acknowledged or not? How can this be denied? Just the same holds if science adopts and implements Popper's methodological principle of simplicity.

Popper might have tried to wriggle out of accepting this conclusion by pointing to the fact that he only declared that a new theory, in order to be acceptable, "should" proceed from some simple, unifying idea. It is desirable, but not essential, that new theories should satisfy this principle. The principle is relevant to the context of discovery, perhaps, but not to the context of acceptance and rejection. (It is a heuristic principle, not a methodological one.) But if Popper's doctrine is interpreted in this way, it immediately fails to overcome the objections spelled out in section 2.3 above. Either falsificationism adopts Popper's principle of simplicity as a *methodological* principle, or it does not. If it does, it encounters the objections just indicated; if it does not, it encounters the objections of section 2.3.

2.5 From falsificationism to aim-oriented empiricism

The conclusion to be drawn from the argument of the last two sections is that science is more rational, more intellectually rigorous, if it makes explicit, as a criticizable tenet of (conjectural) scientific knowledge, that substantial, influential and problematic metaphysical thesis which is implicit in the way physics persistently rejects *ad hoc* theories, however empirically successful they may be.[13] At once two important new problems leap to our attention. What, precisely, does this metaphysical thesis assert? And on what grounds is it to be (conjecturally) accepted as a part of scientific knowledge?

As far as the first of the above two problems is concerned, a wide range of metaphysical theses are available. As I indicated in section 2.3 above, *ad hoc* theories range from the utterly crackpot and silly, to theories that are only somewhat lacking in simplicity or unity. At one extreme, we might adopt a metaphysical thesis that excludes only utterly silly theories; at the other extreme, we might adopt the thesis that the universe is physically comprehensible in the sense that it has a unified dynamic structure, some yet-to-be-discovered unified physical "theory of everything" being true – a thesis that I shall call "physicalism". We might even adopt some specific version of physicalism, which asserts that the underlying physical unity is of a specific type: it is made up of a unified field perhaps, or a quantum field, or empty topologically complex curved space-time, or a quantum string field. Other things being equal, the more specific the thesis (and thus the more it excludes) so the more likely it is to be false, whereas the more unspecific it is so the more likely it is to be true. It is not, it seems, at all clear what metaphysical thesis we should take science to be presupposing.

As far as the second of the above two problems is concerned, it could be argued that grounds for accepting the metaphysical conjecture of physics (whatever precisely it may be) come from the fact that physics inevitably accepts this conjecture in persistently accepting unified theories in preference to empirically more successful disunified rivals, and it is more rigorous to acknowledge the conjecture than to disavow it. No more is required to render acceptance of the conjecture rational. But this is hardly satisfactory. It would always be possible to formulate an aberrant version of the metaphysical conjecture – a version which asserts that the universe exhibits lawful unity until the year 2090, let us say, when quite different physical laws will begin to operate. If this conjecture is accepted, physical theories would be accepted which would be quite different from the ones we do accept, but nevertheless just as empirically successful (until 2090 at least). There does need to be some reason for accepting the unified metaphysical conjecture we do implicitly accept, in preference to aberrant versions of this conjecture – a reason more substantial than "this conjecture is implicit in what theoretical physicists actually do".

We cannot hope to provide an argument that establishes that the unified metaphysical conjecture of physics is true, or probably true. Any such argument that ignores experience and is entirely *a priori* is surely entirely impossible. What could the premises of such an argument be? On the other hand, any such argument that appeals to experience, to empirical science, also seems impossible. Any attempt to establish the truth of the metaphysical conjecture of lawful unity by an appeal to the success of science can always be rebutted by the counter-claim that the aberrant version of the metaphysical conjecture (which postulates radical change in 2090) would receive just as much support from an empirically equally successful aberrant science.

There does not seem to be much hope either for a Kantian argument along the lines that experience cannot refute order in the universe because experience, in order to be conscious, must exhibit some order. We can easily imagine a universe that is sufficiently orderly for conscious experience, and life, to be possible, but in which only disunified, aberrant laws, of one kind or another, hold. Arguments along Kantian lines do not provide grounds for accepting a metaphysical conjecture sufficiently contentful to exclude empirically successful aberrant physical theories.

One possibility is to argue that it is rational to accept a metaphysical conjecture which is such that the truth of the conjecture is required for the pursuit of knowledge to be possible at all. Even though there would be no argument for the *truth* of the thesis, there would be an argument for *accepting* the thesis as a part of knowledge since, if the thesis is false, we cannot acquire knowledge whatever we assume. Accepting the thesis as a part of scientific knowledge cannot imperil or adversely affect science whatever the universe is like.

The problem with this argument is that it does not provide grounds for accepting a metaphysical conjecture that has sufficient content to exclude empirically successful aberrant physical theories of the kind that physics does, in practice, exclude (or just ignore). We can easily imagine a universe which is such that there is no underlying unity of physical law, and yet human life is possible, and new knowledge can be acquired.

A rather more Popperian argument would be that we should accept that metaphysical conjecture which holds out the greatest promise of scientific progress. But any such argument faces the difficulty that there will always be equally valid arguments for aberrant metaphysical conjectures that promise success for aberrant science (science with a succession of empirically successful aberrant theories).

In attempting to justify acceptance of a metaphysical conjecture of science, there are four considerations that we may appeal to, three of which are wholly Popperian in spirit if not in the letter of Popperian doctrine:

(1) If some metaphysical thesis, M, is implicit in some scientific methodological practice, then science is more rigorous if M is made explicit, since this facilitates criticism of it, the consideration of alternatives.

(2) A metaphysical thesis may be such that its truth is a necessary condition for it to be possible for us to acquire knowledge: if so, accepting the thesis can only help, and cannot undermine, the pursuit of knowledge of truth.

(3) Given two rival metaphysical theses, M_1 and M_2, it may be the case that M_1 supports an empirical scientific research programme that has apparently met with far greater empirical success than any rival empirical research programme based on M_2: in this case we may favour M_1 over M_2, at least until M_2, or some third thesis, M_3, shows signs of supporting an even more empirically progressive research programme.[14]

(4) M_1 may be preferred to M_2 on the grounds that it gives greater promise of supporting an empirically progressive research programme.

The above discussion has revealed that these four considerations fail to provide a justification for accepting the metaphysical thesis of physics.

To sum up. Intellectual rigour requires that physics acknowledges that there is a substantial metaphysical thesis implicit in its persistent acceptance of unified theories even though endlessly many empirically more successful disunified rivals are always potentially available. Two problems arise, however, once this metaphysical thesis of physics is acknowledged. What should we take this metaphysical thesis to assert? And what grounds are there for accepting it as a part of scientific knowledge? So far, no satisfactory solution to these two problems has been forthcoming.

These two problems can be solved, however, if physics is construed as adopting, not just one metaphysical conjecture, but a hierarchy of such conjectures concerning the comprehensibility and knowability of the universe, these conjectures becoming more and more insubstantial as one ascends the hierarchy, more and more likely to be true (see Figure 2.1). We need, in short, to adopt the hierarchical view sketched in section 2.1. This hierarchical view of aim-oriented empiricism (AOE) is a radical improvement over Popper's falsificationism. In this section I expound AOE (in a little more detail than the introductory exposition of section 2.1) and indicate how it solves the two problems just mentioned; I indicate further how it solves the methodological problem of induction and the related problem of simplicity, and then consider possible objections.

At level 7 there is the thesis that the universe is such that we can continue to acquire knowledge of our local circumstances, sufficient to make life possible. At level 6 there is the more substantial thesis that there is some rationally discoverable thesis about the nature of the universe which, if accepted, makes it possible progressively to improve methods for the improvement of knowledge. "Rationally discoverable", here, means at least that the thesis is not an arbitrary choice from infinitely many analogous theses. At level 5 we have the even more substantial thesis that the universe is *comprehensible* in some way or other, whether physically or in some other way. This thesis asserts that the universe is such that there is *something* (God, a tribe of gods, a cosmic goal, a physical entity, a cosmic programme or whatever), which exists everywhere in an unchanging form and which, in some sense, determines or is responsible for everything that changes (all change and diversity in the world in principle being explicable and understandable in terms of the underlying unchanging *something*). A universe of this type deserves to be called "comprehensible" because it is such that everything that occurs, all change and diversity, can in principle be explained and understood as being the outcome of the operations of the one underlying *something*, present throughout all phenomena. At level 4 we have the still more substantial thesis that the universe is *physically* comprehensible in some way or other (a thesis I shall call *physicalism*[15]). This asserts that the universe is made up of one unified self-interacting physical entity (or one kind of entity), all change and diversity being in principle explicable in terms of this entity. What this amounts to is that the universe is such that some yet-to-be-discovered unified physical theory of everything is true. At level 3 we have an even more substantial thesis, the best currently available specific idea as to how the universe is physically comprehensible. This asserts that everything is made of some specific kind of physical entity: corpuscle, point-particle, classical field, quantum field, convoluted space-time, string or whatever. Because the thesis at this level is so specific, it is almost bound to be false (even if the universe is physically comprehensible in some way or other). Here, ideas evolve with evolving knowledge. At level 2 we have our best fundamental physical theories, currently general relativity and the so-called standard model, and at level 1 we have empirical data (low-level experimental laws).

The thesis at the top of the hierarchy, at level 7, is such that, if it is false, knowledge cannot be acquired whatever is assumed. This thesis is, quite properly, accepted as a permanent part of scientific knowledge, even though we have no reason to suppose that it is true, since accepting it can only help, and cannot hinder, the acquisition of knowledge whatever the universe is like.

I have two arguments for the acceptance of the thesis of meta-knowability, at level 6:

(i) Granted that there is *some* kind of general feature of the universe which makes it possible to acquire knowledge of our local environment (as guaranteed by the thesis at level 7), it is reasonable to suppose that we do not know all that there is to be known about the *nature* of this general feature. It is reasonable to suppose, in other words, that we can improve our knowledge about the nature of this general feature, thus improving methods for the improvement of knowledge. Not to suppose this is to assume, arrogantly, that we already know all that there is to be known about how to acquire new knowledge. Granted that learning is possible (as guaranteed by the level 7 thesis), it is reasonable to suppose that, as we learn more about the world, we will learn more about how to learn. Granted the level 7 thesis, in other words, meta-knowability is a reasonable conjecture.

(ii) Meta-knowability is too good a possibility, from the standpoint of the growth of knowledge, not to be accepted initially, the idea only being reluctantly abandoned if all attempts at improving methods for the improvement of knowledge fail.

These two arguments for accepting meta-knowability are, admittedly, weak. It is crucial, however, that these two arguments make no appeal to the success of science, for a reason that will become apparent in a moment.

The thesis that the universe is comprehensible, at level 5 is accepted because no rival thesis, at that level, has been so fruitful in leading to empirically progressive research programmes. It is hardly an exaggeration to say that all empirically successful research programmes into natural phenomena have been organized around the search for explanatory theories, of one kind or another. Aberrant rivals to the thesis of comprehensibility, which might be construed as supporting aberrant empirically successful research programmes, are rejected because of incompatibility with the thesis of meta-knowability at level 6. Such rival ideas are not "rationally discoverable" in that each constitutes an arbitrary choice from infinitely many equivalent rivals.

Physicalism at level 4 is accepted because it is by far the most empirically fruitful thesis at that level that is compatible with the thesis of comprehensibility, at level 5.

Since the scientific revolution of the seventeenth century, all new fundamental physical theories have enhanced the overall unity of theoretical physics. Thus Newtonian theory (NT) unifies Galileo's laws of terrestrial motion and Kepler's laws of planetary motion (and much else besides). Maxwellian classical electrodynamics (CEM) unifies electricity, magnetism and light (plus radio, infrared, ultraviolet, X-rays and gamma rays). Special relativity (SR) brings greater unity to CEM (in revealing that the way one divides up the electromagnetic field into the electric and magnetic fields depends on one's reference frame). SR is also a step towards unifying NT and CEM in that it transforms space and time so as to make CEM satisfy a basic principle fundamental to NT, namely the (restricted) principle of relativity. SR also brings about a unification of matter and energy, via the most famous equation of modern physics, $E = mc^2$, and partially unifies space and time into Minkowskian space-time. General relativity (GR) unifies space-time and gravitation, in that, according to GR, gravitation is no more than an effect of the curvature of space-time. Quantum theory (QM) and atomic theory unify a mass of phenomena having to do with the structure and properties of matter, and the way matter interacts with light. Quantum electrodynamics unifies QM, CEM and SR. Quantum electroweak theory unifies (partially) electromagnetism and the weak force. Quantum chromodynamics brings unity to hadron physics (via quarks) and brings unity to the eight kinds of gluon of the strong force. The standard model unifies to a considerable extent all known phenomena associated with fundamental particles and the forces between them (apart from gravitation). The theory unifies to some extent its two component quantum field theories, in that both are locally gauge invariant – the symmetry group being $U(1) \times SU(2) \times SU(3)$. String theory, or M-theory, holds out the hope of unifying all phenomena. All these theories have been accepted because they progressively (a) *increase* the overall unity of theoretical physics and (b) *increase* the predictive power of physical theory, (a) being as important as (b). Physicalism is the key, persisting thesis of the entire research programme of theoretical physics since Galileo, and no obvious rival thesis, at that level of generality, can be substituted for physicalism in this research programme.

It may be asked how this succession of theories can reinforce physicalism when the totality of physical theory has always, up until now, clashed with physicalism. The answer: if physicalism is true, then all physical theories that only unify a restricted range of phenomena, must be false. Granted the truth of physicalism, and granted that theoretical physics advances by putting forward theories of limited but ever increasing empirical scope, then it follows that physics will advance from one false theory to another (as it has done: see point 7 of section 2.6 below), all theories being false until a unified theory of everything is achieved (which just might be true). The successful pursuit of physicalism requires progressive increase in both empirical scope and unity of the totality of fundamental physical theory. It is just this which the history of physics, from Galileo to today, exemplifies – thus demonstrating the unique fruitfulness of physicalism.

At level 3 that metaphysical thesis is accepted which is the best specific version of physicalism available, that seems to do the best justice to the evolution of physical theory. Two considerations govern acceptance of testable fundamental dynamical physical theories. Such a theory must be such that (i) it, together with all other accepted fundamental physical theories, exemplifies, or is a special case of, the best available metaphysical blueprint (at level 3), and physicalism (at level 4) to a sufficiently good extent, and (ii) it is sufficiently successful empirically (where empirical success is to be understood, roughly, in a Popperian sense).[16]

How does this hierarchical view of AOE overcome the problems and difficulties, indicated above, that confront any view which holds that science makes just one, possibly composite metaphysical assumption, at just one level? Given the one-thesis view, it must remain entirely uncertain as to what the one thesis should be. If it is relatively contentful and precise, more or less equivalent to the current level 3 thesis of AOE, then it is all too likely that this is false, and will need to be replaced in the future. If it is relatively contentless and imprecise, equivalent to theses at levels 7 or 6, this will not be sufficiently precise to exclude empirically successful but grossly *ad hoc*, aberrant theories. Even the level 4 thesis of physicalism is both too contentful and precise, and not contentful and precise enough. Physicalism may be false, and may need to be revised. At the same time, physicalism lacks the potential heuristic power to suggest good new fundamental theories which the more precise and contentful theses at level 3 possess. All these difficulties are avoided by the hierarchical view of AOE, just because of the hierarchy of assumptions, graded from the relatively contentless, imprecise and permanent at the top, to the relatively contentful, precise and impermanent (but methodologically and heuristically fruitful) at the bottom.

Any one-thesis view faces the even more serious problem of how this one thesis is to be critically assessed, revised and improved. The hierarchical view of AOE overcomes this problem by providing severe constraints on *what* is to be revised, and *how* this revision is to proceed. In the first instance, and even then in quite exceptional circumstances, only the current level 3 thesis can be revised. This revision must proceed, however, within constraints provided by the level 4 thesis of physicalism, on the one hand, and accepted, empirically successful level 2 theories, on the other hand. In a really exceptional situation, scientific progress might require the revision of the level 4 thesis of physicalism, but this too would proceed within the constraints of the thesis at level 5, and empirically successful theories at level 2, or empirically progressive research programmes at levels 2 and 3. The great merit of AOE is that it separates out what is most likely to be true from what is most likely to be false in the metaphysical assumptions of physics, and employs the former to assess critically, and to constrain, theses that fall into the latter category. It concentrates criticism and innovation where it is most likely to promote scientific progress.

Finally, any one-thesis view cannot, as we have seen, simultaneously call upon principles (1) to (4) above to justify acceptance of the single thesis, whatever it may be. The hierarchical view of AOE is able to do just that. It can appeal to *different* principles, (1) to (4) above, to justify (to provide a rationale for) acceptance of the *different* theses at the different levels of the hierarchy of AOE.[17] Thus acceptance of the thesis at level 7 is justified by an appeal to (2); acceptance of theses at levels 3 to 5 are accepted as a result of (a) an appeal to (3), and (b) compatibility with the thesis above in the hierarchy. The thesis at level 6 is accepted as a result of an appeal to (4). Aberrant rivals to theses accepted at levels 3 to 5 (which might be construed to support aberrant, rival empirically progressive research programmes) are excluded on the grounds that these clash with the thesis at level 6.[18]

It may be objected that AOE suffers from vicious circularity, in that acceptance of physical theories is justified by (in part) an appeal to physicalism, the acceptance of which is justified, in turn, by the empirical success of physical theory. My reply to this objection is that the level 6 thesis of meta-knowability asserts that the universe is such that this kind of circular methodology, there being positive feedback between metaphysics, methods and empirically successful theories, is just what we need to employ in order to improve our knowledge. The thesis of meta-knowability, if true, justifies implementation of AOE. This response is only valid, of course, if reasons for accepting the level 6 thesis of meta-knowability do not themselves appeal to the success of science (which would just reintroduce vicious circularity at a higher level). As I made clear above, the two arguments given for accepting meta-knowability make no appeal to the success of science whatsoever.[19]

A basic idea of AOE is to channel or direct criticism so that it is as fruitful as possible, from the standpoint of aiding progress in knowledge. The function of criticism within science is to promote scientific progress. When criticism demonstrably cannot help promote scientific progress, it becomes irrational (the idea behind (2) above). In an attempt to make criticism as fruitful as possible, we need to try to direct it at targets which are the most fruitful, the most productive, to criticize (from the standpoint of the growth of knowledge). This is the basic idea behind the hierarchy of AOE. Conjectures at all levels remain open to criticism. But, as we ascend the hierarchy, conjectures are less and less likely to be false; it is less and less likely that criticism, here, will help promote scientific knowledge. The best currently available level 3 conjecture is almost bound to be false: at this level, the history of physics reveals, as I have indicated above, that a number of different conjectures have been adopted and rejected in turn. Here, criticism, the activity of developing alternatives (compatible with physicalism), is likely to be immensely fruitful for progress in theoretical physics. Indeed, elsewhere I have argued that this provides physics with a rational, though fallible and non-mechanical method for the discovery of new fundamental physical theories,[20] a method invented and exploited by Einstein in discovering special and general relativity (see the next chapter) something which Popper has argued is not possible (see Popper, 1959, pp. 31–2). Criticizing physicalism, at level 4, may also be fruitful for physics,[21] but (the conjecture of AOE is that) this is not as likely to be as fruitful as criticism at level 3. And, as we ascend the hierarchy (so AOE conjectures), criticism becomes progressively less and less likely to be fruitful. Against that, it must be admitted that the higher in the hierarchy we need to modify our ideas, so the more dramatic would be the intellectual revolution that this would bring about. If physicalism is rejected altogether, and some quite different version of the level 5 conjecture of comprehensibility is adopted instead, the whole character of natural science would change dramatically; physics, as we know it, might even cease to exist.

The biggest change, in moving from falsificationism to AOE, has to do with the role of metaphysics in science, and the scope of scientific knowledge. According to falsificationism, untestable metaphysical theses may influence scientific research in the context of discovery, and may even lead to metaphysical research programmes; they cannot, however, be a part of scientific knowledge itself. But according to AOE, the metaphysical theses at levels 3 to 7 are all part of current (conjectural) scientific knowledge. This is the case, in particular, as far as physicalism is concerned. According to AOE, it is a part of current scientific knowledge that the universe is physically comprehensible – certainly not the case granted falsificationism.

Another important change has to do with the relationship between science and the philosophy of science. Falsificationism places the study of scientific method, the philosophy of science, outside science itself, in accordance with Popper's demarcation principle. AOE, by contrast, makes scientific method and the philosophy of science an integral part of science itself. The activity of tackling problems inherent in the aims of science, at a variety of levels, and of developing new possible aims and methods, new possible more specific or less specific philosophies of science (views about what the aims and methods of science ought to be) is, according to AOE, a vital research activity of science itself. But this is also philosophy of science, being carried on within the framework of AOE.[22]

AOE differs in many other important ways from Popper's falsificationism, whether bare or dressed (see Maxwell, 1998). Nevertheless the impulse, the intellectual aspirations and values, behind the hierarchical view of AOE are, as I have tried to indicate, thoroughly Popperian in character and spirit. The whole idea is to turn implicit assumptions into explicit conjectures in such a way that criticism may be directed at what most needs to be criticized from the standpoint of aiding progress in knowledge, so that conjectures may be developed and adopted that are the most fruitful in promoting scientific progress, at the same time no substantial conjecture, implicit or explicit, being held immune from critical scrutiny.

2.6 Aim-oriented empiricism: an improvement over falsificationism

AOE is also, in a number of ways, a considerable improvement over Popper's falsificationism:

1. *Consistency*. Bare falsificationism fails dramatically to do justice to scientific practice, and is an inherently unworkable methodology, in any case. (In what follows I shall mostly ignore bare falsificationism as obviously untenable, and concentrate on comparing dressed falsificationism and AOE.) Dressed falsificationism does better justice to scientific practice, but at the cost of consistency; persistent rejection of empirically successful theories that do not "proceed from some simple ... unifying idea" commits science to accepting a metaphysical thesis of simplicity as a part of scientific knowledge (though this is not recognized); this contradicts Popper's demarcation principle. AOE is free of such lethal defects.

2. *Criticism*. Pursuing physics in accordance with dressed falsificationism protects the implicit metaphysical thesis of simplicity from criticism within science itself, just because this thesis is metaphysical (and therefore not a part of science) and implicit (and therefore not available for sustained, explicit critical scrutiny). AOE, by contrast, is specifically designed to provide a framework of metaphysical assumptions and corresponding methodological rules within which level 3 metaphysical blueprints may be developed, and critically assessed, within science.

3. *Rigour*. Science pursued in accordance with AOE is more rigorous than science pursued in accordance with falsificationism. An elementary, but important requirement for rigour is that assumptions that are substantial, influential, problematic and implicit need to be made explicit so that they can be criticized, and so that alternatives can be considered.

If the attempt is made to do science in accordance with falsificationism, bare or dressed, one substantial, influential and problematic assumption must remain implicit (as we have just seen), namely the metaphysical assumption that nature behaves as if simple or unified, no *ad hoc* theory being true. This is implicit in the adoption of the methodological simplicity principle of dressed falsificationism. AOE, by contrast, makes this implicit assumption explicit, and provides a framework within which rival versions can be proposed and critically assessed.

4. *Simplicity*. Falsificationism fails to say what the simplicity of a theory *is*. Bare falsificationism provides an account of simplicity in terms of falsifiability, but we have already seen that this account is untenable. Popper's (1963) "requirement of simplicity" appeals to a conception of simplicity or unity that is wholly in addition to falsifiability, but does not explain what the simplicity or unity of a theory is. It fails to explain how the simplicity of a theory can possibly be methodologically or epistemologically significant when a simple theory can always be made complex by a suitable change of terminology, and vice versa. Popper himself recognized the inadequacy of his simplicity requirement when he called it "a bit vague", said that "it seems difficult to formulate it very clearly" and acknowledged that it threatened to involve one in an infinite regress (Popper, 1963, p. 241). By contrast, AOE solves the problems of explaining what the simplicity or unity of a theory is without difficulty, as we shall see in Chapter 4. Put very briefly, in order to be unified, a physical theory must be such that its *content*, what it asserts about the world, must be the same throughout all the phenomena to which the theory applies. If the content of a physical theory, T, splits into N distinct regions, so that the content of any one region is different from what it is in all the others, then T is disunified to degree N. For perfect unity we require $N = 1$. Because what matters is content, not form, the way T is *formulated* is irrelevant to this way of assessing simplicity or unity. Falsificationism cannot avail itself of this way of assessing unity because to do so makes it abundantly clear that acceptance of unified theories only, in this sense of "unified", when endlessly many empirically more successful disunified rivals are available, involves making a persistent metaphysical assumption about the nature of the universe. Within AOE, there is a second way in which the unity of T may be assessed: in terms of the extent to which the content of T exemplifies the best available level 3 metaphysical blueprint. This second conception of simplicity or unity evolves with the evolution of level 3 ideas. As we improve our ideas about how the universe is unified, with the advance of knowledge in theoretical physics, so

non-empirical methods for selecting theories on the basis of simplicity or unity improve as well. Thus current symmetry principles of modern physics, such as Lorentz invariance and gauge invariance, which guide acceptance of theory, are an advance over simplicity criteria upheld by Newton.[23] Ultimately, as we shall see in Chapter 4, requirements of theoretical unity, as these apply in physics, need to be applied to the totality of fundamental physical theory in physics, and not to individual theories.

5. *Scientific method*. Dressed falsificationism acknowledges (correctly) that two considerations govern selection of theory in science, namely considerations that have to do with (a) evidence, and (b) simplicity. But because it cannot solve the problem of what simplicity *is*, dressed falsificationism cannot, with any precision, specify what methods are involved when theories are selected on the basis of simplicity. Nor can the view do justice to the way in which the methods of physics evolve with evolving knowledge, especially methods that assert that acceptable theories must satisfy this or that *symmetry*. In other words, falsificationism fails to solve what may be called the "methodological" problem of induction: the problem of specifying, merely, what the methods are that are employed by science in accepting and rejecting theories (leaving aside the further problem of justifying these methods given that the aim is to acquire knowledge). AOE, by contrast, solves the problem of simplicity, and thus can specify precisely what methods are involved when theories are selected on the basis of simplicity or unity. Furthermore, AOE can do justice to *evolving* criteria of simplicity (as we have just seen), and hence evolving methods. According to AOE, the totality of fundamental physical theory, T, can be assessed (i) by considering the extent to which its content is unified, and thus exemplifies the fixed level 4 thesis of physicalism, or (ii) by considering the extent to which its content exemplifies the evolving, best available level 3 thesis. Whereas (i) constitutes a fixed criterion of unity (as long as physicalism is not modified), (ii) constitutes an evolving criterion, a criterion of unity that improves with improving knowledge.

6. *Evolving aims and methods*. A point, briefly alluded to in 4 and 5 above, deserves further emphasis. As physics has evolved, from Newton's time to today, non-empirical methods, determining what theories will be accepted and rejected, have evolved as well. Newton, in his *Principia*, formulated four rules of reasoning, three of which are concerned with simplicity (Newton, 1962, vol. 2, pp. 398–400). Principles that have been proposed since include: invariance with respect to position, orientation,

time, uniform velocity, charge conjugation, parity and time-reversal; principles of conservation of mass, momentum, angular momentum, energy and charge; Lorentz invariance; Mach's principle; the principle of equivalence; principles of gauge invariance, global and local; supersymmetry; duality principles; the principle that different kinds of particle should be reduced to one kind, and different kinds of force should be reduced to one kind; and the principle that space-time on the one hand, and particles and forces on the other, should be unified. All of these principles can be interpreted as methodological rules which specify requirements theories must meet in order to be accepted. They can also be interpreted as physical principles, making substantial assertions about such things as space, time, matter and force. Some, such as conservation of mass, parity and charge conjugation, have been shown to be false; others, such as Mach's principle, have never been generally accepted; still others, such as supersymmetry, remain speculative.

Principles such as these, which can be interpreted either as physical assertions or as methodological principles, which are made explicit, developed, revised and, on occasions, rejected or refuted, are hard to account for within the framework of falsificationism. It is especially difficult, within this framework, to account for principles which (a) have a quasi *a priori* role in specifying requirements theories must satisfy in order to be accepted, but which at the same time (b) make substantial *physical* assertions about the nature of the universe. AOE, on the other hand, predicts the existence of such principles, with just the features that have been indicated. Accepted principles are components of the currently accepted level 3 blueprint. As the accepted blueprint evolves, these principles, interpreted either as physical or methodological principles, evolve as well. Indeed, according to AOE, these principles, and associated blueprints, do not just evolve, they are *improved* with improving theoretical knowledge. AOE provides a more or less fixed framework of relatively unproblematic assumptions and associated methods (at level 4 or above) within which highly problematic level 3 assumptions and associated methods may be improved in the light of the empirical success and failure of rival research programmes (which adopt rival level 3 assumptions and associated methods).

This can be reformulated in terms of *aims* and *methods* of physics. A basic aim of theoretical physics is to discover the true theory of everything. This aim can be characterized in a range of ways, depending on how broadly or narrowly "theory of everything" is construed, what degree of unity such a theory must have in order to be a theory at all, and thus how much metaphysics is built into, or is presupposed by, the aim so characterized. The aim might be construed in such a way that no more than the truth of the thesis at level 7, or at level 6, is presupposed. Or, more specifically, the truth of the thesis at level 5 might be presupposed, or even more specifically, the truth of physicalism at level 4; or a range of increasingly specific blueprints at level 3 might be presupposed. Corresponding to these increasingly specific *aims* there are increasingly restrictive *methods*. As the aim becomes more specific, so it becomes more problematic, in that the presupposed metaphysics becomes increasingly likely to be false, which would make the corresponding aim unrealizable. AOE can thus be construed as providing a kind of nested framework of aims and methods, the aims becoming, as one goes down the hierarchy, increasingly problematic, and vulnerable to being unrealisable in principle, because the presupposed metaphysics is false. Within the framework of relatively unspecific, unproblematic, permanent aims and methods (high up in the hierarchy) much more specific, problematic, fallible aims and methods (low down in the hierarchy) can be revised and improved in the light of improving knowledge. There is, as I have already in effect said, something like positive feedback between improving scientific knowledge and improving aims and methods. As knowledge improves, knowledge-about-how-to-improve-knowledge improves as well. This capacity of science to adapt itself – its aims and methods (its philosophy of science) – to what it finds out about the universe is, according to AOE, the methodological key to the astonishing progressive success of science. Falsificationism, with its fixed aim and fixed methods, is quite unable to do justice to this positive feedback, meta-methodological feature of science, this capacity of science to learn about learning as it proceeds.

7. *Verisimilitude*. The so-called problem of verisimilitude arises because physics usually proceeds from one false theory to another, thus rendering obscure what it can mean to say that science makes progress. Popper (1963, ch. 10 and addenda) tried to solve this problem within the framework of falsificationism but, as Miller (1974) and Tichy (1974) have shown, this attempted solution does not work. Not only does falsificationism fail to specify properly the methods that make progress in theoretical physics possible, it fails even to say what progress in theoretical physics *means*.

AOE solves the problem without difficulty. First, the fact that physics does proceed from one false theory to another, far from undermining physicalism, and hence AOE as well, is just the way theoretical physics must proceed, granted physicalism (as I have already indicated). For, granted physicalism, any theory, T*, which captures precisely how phenomena evolve in some restricted domain, must be generalizable to cover *all* phenomena. If T* cannot be so generalized then, granted physicalism, it cannot be precisely true. In so far as physics proceeds by developing theories which apply to restricted, but successively increasing, domains of phenomena, it is bound (granted physicalism) to proceed by proposing one false theory after another.

Second, AOE solves the problem of what it can *mean* to say that theories, $T_0, \ldots T_N$, get successively closer and closer to the true theory-of-everything, T, as follows. For this we require that T_N can be "approximately derived" from T (but not vice versa), T_{N-1} can be "approximately derived" from T_N (but not vice versa), and so on down to T_0 being "approximately derivable" from T_1 (but not vice versa).

The key notion of "approximate derivation" can be indicated by considering a particular example: the "approximate derivation" of Kepler's law that planets move in ellipses around the sun (K) from Newtonian theory (NT). The "derivation" is done in three steps. First, NT is restricted to N body systems interacting by gravitation alone within some definite volume, no two bodies being closer than some given distance, r. Second, keeping the mass of one object constant, we consider the paths followed by the other bodies as their masses tend to zero. According to NT, in the limit, these paths are precisely those specified by K for planets. In this way we recover the *form* of K from NT. Third, we reinterpret this "derived" version of K so that it is now taken to apply to systems like that of our solar system. (It is of course this *third* step of reinterpretation that introduces error: mutual gravitational attraction between planets, and between planets and the sun, ensures that the paths of planets, with masses greater than zero, must diverge, however slightly, from precise Keplerian orbits.)

Quite generally, we can say that T_{r-1} is "approximately derivable" from T_r if and only if a theory empirically equivalent to T_{r-1} can be extracted from T_r by taking finitely many steps of the above type, involving (a) restricting the range of application of a theory, (b) allowing some combination of variables of a theory to tend to zero, and (c) reinterpreting a theory so that it applies to a wider range of phenomena.[24]

This solution to the problem of what progress in theoretical physics *means* requires AOE to be presupposed; it does not work if falsificationism is presupposed. This is because the solution requires one to assume (a) that the universe is such that a yet-to-be-discovered, true theory of everything, T, exists, and (b) current theoretical knowledge can be approximately derived from T. Both assumptions, (a) and (b), are justified granted AOE; neither assumption is justifiable granted falsificationism.[25]

8. *Discovery of new fundamental theories.* Given falsificationism, the discovery of new fundamental physical theories that turn out, subsequently, to meet with great empirical success, is inexplicable. (One thinks here of Newton's discovery of his mechanical theory and theory of gravitation, Maxwell's discovery of classical electromagnetism, Einstein's discovery of the special and general theories of relativity, Bohr's discovery of "old" quantum theory, Heisenberg's and Schrödinger's discovery of "new" quantum theory, Dirac's discovery of the relativistic quantum theory of the electron and, in more recent times, the discovery of quantum electrodynamics, the electroweak theory, quantum chromodynamics and the standard model.) Granted that a new theory is required to explain a range of phenomena, there are, on the face of it, infinitely many possibilities. In the absence of rational guidance towards good conjectures, it would seem to be infinitely improbable that anyone should, in a finite time, be able to come up with a theory that successfully predicts new phenomena. The only guidance that falsificationism can provide is to think up new theories that "proceed from some simple, new, and powerful, unifying idea", in accordance with Popper's (1963) requirement of simplicity, but this is so vague and ambiguous as to be almost useless. Famously, Popper explicitly denied that a rational method of discovery is possible at all (see Popper, 1959, p. 31). But if discovery is not rational, it becomes miraculous that good new theories are ever discovered. Scientific progress becomes all but inexplicable.

AOE, by contrast, provides physics with a rational, if fallible and non-mechanical, method for the discovery of new fundamental physical theories. This method involves modifying the current best level 3 blueprint so that:

(a) the new blueprint exemplifies physicalism better than its predecessor;
(b) the new blueprint promises, when made sufficiently precise to become a testable theory, to unify clashes between predecessor theories;
(c) the new theory promises to exemplify the new blueprint better than the predecessor theories exemplify the predecessor blueprint.

(a), (b) and (c) provide means for assessing how good an idea for a new theory is which do not involve empirical testing (which is brought in once the new theory has been formulated). The level 4 thesis of physicalism provides continuity between the state of knowledge before the discovery of the new theory, and the state of knowledge after this discovery. Modifying the current level 3 blueprint ensures that the new theory will be incompatible with its predecessors; it will postulate new kinds of entities, forces, space-time structure, and will exhibit new symmetries. In other words, because of the hierarchical structure of AOE, there is (across revolutions) both continuity (at level 4) and discontinuity (at levels 2 and 3), something that is not possible given falsificationism. AOE provides physics with specific non-empirical tasks to perform, specific non-empirical problems to be solved, and non-empirical methods for the assessment of ideas for new theories, all of which adds up to a rational, if fallible, method of discovery. It all stems from recognizing that physicalism is a part of current scientific knowledge. The discovery of new fundamental physical theories then ceases to be inexplicable. None of this is possible granted falsificationism.[26]

The fact that AOE is able to provide a rational method of discovery, while falsificationism is not, is due to the greater rigour of AOE (a point mentioned in 3 above). AOE has greater rigour because AOE acknowledges, while falsificationism denies, metaphysical assumptions implicit in persistent scientific preference for simple, explanatory theories. It is precisely the explicit acknowledgement of these metaphysical assumptions which makes the rational method of discovery of AOE possible.

9. *Diversity of scientific method*. One striking feature of natural science, often commented on, is that different branches of the natural sciences have somewhat different methods. Experimental and observational methods, and methods or principles employed in constructing and assessing theories, vary as one moves from theoretical to phenomenological physics, from physics to chemistry, from astronomy to biology, from geology to ethology. Falsificationism can hardly do justice to this striking diversity of method within the natural sciences. Popper, indeed, tends to argue that there is unity of method, not only in natural science, but across the whole of science, including social science as well (see Popper, 1961). AOE, by contrast, predicts diversity of method throughout natural science, overlaid by unity of method at a meta-methodological level. AOE can do justice to the diversity of methods to be found in diverse sciences, without underlying unity and rationality being sacrificed.

It is important to appreciate, first, that different branches of the natural sciences are not isolated from one another: they form an interconnected whole, from theoretical physics to molecular biology, neurology and the study of animal behaviour. Different branches of natural science, even different branches of a single science such as physics, chemistry or biology, have, at some level of specificity, different aims, and hence different methods. But at some level of generality all these branches of natural science have a common aim, and therefore common methods: to improve knowledge and understanding of the natural world. All (more or less explicitly) put AOE into practice, but because different scientific specialities have different specific aims,[27] at the lower end of the hierarchy of methods different specialities have somewhat different methods, even though some more general methods are common to all the sciences. Furthermore, all natural sciences apart from theoretical physics presuppose and use results from other scientific specialities, as when chemistry presupposes atomic theory and quantum theory, and biology presupposes chemistry. The results of one science become a part of the presuppositions of another, implicit in the aims of the other science (equivalent to the level 3 blueprint of physics, or the level 4 thesis of physicalism). This further enhances unity throughout diversity, and helps explain the need for diversity of method.

A key feature of AOE is that methods depend somewhat on aims, methods varying somewhat as aims vary. When the hierarchical structure of aims and methods of AOE, depicted in Figure 2.1, is applied to specific branches of natural science – geology, evolutionary biology, neuroscience, organic chemistry – the various spherical discs of Figure 2.1 will need to be reinterpreted so that they come to represent the different more or less specific aims of the various branches of natural science. What the various discs of the figure depict will vary as we move from geology to neuroscience and so on – even the number of levels required will vary. What remains constant throughout all these diverse applications of AOE, however, is the hierarchical structure of aims and methods, aims becoming increasingly insubstantial and unproblematic as one goes up the hierarchy in each case.

In short, in order to exhibit the rationality of the diversity of method in natural science – apparent in the evolution of methods of a single science, and apparent as one moves, at a given time, from one scientific speciality to another – it is essential to adopt the meta-methodological, hierarchical standpoint of AOE, which alone enables one to depict methodological unity (high up in the hierarchy) throughout methodological diversity (low down in the hierarchy).[28] Figure 2.1, in depicting the aims and methods of theoretical physics, depicts what is common to all of natural science (physics being the fundamental to all of natural science). In addition, the hierarchical structure of Figure 2.1 forms a kind of template, a paradigm, for many similarly structured but diversely interpreted depictions of aims and methods of the many diverse specialized disciplines of natural science. It is in this way that the general AOE idea can do justice both to the diverse aims and methods of different sciences at any one moment, *and* to the evolving aims and methods of any one science over time. At the same time, it depicts what is common to all this diversity of aims and methods: it is to be found at the metalevel, or the meta-meta-metalevel, and the common hierarchical structure. Furthermore, and crucially, AOE is able to provide a *rationale* for all this diversity of aims and methods: it does this at the metalevel, meta-methods having the role of assessing methods one step down in the hierarchy.

Falsificationism, lacking this hierarchical structure, cannot begin to do justice to this key feature of scientific method, diversity at one level, unity at another; nor can it begin to do justice to the rational *need* for this feature of scientific method.[29]

There is a further, important point. Any new conception of science which improves our understanding of science ought to enable us to improve scientific practice. It would be very odd if our ability to do science well were wholly divorced from our understanding of what we are doing. A test for a new theory of scientific method ought to be, then, that it improves scientific practice, and does not merely accurately depict current practice. AOE passes this test. In providing a framework for the articulation and scrutiny of level 3 metaphysical blueprints, as an integral part of science itself, and thus providing a rational means for the development of new non-empirical methods, new symmetry principles and new theories, AOE advocates, in effect, that current practice in theoretical physics be modified. AOE makes explicit what is at present only implicit. And more generally, in depicting scientific method in a hierarchical, meta-methodological fashion, AOE has implications for method throughout the natural sciences, and not just for theoretical physics.

In case it should seem miraculous that science has made progress without AOE being generally understood and accepted, I should add that good science has always put something close to AOE into practice in an implicit, somewhat covert way, and it is this which has made progress possible.

2.7 Thomas Kuhn

As I remarked in section 2.1 above, the main difference between Kuhn's (1970a) picture of science and Popper's is that, whereas Kuhn stresses that, within normal science, paradigms are dogmatically protected from refutation, from criticism, Popper holds that theories must always be subjected to severe attempted refutation. AOE is even more Popperian than Popper's falsificationism, in that AOE exposes to criticism assumptions that falsificationism denies, and thus shields from criticism. One might think, therefore, that AOE would differ even more from Kuhn's picture of science than falsificationism does.

It is therefore rather surprising that exactly the opposite is the case: in some important respects, AOE is closer to Kuhn than to Popper.

The picture of science that emerges from Kuhn (1970a) may be summarized like this. There are three stages to consider. First, there is a pre-scientific stage: the discipline is split into a number of competing schools of thought which give different answers to fundamental questions. There is debate about fundamental questions between the schools, but no overall progress, and no science.

Second, the ideas of one such school begin to meet with empirical success; these ideas become a "paradigm", and the pre-scientific school becomes normal science (competing schools withering away). Within normal science, no attempt is made to refute the paradigm (roughly, the basic theory of the science); indeed, the paradigm may be accepted even though there are well-known apparent refutations. When the paradigm fails to predict some phenomenon, it is not the paradigm, but the skill of the scientist, that is put to the test. The task of the normal scientist is to solve puzzles, rather than problems. The paradigm specifies what is to count as a solution, specifies what methods are to be employed in order to obtain the solution and guarantees that the solution exists: these are all characteristics of puzzles rather than open-ended problems. The task is gradually to extend the range of application of the paradigm to new phenomena, textbook successes being taken as models of how to proceed. Methods devolve from paradigms.

Third, the paradigm begins to accumulate serious failures of prediction; these resist all attempts at resolution, and some scientists lose faith in the capacity of the paradigm to overcome these "anomalies". A new paradigm is proposed, which does resolve these recalcitrant anomalies, but which may not, initially, successfully predict all that the old paradigm predicted. Empirical considerations do not declare that the new paradigm is, unequivocally, better than the old. Normal science gives way to a period of revolutionary science. Scientists again debate fundamentals, arguments for and against the rival paradigms, often presupposing what they seek to establish. Rationality breaks down. If the revolution is successful, the new paradigm wins out, and becomes the basis for a new phase of normal science. Many old scientists do not accept the new paradigm; they die holding on to their convictions.

Kuhn argues that the dogmatic attitude inherent in normal science is necessary if science is to make progress. Applying a paradigm to new phenomena, or to old phenomena with increasing accuracy, is often extremely difficult. If every failure was interpreted as a failure of the paradigm, rather than of the scientist, paradigms would be rejected before their full range of successful application had been discovered. By refusing to reject a paradigm until the limits of its successes have been reached, scientists put themselves into a much better position to develop and apply a new paradigm. For reasons such as these, normal science, despite being ostensibly designed to discover only the expected, is actually uniquely effective in disclosing novelty. Popper (1970), in criticizing Kuhn on normal science, ignored these arguments in support of the necessity of normal science for scientific progress.

AOE holds that much scientific work ought indeed to resemble Kuhn's normal science, in part for reasons just indicated. But there are even more important considerations. According to AOE, and in sharp contrast with falsificationism, theoretical physics accepts a level 3 metaphysical blueprint, which exercises a powerful constraint on what kind of new theories physicists can try to develop, consider or accept. The blueprint has a role reminiscent, in some respects, of Kuhn's paradigm, and theoretical physics, working within the constraints of the blueprint, its non-empirical methods set by the blueprint, has some features of Kuhn's normal science.

Furthermore, according to AOE, other branches of natural science less fundamental than theoretical physics invariably presuppose relevant parts of more fundamental branches. Thus chemistry presupposes relevant parts of atomic theory and quantum theory, biology relevant parts of chemistry, astronomy relevant parts of physics. Such presuppositions of a science have a role, for that science, that is analogous to the role that the current level 3 blueprint, or the level 4 thesis of physicalism, has for theoretical physics. The presuppositions act as a powerful constraint on theorizing within the science. They set non-empirical methods for that science. Such presuppositions have a role, in other words, which is similar, in important respects, to Kuhn's paradigms. Viewed from an AOE perspective, one can readily see how and why much of science is Kuhnian puzzle solving rather than Popperian problem solving.

There are also, it must be emphasized, major differences between Kuhn and AOE. The chief difference is that, according to AOE, science has a paradigm for paradigms – to put it in Kuhnian terms.[30] In order to be acceptable, level 3 blueprints must exemplify the level 4 thesis of physicalism (which in turn must exemplify the level 5 thesis of comprehensibility and so on, up to level 7). This means that, as long as physicalism continues to be accepted as the best available level 4 thesis for science, metaphysical blueprints can be assessed in a quasi non-empirical way, in terms of how well they accord with physicalism. Natural science is, according to AOE, one sustained, gigantic chunk of normal science, with physicalism as its paradigm. In this respect, AOE is more Kuhnian than Kuhn (in addition to being more Popperian than Popper!).

Like falsificationism, Kuhn's picture of science is hardly tenable. In the first place, it does not fit scientific practice very well. Normal science undoubtedly exists, as even Popper recognized; it may well be that most scientific activity has the character of Kuhn's normal science. But even when a discipline seems most like normal science, almost always there are a few scientists actively engaged in developing alternatives to the reigning paradigm. And on occasions, it is from the work of these few that a new paradigm, and a new phase of normal science, springs, often in a way that is quite different from Kuhn's account. It is not obvious that accumulation of anomalies, resulting in a crisis in biology, led to Darwin's theory of evolution. Quantum theory did not emerge, initially, from a crisis in classical physics. Planck's work around 1900 on blackbody radiation engendered the quantum revolution. It is true that classical physics, applied to a so-called black body emitting electromagnetic radiation, made a drastically incorrect prediction, but no one, not even Planck, thought that this posed a serious problem for classical physics. The fallacious prediction of classical physics was dubbed "the ultraviolet catastrophe", but this phrase was coined by Ehrenfest, after the quantum revolution was under way, around 1911, as propaganda for the new theory. It was Einstein who first recognized that Planck's work spelled the downfall of classical physics; but general recognition of this only came later, probably with Bohr's quantum theory of the atom, around 1913. Again, Einstein's general theory of relativity emerged, not because Newton's theory had accumulated anomalies and was in a state of crisis, but because it contradicted special relativity. Einstein sought a theory of gravitation compatible with special relativity, and it was this that led him to general relativity. These three revolutions, resulting in Darwinian theory, quantum theory and general relativity, are among the biggest and most important in the history of science; and yet they do not fit Kuhn's pattern.

Failure to fit scientific practice in detail does not, however, provide decisive grounds for rejecting a normative account of scientific method. One can always reply that the account specifies how science *ought* to proceed, not how it has in fact proceeded. Much more serious are the objections of principle to Kuhn's account. Kuhn, like Popper, provides no account of the creation of new paradigms. And given Kuhn's insistence that a new paradigm, after a successful revolution, is incommensurable with its pre-revolutionary predecessor, it would seem impossible to provide rational (if fallible) procedures for the creation of good new paradigms while maintaining consistency with the rest of Kuhn's views. Kuhn does allow that non-empirical criteria, or values, such as consistency and simplicity, are employed by science permanently (and therefore, presumably, across revolutions) to assess theories or paradigms; but Kuhn also emphasizes that these criteria are flexible and open to different interpretations (Kuhn, 1970a, p. 155; 1977, ch. 13). There is no account of what simplicity is, and no advance over Popper's "requirement of simplicity". Furthermore, Kuhn's appeal to simplicity faces the same difficulty we have seen arising in connection with Popper's appeal to simplicity. If "simplicity" is interpreted in such a way that it has real content, and is capable of excluding "complex" or disunified and aberrant theories or paradigms from science, then its permanent employment by science commits science to a permanent metaphysical assumption that persists through revolutions, something Kuhn explicitly rejects (and could not, in any case, provide a rationale for). If "simplicity" is interpreted sufficiently loosely and flexibly to ensure that no such metaphysical thesis is involved, invoking simplicity must fail to exclude complex, disunified, aberrant paradigms from science. Any Kuhnian requirement of simplicity, in short, must either be incompatible with the rest of Kuhn's views, or toothless and without content. Either way, Kuhn has no consistent method for excluding complex, aberrant paradigms from consideration. It should be noted that Kuhn is emphatic that no sense can be made of the idea that there is progress in knowledge across revolutions, the new paradigm being better, closer to the truth, than the old one (see Kuhn 1970a, ch. 13). But this is a disaster for Kuhn's whole view. Why engage in normal science if the end result is the rejection of all that has been achieved, all the progress in knowledge of that period of normal science being sacrificed when the science adopts a new paradigm? Kuhn's arguments for the progressive character of normal science, indicated above, are all defeated.

Perhaps the most serious objection to Kuhn's picture of science is the obvious basic unintelligence of its prescriptions for scientific research. Suppose we have the task of crossing on foot difficult terrain, containing ravines, cliffs, rivers, swamps, thickets. Kuhn's view, applied to this task, would be as follows. After debate about which route to follow (pre-science), one particular route is chosen and then followed with head down, no further consideration being given to changing the route (normal science). Eventually, this leads to an impasse: one comes face to face with an unclimbable cliff, or finds oneself waist-deep in a swamp and in danger of drowning (crisis). Finding oneself in these dire circumstances, a new route is taken (new paradigm), and again, with head down, this new route is blindly followed (normal science) until, again, one finds oneself unable to proceed, about to drown in a river or tumble into a ravine.

This is clearly a stupid way to proceed. It would be rather more intelligent if, as one tackles immediate problems of wading through this stream, climbing down this scree (puzzle-solving of normal science), one looks ahead, whenever possible, and reconsiders, in the light of the terrain that has been crossed, what adjustments one needs to make to the route one has opted to follow. Exactly the same point holds for science. There can be division of labour. Even if a majority of scientists tackle the multitude of puzzles that go to make up normal scientific research, taking the current theory, or paradigm, for granted, there ought also to be some scientists who are concerned to look ahead, consider more fundamental problems, explore alternatives to the current paradigm. In this way new paradigms may be developed before science plunges deep into crisis. And just this does go on in scientific practice, as I have already indicated in the brief discussion of the work of Darwin and Einstein (and somewhat less convincingly, Planck). Another example of a new, revolutionary theory or paradigm being proposed in the absence of crisis is Wegener's advocacy of the movement of continents, anticipating the plate tectonic revolution by decades. Science is, in practice, more intelligent than Kuhn allows.

In sharp contrast to Kuhn, AOE does not merely stress the importance of "looking ahead", of trying to develop new theories, new paradigms, before science has plunged into crisis; even more important, AOE provides a framework for theoretical physics (and therefore, in a sense, for the whole of natural science) within which ideas for fundamental new theories may be developed and assessed.

According to Kuhn, successful revolutions mark radical discontinuities in the advancement of science, to the extent, indeed, that old and new paradigms are "incommensurable" (i.e. so different that they cannot be compared). This Kuhnian view is most likely to be correct when applied to revolutions in fundamental theoretical physics, where radical discontinuity seems most marked. But it is precisely here that Kuhn's claim turns out to be seriously inadequate. As I have already emphasized, all revolutions in theoretical physics, despite their diversity in other respects, reveal one common theme: they are all gigantic steps in unification. From Newton, via Maxwell, Einstein, Bohr, Schrödinger and Dirac, to Salam, Weinberg and Gell-Mann, all new revolutionary theories in physics bring greater unity to physics. (And Darwinian theory, one might add, brings a kind of unity to the whole of biology.) The very phenomenon that Kuhn holds to mark discontinuity, namely revolution, actually also reveals continuity – continuity of the search for, and the successful discovery of, underlying theoretical unity.[31]

This aspect of natural science, to which Kuhn fails entirely to do justice, is especially emphasized by AOE. According to AOE, revolutions in theoretical physics mark discontinuity at the level of theory, at level 2, and even discontinuity at level 3, but continuity at level 4. Physicalism, which asserts that underlying dynamic unity exists in nature, persists through revolutions – or, at least, has persisted through all revolutions in physics since Galileo. In order to make rational sense of natural science, we need to interpret the whole enterprise as seeking to turn physicalism, the assertion of underlying dynamic unity in nature,[32] into a precise, unified, testable, physical "theory of everything". That, in a sentence, is what AOE asserts. Physicalism, according to AOE, despite its metaphysical (untestable) character, is the most secure item of theoretical knowledge in science; it is *the* most fruitful idea that science has come up with, at that level in the hierarchy of assumptions.

Because of its recognition that, despite the discontinuity of revolutions at levels 2 and 3, there is the continuity of the persistence of physicalism at level 4 (and of other theses at levels higher up in the hierarchy), AOE is able to resolve problems concerning the discovery and assessment of paradigms which Kuhn's view is quite unable to solve. Both fundamental physical theories and level 3 blueprints can be partially ordered with respect to how well they exemplify physicalism, entirely independent of ordinary empirical assessment. Assessing progress through revolution poses no problem for AOE. As we have seen, AOE solves the problem of verisimilitude.

I have already mentioned that AOE does not merely describe scientific practice; it carries implications as to how scientific practice can be improved. One such implication concerns scientific revolutions. Kuhn (1970a) gives a brilliant description of the way, during a scientific revolution, there is a breakdown of rationality, competing arguments for the rival paradigms being circular, and each presupposing what is being argued for. This is a feature of actual science. Scientists do not know how to assess competing theories objectively when empirical considerations are inconclusive. But all this can be seen to be a direct consequence of trying to do science without the explicit acknowledgement of persisting metaphysical assumptions concerning the comprehensibility of the universe, there thus being nothing available to constrain acceptance of theories when empirical considerations are inconclusive. Consider Kuhn's breakdown of rationality. A substantial revolution will involve, not just two rival paradigms or theories, T_1 and T_2, but two rival blueprints, B_1 lurking behind T_1, and B_2 lurking behind T_2. Granted B_1, T_1 is far more acceptable than T_2, but the reverse holds granted B_2. But B_1 and B_2, being untestable, metaphysical theses, they are not explicitly discussable, and objectively assessable, within science; so they are more or less repressed, excluded from discussion. Nevertheless, scientists do think in terms of B_1 and B_2. Kuhn's Gestalt switch, involved in switching allegiance from T_1 to T_2, can be pinpointed as the act of abandoning the old blueprint and adopting the new one. Non-empirical arguments in favour of T_1 or T_2 can only take the form of an appeal to B_1 or B_2, in however a muffled a way (due to the point that blueprints are not open to explicit discussion). Such arguments will be circular, and entirely unconvincing to the opposition, in just the way described by Kuhn. Accept B_1, and T_1 becomes the only possible choice; accept B_2 and T_2 is the only choice. Each side in the dispute is convinced that the other side is wrong, even incoherent. What needs to be done, and cannot be done, of course, is to discuss the relative merits of B_1 and B_2. Just this can be done, granted AOE. T_1, B_1, T_2 and B_2 can all be assessed from the standpoint of adequacy in exemplifying physicalism. When the scientific community adopts AOE, the Kuhnian irrationality of revolutions will disappear from science.

It may be asked how it is possible for AOE to be *both* more Popperian than Popper and more Kuhnian than Kuhn. The answer is that AOE is more Popperian that Popper in making explicit, and so criticizable, metaphysical theses which falsificationism denies, and thus leaves implicit and uncriticizable within science. But AOE is also more Popperian than Popper in insisting we need to exploit criticism *critically*, so that it furthers, and does not sabotage, the growth of knowledge. Criticism needs to be marshalled and directed at that part of our conjectural knowledge which it is, we conjecture, the most fruitful to criticize. This means directing critical fire at level 2 theories and level 3 blueprints, it being less likely, though still possible, that criticism of the level 4 thesis of physicalism will aid the growth of empirical knowledge. Physicalism has played an extraordinarily fruitful role in the advancement of scientific knowledge; it should not be abandoned unless an even more apparently fruitful idea is forthcoming, or unless the empirical and explanatory success that physicalism appears to have engendered turns out to be illusory. It is this persistence of physicalism, for good Popperian reasons, which gives to theoretical physics, and indeed to the whole of natural science, something of the character of Kuhn's normal science, with physicalism as its quasi-permanent "paradigm".

2.8 Imre Lakatos

Lakatos sought to reconcile the very different views of science held by Popper and Kuhn. According to Kuhn, far from seeking falsifications of the best available theory, as Popper held, scientists protect the accepted theory, or "paradigm", from refutation for most of the time, the task being to fit recalcitrant phenomena into the framework of the paradigm. Only when refutations become overwhelming, does crisis set in; a new paradigm is sought for and found, a revolution occurs, and scientists return to doing "normal science", to the task of reconciling recalcitrant phenomena with the new paradigm. Lakatos sought to reconcile Popper and Kuhn by arguing that science consists of competing fragments of Kuhnian normal science, or "research programmes", to be assessed, eventually, in terms of their relative empirical success and failure. Instead of research programmes running in series, one after the other, as Kuhn thought, research programmes run in parallel, in competition, this doing justice to Popper's demand that there should be competition between theories (a point emphasized especially by Feyerabend).[33] Lakatos became so impressed with the Kuhnian point that theories always face refutations, the empirical successes of a theory being a far more important guide to scientific progress than refutation, that he finally came to the conclusion that Popper's philosophy of science was untenable.

AOE has a number of features in common with Lakatos's methodology of scientific research programmes. AOE makes extensive use of the notion of scientific research programme. Like Lakatos's view, AOE exploits the idea that such research programmes can, sometimes, be compared with respect to how empirically progressive they are. AOE, again like Lakatos's view, sees the whole of science as a gigantic scientific research programme. And, like Lakatos's view, AOE can be construed as synthesizing Popper's and Kuhn's views.

But there are also striking differences. There are differences in the way scientific research programmes are conceived, especially research programmes in fundamental physics. For Lakatos, the main components of a research programme are the "hard core" (corresponding to Kuhn's "paradigm"), and the "protective belt" of "auxiliary hypotheses", which facilitate the application of the hard core to empirical phenomena. The main business of a research programme is to develop the protective belt, thus extending, and making more accurate, the empirical predictions of the hard core. The hard core is a testable theory rendered metaphysical by the methodological decision not to allow it to be refuted, refutations being directed at the protective belt rather than the hard core.

According to AOE, by contrast, the metaphysical kernel of a research programme is not a testable theory but rather a thesis that is genuinely metaphysical (i.e. more or less unspecific, and usually untestable) – a thesis such as the corpuscular hypothesis, Boscovich's point-atom blueprint, Einstein's unified field blueprint and so on. The basic aim of the programme is to turn the relatively unspecific blueprint into a precise, testable (and true) physical theory. The research programme thus consists of a succession of theories, $T_1, T_2 \ldots T_n$, which can be compared, not only with respect to empirical success, but also with respect to how adequately each theory encapsulates, or exemplifies, the blueprint of the programme. (The latter is not possible within a Lakatosian programme.) Whereas a Lakatosian programme has a fixed basic theory (or hard core), and seeks to improve auxiliary hypotheses (the protective belt), an AOE programme strives to capture the blueprint more and more adequately by means of testable physical theories.

Both Lakatos's view and AOE permit one to see natural science as one gigantic research programme, but how this programme is construed is very different. For Lakatos, "science as a whole can be regarded as a huge research programme with Popper's supreme heuristic rule: 'devise conjectures with more empirical content than their predecessors'" (1970, p. 132). The huge research programme of natural science has, for Lakatos, no hard core; to this extent, Lakatos's view is a variant of Popper's.[34] According to AOE, however, if natural science is viewed as one gigantic research programme, then it does have something like a hard core. First, there is physicalism at level 4, a metaphysical but nevertheless substantial thesis about the nature of the universe. And then there is the current blueprint at level 3, an even more substantial metaphysical thesis about the nature of the universe. These provide severe constraints on what theories are acceptable that are not straightforwardly empirical,[35] something that is not possible given the views of Popper or Lakatos (or even Kuhn).[36]

Lakatos and AOE have very different motivations for taking scientific research programmes so seriously. For Lakatos, the motivation comes from appreciating that a scientific theory, T, cannot be decisively refuted at an instant, as it were, partly because auxiliary hypotheses can always be invented to salvage T from a refutation, partly because early applications of a new theory, such as Newton's, may make simplifying assumptions which may well lead to false predictions (the fault lying with the simplifying, auxiliary hypotheses rather than the basic theory). Only by looking at a series of theories, a given T_1 (the hard core) plus changing auxiliary hypotheses (the protective belt), and comparing this with a rival series based on a different hard core, T_2, and comparing the extent to which the two series are empirically progressive or degenerating, can one assess the relative empirical merits of T_1 and T_2. For AOE, the situation is very different. A research programme in theoretical physics consists of a blueprint, B, and a succession of theories, $T_1, T_2 \ldots T_n$ (each equivalent to a Lakatosian hard core), which are successive attempts to capture B as a testable theory. If $T_1, T_2 \ldots T_n$ are increasingly empirically successful (in a roughly Popperian sense) and also increasingly successful at capturing B, then this means that B is empirically fruitful. A rival blueprint, B*, might be such that the series $T_1, T_2 \ldots T_n$ moves further and further away from B*: this would mean that B* is empirically sterile. A major part of the point of research programmes, for AOE, is to assess the relative empirical fruitfulness of rival metaphysical theses, at levels 3 and 4 (and above, if necessary). Though mostly untestable, nevertheless metaphysical theses can be assessed in a quasi-empirical way, in terms of the empirical progressiveness or degeneracy of the research programmes with which they are associated (or can be regarded as being associated).[37] This is, according to AOE, a key feature of scientific method, one which makes scientific progress possible. It makes it possible for improving theoretical knowledge to lead to a reassessment of what is the best available blueprint, which in turn leads to a reassessment of the best available non-empirical methodological rules, such as symmetry principles. In other words, it makes it possible for there to be positive feedback between improving knowledge and improving aims and methods (improving knowledge-about-how-to-improve-knowledge), a vital feature of scientific rationality according to AOE.

The differences indicated enable AOE to overcome problems which Lakatos's view cannot solve. Lakatos insists that there is no such thing as instant rationality: however apparently decisive the refutation of a theory may be, it is always possible to salvage it from refutation in a content-increasing way by the invention of an appropriate auxiliary hypothesis. It is this consideration which leads Lakatos to argue that only series of theories, competing research programmes, can be assessed rationally, in terms of relative empirical progressiveness. But in practice in science there do seem to be instant refutations. A famous example is the refutation of parity. This is a symmetry which declares, roughly, that if a process can occur, then so can its mirror image. This was decisively refuted by Wu et al. (1957), by means of an experiment which showed that electrons were emitted in a preferential direction from cobalt nuclei undergoing radioactive decay in a magnetic field. Parity conservation implied that this would not occur. Strictly speaking, it was not parity conservation on its own that was refuted, but parity plus quantum theory plus the theory of weak interactions plus the theory of nuclear structure plus a highly theoretical description of the experiment. One would think there was plenty of scope, here, for auxiliary hypotheses to be invented to salvage parity from refutation. No such hypothesis was forthcoming; the refutation of parity conservation was accepted immediately by the physics community, despite strong resistance to accepting such a conclusion (because of the implausibility of supposing that nature distinguishes between left-handedness and right-handedness at the level of fundamental physical theory). Allan Franklin, who has produced what is probably the best account of the downfall of parity conservation, has put the matter like this: "It is fair to say that as soon as any physicist saw the experimental result they were convinced that parity was not conserved in the weak interactions" (Franklin, 1990, p. 66).[38] Scientific practice seems almost to refute Lakatos's view.

But it does not refute AOE. According to Lakatos, in the end only empirical considerations, plus considerations of empirical content, restrict choice of theory; few restrictions are placed on how a body of theory may be modified to salvage it from refutation. AOE places much more severe restrictions on choice of theory. In addition to those that it has in common with Lakatos's view, AOE demands of a fundamental physical theory that it, together with other such theories, exemplifies physicalism, to a sufficient degree. This makes it very much more difficult to modify a body of theory so as to salvage it from refutation. Instant refutation is not surprising, granted AOE.[39]

Lakatos's view requires that science consists of competing research programmes. Unquestionably, the history of science reveals that competing research programmes have, on occasions existed. But it is not clear that all science has this character, as Lakatos's view would seem to require. After Heisenberg and Schrödinger had developed quantum theory in the mid 1920s, there continued to be debate about how the new theory should be interpreted, and whether the new theory, interpreted along the orthodox lines advocated by Bohr, Heisenberg and others, was ultimately acceptable. But there was nothing like a competing research programme. Viewed from the perspective of AOE, all this makes perfect sense. There were indeed serious grounds for regarding the new theory as unsatisfactory (see Chapter 3; see also Maxwell, 1998, ch. 7). But the new theory had achieved such striking successes, it was rational to conjecture that progress lay in developing the new theory, applying it to new phenomena, reconciling it with special relativity – in doing something like Kuhnian normal science, in other words – rather than in trying to develop a rival theory, a rival research programme. (To say this is not to say that serious attention should not have been given to the theoretical defects of orthodox quantum theory.) Not only does the history of science fail to reveal that there are always competing research programmes, whenever a new theory arrives on the scene that meets with extraordinary empirical success and no refutation, no good rationale may exist for inventing a rival research programme. (As we have seen, unlike Popper's falsificationism and Lakatos's research programme view, AOE holds that something like Kuhn's normal science may well be rational, as long as it is accompanied by some sustained tackling of problems associated with the currently accepted blueprint. This may, eventually, but not immediately, lead to the development of a new fundamental theory, a new research programme.)

There are other, much more decisive ways in which AOE is an improvement over Lakatos's view. Lakatos's methodology of research programmes inherits a number of unsolved problems from its two sources, Popper and Kuhn. Like Popper and Kuhn, Lakatos has no solution to the problem of what the simplicity, unity or explanatory character of a theory, or hard core, is; AOE, as I have indicated briefly above, solves the problem without difficulty. In failing to say what simplicity *is*, Lakatos also fails to articulate with any precision that part of scientific method concerned with simplicity; AOE faces no difficulty here either. Like Popper and Kuhn, Lakatos can say nothing useful about how new theories, new hard cores, are created or discovered; AOE, as a result of including levels 3 and 4 within the domain of scientific knowledge, is able to specify a rational, if fallible and non-mechanical, method for the creation of new theories, even new fundamental theories of physics. Finally, Lakatos's view fails to solve the problem of verisimilitude, a problem which can be readily solved granted AOE.

Popper, Kuhn and Lakatos, despite their differences, have one big failure in common (the source of almost all the others). All three take for granted that:

(A) In science no untestable but nevertheless substantial thesis about the world can be accepted as a part of scientific knowledge in such a firm way that theories which clash with it, even if highly successful empirically, are nevertheless rejected.

Popper accepts (A) in that, for him, untestable theses are metaphysical, and therefore not a part of scientific knowledge. Kuhn holds it, because, for Kuhn, nothing theoretical survives a revolution. Kuhn's acceptance of (A) is also apparent in his whole treatment of revolutions: precisely because Kuhn accepts (A), Kuhn cannot invoke anything like the level 4 thesis of physicalism to assess rival paradigms during a revolution, when empirical considerations are inconclusive. The Kuhnian irrationality of revolutions is a consequence of scientists accepting (A); and in so far as Kuhn thinks this irrationality is inevitable, Kuhn accepts (A) as well.

A case could be made out for saying that Lakatos came near to rejecting (A) in arguing for the need for science to adopt a conjectural metaphysical inductive principle which, if true, would more or less guarantee that Popperian, or rather Lakatosian, methods deliver authentic theoretical knowledge.

But Lakatos here missed the fundamental point, central to AOE, and highly Popperian in spirit, that our current methods are all too likely to be more or less the *wrong* methods to adopt, the metaphysics implicit in these methods being *false*. There is thus a vital need, for the sake of scientific progress, to make this false metaphysics explicit so that it can be criticized, so that alternatives can be developed and considered, leading to *improved* metaphysics and methods. In order to do this in the best possible way, we need to develop a hierarchy of metaphysical theses to form a framework of relatively unproblematic theses, at the top of the hierarchy, below which more specific and problematic theses may be developed and assessed, especially in the light of their potential and actual empirically progressive character.

Interestingly enough, Lakatos himself was aware of this deficiency in his "plea to Popper for a whiff of 'inductivism'" (1978, p. 159). Discussing his proposal that one should appeal to a metaphysical inductive principle as a conjecture as a part of the solution to the problem of induction, Lakatos says:

> Alas, a solution is interesting only if it is embedded in, or leads to, a major research programme; if it creates new problems – and solutions – in turn. *But this would be the case only if such an inductive principle could be sufficiently richly formulated so that one may, say criticize our scientific game from its point of view.* My inductive principle tries to explain why we "play" the game of science. But it does so in an *ad hoc*, not in a "fact-correcting" (or, if you wish, "basic value judgment correcting") way. (Lakatos, 1978, p. 164)

Lakatos highlights, here, the difference between his own position and that of AOE. The (revisable) AOE thesis of physicalism is indeed "sufficiently richly formulated so that one may ... criticize our scientific game from its point of view". AOE not only offers a new research programme for the philosophy of science, it modifies the research programme of science, one modification being that the philosophy of science becomes an integral part of science itself. The passage above makes me wonder whether Lakatos might not have gone on to develop or endorse AOE if he had lived.

3
Einstein, aim-oriented empiricism, and the discovery of special and general relativity*

3.1 Einstein's new method of discovery

According to Popper, Einstein is a falsificationist. Thus Popper declares: "Einstein *consciously seeks error elimination*. He tries to kill his theories: he is *consciously critical* of his theories" (Popper, 1972, p. 25). And elsewhere Popper declares: "what I have done is mainly to make explicit certain points which are implicit in the work of Einstein" (Whitrow, 1973, p. 23). Paul Feyerabend, on the other hand, holds Einstein to be a methodological "opportunist or cynic" or, in other words, a methodological anarchist (Feyerabend, 1978, p. 213n; see also p. 18 and pp. 56–7 and note). For Arthur Fine, Einstein adopts a view close to the natural ontological attitude (NOA). Fine writes: "In its antimetaphysical aspect, NOA is at one with Einstein's motivational realism" (Fine, 1986, p. 9). As far as I know, van Fraassen has not yet claimed that Einstein is a constructive empiricist but, amazingly, the claim has been made on his behalf by Fine, who writes:

> Indeed it would not be too far off if we summarized Einstein's views this way: "Science aims to give us theories which are empirically adequate; and acceptance of a theory involves as belief only that it is empirically adequate" [a straight quote from Van Fraassen's *The Scientific Image* (1980)] ... My argument, then, is that if we understand Einstein in the way that he asks us to, his own realist-sounding language maps out a position closer to constructive empiricism than to either "metaphysical realism" or "scientific realism". (Fine, 1986, p. 108)

* I am grateful to Harvey Brown for critical comments concerning the first draft of this chapter.

The temptation to see one's own view in Einstein's thought is, it seems, all but irresistible. I too, it seems, am unable to resist this temptation. For it is my claim that aim-oriented empiricism – the view which, I argued in the last chapter, emerges as a sort of synthesis of the views of Popper, Kuhn and Lakatos but which, at the same time, greatly improves on all these views – is close to Einstein's mature view of science.

I must confess that I did not arrive at aim-oriented empiricism as a result of reading Einstein. I developed the view during the course of criticizing Popper, and as the key to the solution to the problem of induction.[1] At first I was convinced that the orthodox view of science, which I propose to call *standard empiricism*, had such a dogmatic stranglehold on science that it would be quite impossible for any scientist to uphold aim-oriented empiricism. It did not occur to me that Einstein might be an exception.

By standard empiricism I mean the view that in science theories are accepted and rejected solely on the basis of empirical success and failure. If science gives preference to theories that are simple, unified or explanatory, this must not be done in such a way that the universe itself is permanently assumed to be simple, unified or comprehensible. The crucial tenet of standard empiricism is that science must not make any permanent assumption about the nature of the universe that is upheld independently of empirical considerations, and certainly not in violation of empirical considerations. Science may accept a paradigm or hard core for a time, in the sort of way depicted by Kuhn and Lakatos, but it must not accept any such assumption permanently, independently of evidence.

Aim-oriented empiricism (AOE) is in stark conflict with standard empiricism in that it holds that science does make permanent assumptions about the nature of the universe independent of empirical considerations. Assumptions at levels 7 and 6 are, according to AOE, accepted permanently by science independently of evidence, and the even more substantial assumptions at levels 5 and 4 are accepted pretty permanently as well, and certainly not straightforwardly on the basis of empirical success, as these are untestable *metaphysical* theses. AOE, as a result of being in such stark conflict with standard empiricism, would be, I thought, far too heretical a view to be acceptable to any scientist.

But then it began to occur to me that Einstein, in developing special and general relativity, had made essential use of aim-oriented empiricism – his success owing much to his exploitation of the view in scientific practice.[2] I read everything I could lay my hands on that Einstein had written about the nature of science, and I discovered, so it seemed to me, that Einstein had actually advocated key tenets of aim-oriented empiricism in an increasingly explicit way as the years went by. He had, however, been ignored and misunderstood because of the powerfully prevailing influence of standard empiricism.[3] Here are my reasons for holding this view.

Einstein invented aim-oriented empiricism in scientific practice in order to overcome a severe scientific crisis. The crisis was the demise of classical physics as a result of Planck's 1900 quantum theory of black-body radiation. Initially, it was only Einstein who understood just how grave, how wholesale, the crisis was. In his "Autobiographical Notes" he puts the matter like this:

> it [became] clear to me as long ago as shortly after 1900, i.e. shortly after Planck's trailblazing work, that neither mechanics nor electrodynamics could (except in limiting cases) claim exact validity. By and by I despaired of the possibility of discovering the true laws by means of constructive efforts based on known facts. The longer and the more despairingly I tried, the more I came to the conviction that only the discovery of a universal formal principle could lead us to assured results. The example I saw before me was thermodynamics. The general principle was there given in the theorem: the laws of nature are such that it is impossible to construct a *perpetuum mobile* (of the first and second kind). How, then, could such a universal principle be found? (Einstein, 1969, pp. 51–3)

This, I claim, is the beginning of the explicit employment of aim-oriented empiricism in scientific practice. It is to this that Einstein owed his extraordinary success in discovering special and general relativity. Soon after 1900, Einstein found himself bereft of guidelines as to how to proceed, because Planck's "trailblazing" result had cast into doubt the whole of classical physics. Ordinarily a theoretical physicist can proceed by applying, extending, modifying or reinterpreting existing established physical theory. This is how classical physics had developed so far, after Newton. Einstein, however, found himself in what seemed an unprecedented situation. Existing physical theory – especially Newtonian mechanics and Maxwell-Lorentzian electrodynamics – had to be fundamentally wrong, given Planck's result. A fundamentally new kind of theory was needed to stand in their stead. But, in order to discover this new theory, it would be useless to try to extend or modify existing physical theories, in the ordinary manner, since it was just these theories which were fundamentally wrong. In order to proceed, Einstein was obliged to invent a new method of discovery for theoretical physics – a rational method capable of leading to the discovery of fundamentally new kinds of theories.

Within the framework of standard empiricism there can be no such rational method of discovery. If the *only* way in which theories can be rationally assessed in physics is by means of empirical success and failure, there can be no rational method for the invention of good, radically new physical theories which are incompatible with existing theories.

Popper (1959, 1963), Kuhn (1970a) and Lakatos (1970), all of whom defend versions of standard empiricism, not surprisingly all deny the possibility of there being a rational method of discovery of fundamentally new theories or paradigms – theories whose invention and acceptance constitute a "scientific revolution". Kuhn even denies that there can be rational assessment of a revolutionary new theory (with respect to its predecessor). The problem of how to proceed when confronted by wholesale scientific crisis, the breakdown of all existing theoretical knowledge, which Popper, Kuhn and Lakatos failed to solve in principle in the 1930s, 1960s and 1970s, Einstein had already solved in successful scientific practice by the year 1905. He solved it by inventing special relativity.

What, then, is Einstein's new rational method of discovery, which led to the discovery of special and general relativity? It can be put, quite simply, like this. Choose two of the most fundamental physical theories, T_1 and T_2, say, which are a part of "scientific knowledge" but which contradict each other. Discard everything about T_1 and T_2 that does not seem relevant to the contradiction until two mutually contradictory principles, P_1 and P_2, are arrived at, P_1 from T_1 and P_2 from T_2, thus arriving, it is hoped, at the nub of the contradiction between T_1 and T_2. Modify P_1 or P_2 (or both) or relevant background assumptions to resolve the contradiction into a new unified principle, P_3 (a synthesis of a transformed P_1 and P_2). Take P_3 as the basis for a new theory, T_3, which unifies T_1 and T_2.

In order for this method of discovery to be a rational one to adopt, one crucial assumption must be made: the universe has some kind of discoverable unified structure, of which our present fundamental physical theories give us limited, approximate (and incompatible) glimpses. Given the truth of this assumption, we have rational grounds for holding that the method can lead to success. If the assumption is false, we have no such grounds. As we shall see, Einstein seems only to have fully understood this point after the discovery of general relativity.

3.2 The discovery of special relativity

As far as the discovery of special relativity is concerned, Einstein used the above method in the following way. The two fundamental physical theories that he takes as his starting point (T_1 and T_2) are Newtonian mechanics (NM) and Maxwellian electrodynamics (ME). These two theories are incompatible, fundamentally because, given their most natural interpretation, NM is about forces-at-a-distance between point particles with mass, whereas ME is about one entity, the continuous electromagnetic field. More specifically, however, there is the following contradiction. NM asserts that forces affect *accelerations*, not *velocities*. Dynamic laws (laws concerning forces and their affects), formulated within the framework of NM, do not pick out any special velocity any more than they pick out some special place or time. ME does, however, pick out a special velocity: the velocity of light, the velocity at which, according to ME, vibrations in the field strengths of the electromagnetic field travel through space.

Both points are absolutely fundamental to the two theories. It is fundamental to the whole structure of NM that forces affect accelerations, not velocities (there thus being no role for absolute velocity within the theory). And it is fundamental to ME that influences should spread through the field at some fixed, *finite* velocity: for it is this which creates the need for a field theory in the first place. (Because gravitational influences, in Newton's theory of gravitation, spread at infinite velocity, instantaneous physical states can be specified in terms of point particles. When influences travel at some finite velocity, as in ME, this can no longer be done, as momentum and energy associated with variations in the force travelling at finite velocity through space will not be specified.)

One way in which the clash between NM and ME may be resolved is to interpret ME as a theory which presupposes the existence of the aether, states of the electromagnetic field being states of the aether. In this case, it is reasonable to hold that light has a constant velocity with respect to the aether, and the clash with NM disappears (the constancy of the velocity of light being as unproblematic as the constancy of the velocity of sound with respect to air). In his 1905 paper expounding special relativity, Einstein gave two reasons for rejecting this approach. First, it introduces an implausible asymmetry in the explanation of electromagnetic induction, implausible because of the symmetry in the phenomena to be explained. The theoretical explanation for the current in a conductor moving near a magnet at rest is strikingly different from the explanation of the current if the conductor is at rest and the magnet moves, even though all that matters is the relative motion as far as the effect is concerned. Second, it runs into empirical difficulties in that all attempts to detect the motion of the earth relative to the "light medium" – the aether – have failed. Einstein concluded that "the phenomena of electrodynamics as well as of mechanics possess no properties corresponding to the idea of absolute rest" (Einstein, 1905; translated in Einstein et al., 1952, p. 37).

Einstein was of course well aware that the null result of the Michelson-Morley experiment does not decisively demolish the aether; he knew of Lorentz's efforts to employ the FitzGerald contraction hypothesis to develop a version of electrodynamics which both presupposes the aether and is compatible with observation. In a paper published in 1907, however, Einstein remarked of the FitzGerald-Lorentz approach (surely with some justice) that it is "*ad hoc*" and "artificial" (Holton, 1973, p. 334) – although, as Grünbaum and Zahar remind us, this approach is not as grossly *ad hoc* as some have supposed (Grünbaum, 1963, pp. 386–94; Zahar, 1973).

We know that during the decade before 1905, Einstein took the aether hypothesis sufficiently seriously to wonder how motion through the aether might be detected (Pais, 1982, pp. 130–2). Nevertheless, it seems that, early on, Einstein was drawn to what may be called the "Faraday interpretation" of electromagnetism, according to which, instead of seeking to interpret electromagnetism in terms of some more fundamental kind of aetherial matter, one should, on the contrary, seek to understand matter in terms of electromagnetism, which is to be regarded as fundamental (the whole idea of the aether being a mistake). This is implicit in the "paradox" that Einstein discovered at the age of sixteen, and which he later saw as the germ from which special relativity grew. In his "Autobiographical Notes", Einstein describes the paradox thus:

> If I pursue a beam of light with the velocity c ... I should observe such a beam of light as a spatially oscillatory electromagnetic field at rest. However, there seems to be no such thing, whether on the basis of experience or according to Maxwell's equations. From the very beginning it appeared to me intuitively clear that, judged from the standpoint of such an observer, everything would have to happen according to the same laws as for an observer who, relative to the earth, was at rest. For how, otherwise, should the first observer know, i.e. be able to determine, that he is in a state of fast uniform motion? (Einstein, 1969, p. 53)

This only makes intuitive sense as a paradox in so far as electromagnetism is being conceived of in the absence of the aether.

As I have argued elsewhere,[4] there are strong aim-oriented empiricist, quasi *a priori* grounds for favouring the Faraday interpretation of electrodynamics over the aether interpretation, the view that the aether is required to make electrodynamics intelligible being a sort of metaphysical blunder. And there is an additional consideration. According to aim-oriented empiricism, the acceptability of the aether hypothesis is to be judged in terms of its heuristic and methodological power. But as ME was developed, up to 1905, especially in the hands of Lorentz, the role of the aether seemed to become increasingly tenuous. This, according to aim-oriented empiricism, counts against the aether approach. We may thus detect, in Einstein's adoption of the Faraday interpretation of electrodynamics, and his rejection of the aether interpretation, an instinctive allegiance to aim-oriented empiricism.

There is, however, another approach to resolving the clash between NM and ME. It is possible that the velocity of light is constant with respect to the source. Einstein tried this approach; he tells us that he abandoned it because of the complications to which it led (Shankland, 1963). Evidence against this hypothesis only began to come in later, in 1913, with observations of double stars.

Granted, then, that the above two approaches to resolving the clash between NM and ME are to be rejected, we are left with the following situation: ME appears to be committed to the existence of a fundamental, *absolute* velocity – the velocity of light – just that which NM rules out. We have here, then, two good candidates for P_1 and P_2, extracted from T_1 (NM) and T_2 (ME) in order to highlight the clash between the two theories, namely:

P_1: The laws of nature have the same form with respect to all inertial (non-accelerating) reference frames.

P_2: It is a law of nature that light travels with constant velocity c (in a vacuum).

P_1 and P_2 together form, it would seem, a horrible contradiction. In order for P_1 and P_2 to be compatible, it would be necessary for a beam of light to have the same velocity c with respect to *all* inertial reference frames, even though these are moving with all possible velocities with respect to each other.

Astonishingly, Einstein discovered how to make this apparently blatant absurdity entirely consistent. What we need to do is to modify our ideas about time and space, so that light *does* have the same velocity c in all reference frames. The basic postulates of special relativity are just P_1 and P_2: the many consequences of the theory arise from demanding that P_1 and P_2 be consistent.

More precisely, Einstein took P_1 as one of his basic postulates, but modified P_2 to become:

> P_2^*: It is a law of nature that the velocity of light is a constant c in some "resting" reference frame, and is independent of the velocity of the source.

P_2 is then derived from P_1 and P_2^*. It is entirely understandable that Einstein took P_2^* as his axiom rather than P_2 interpreted to mean: it is law of nature that light has constant velocity c in all inertial reference frames. To adopt this latter postulate is to assume as comprehensible that which only becomes comprehensible with the development of the theory. P_2^* is not initially incomprehensible in this way; on the contrary, P_2^* is a basic tenet of the Lorentzian approach, of the aether approach widely held at the time.

How, then, is the contradiction between P_1 and P_2 to be resolved? Ordinarily we assume that the rate of clocks, and the length of rods, are unaffected by uniform motion, temporal and spatial distances being frame-independent and absolute. Suppose we have two reference frames, R_1 and R_2, with parallel axes, and with origins that coincide at $t_1 = t_2 = 0$, the origin of R_2 travelling along the x-axis of R_1 with velocity v in the +ve direction, and the coordinates of an event P being (x_1, y_1, z_1, t_1) and (x_2, y_2, z_2, t_2) in R_1 and R_2, respectively. In effect, we ordinarily assume that the coordinates are related by the "Galilean" transformations:

$$x_2 = x_1 - vt_1; \quad y_2 = y_1; \quad z_2 = z_1; \quad t_2 = t_1.$$

We assume, that is, that length and time are unaffected by motion, and that if a pulse of light has velocity c along the x-axis in the +ve direction in R_1, then its velocity in R_2 is c – v.

What Einstein realized was that if rates of clocks and lengths of rods *are* affected by relative motion, so that $x_2 = x_1 - vt_1$ and $t_2 = t_1$ are both false, then it is entirely possible that any given pulse of light has the same velocity c in both R_1 and R_2 – indeed, the same velocity c in *all* inertial reference frames.

It turns out that the thesis that light does have the same velocity in all inertial reference frames – which is implied by P_1 plus P_2 – suffices to fix uniquely just how the coordinates of R_1 and R_2 are related. All that we need, in addition, is that the relationship is *symmetric* (which may be said to be inherent in P_1 in any case), *linear* and *isotropic*. With these assumptions it is not hard to show that the coordinates of R_1 and R_2 are related by the following equations, the "Lorentz" transformations:

$$x_2 = \frac{x_1 - vt_1}{(1-v^2/c^2)^{1/2}} \quad y_2 = y_1 \quad z_2 = z_1 \quad t_2 = \frac{t_1 - vx_1/c^2}{(1-v^2/c^2)^{1/2}}$$

According to these equations, all but uniquely determined by P_1 plus P_2, relative motion contracts rods and makes clocks go slow, but in such a way that the velocity of light is c in *all* inertial frames. The miracle of reconciling P_1 and P_2 has been achieved.

Special relativity has a number of startling implications. One is that mass, along with the speed of clocks and length of objects, is affected by uniform motion, so that $m_2 = m_1/(1-v^2/c^2)^{1/2}$, where m_1 is the mass of an object in rest frame R_1, with respect to which the object is at rest, and m_2 is the mass of an object in R_2. Another implication – the most famous of all – is that *mass* is a form of *energy*, in accordance with $E = mc^2$.

From the standpoint of aim-oriented empiricism, special relativity is doubly significant. First, the way in which Einstein *discovered* special relativity exemplifies the method of discovery of aim-oriented empiricism, to the extent that Einstein used the method I have indicated above: namely, creating a new theory as the outcome of resolving a clash between two existing theories – thus creating greater conceptual and theoretical unification. Second, and quite strikingly, special relativity itself exemplifies aim-oriented empiricism, and in an important sense cannot be adequately understood within the framework of standard empiricism. For, as Einstein himself remarks in his "Autobiographical Notes", two pages on from the quotation given above:

> The universal principle of the special theory of relativity is contained in the postulate: The laws of physics are invariant with respect to the Lorentz-tranformations ... This is a restricting principle for natural laws, comparable to the restricting principle of the non-existence of the *perpetuum mobile* which underlies thermodynamics. (Einstein, 1969, p. 57)

Special relativity is thus a law of laws, a meta-law, a guiding principle, a heuristic and methodological rule to be employed in discovering and assessing physical theories – above all, for Einstein in 1905, to be employed as a heuristic tool for the discovery of the new theory to unify classical mechanics and electrodynamics. (When viewed from this perspective, what Einstein did in creating special relativity was to take a basic restricting principle of Newtonian mechanics, namely Galilean invariance – the pre-relativistic way to interpret P_1 – and modify this to make it compatible with P_2, thus forming a new restrictive principle, P_3, i.e. Lorentz invariance.) As a heuristic and methodological principle, special relativity has amply fulfilled Einstein's hopes for it. It played a vital role in the discovery of de Broglie's wave theory of matter, the so-called Klein-Gordon equation (first discovered by Schrödinger), the Dirac equation of the electron, quantum electrodynamics, quantum electro weak theory and quantum chromodynamics. In a modified form, it played a crucial role in the discovery of general relativity; and it continues to be relevant to superstring theory. Here, then, is a heuristic and methodological principle of enormous fruitfulness for all of theoretical physics, which can be formulated as the demand that acceptable theories must be Lorentz invariant. This demand – equivalent to the demand that space-time be Minkowskian (in the formulation of theories) – is not *merely* a methodological principle for, as we have seen, it has substantial physics in it. Special relativity is capable of being falsified and, from the standpoint of general relativity, it is false. All this is very hard to make sense of, or do justice to, within the confines of any version of standard empiricism, precisely because standard empiricism rejects the idea that methodological principles have physics, or metaphysics, built into them – there being, within standard empiricism, no (level 2) *metamethodological* framework within which rival (level 1) methodological principles can be rationally assessed. From the standpoint of aim-oriented empiricism, there is no difference in principle between an *ordinary* methodological rule, such as position invariance (acceptable laws and theories must be invariant with respect to change of position in space), and full Lorentz invariance. To both there correspond substantial *physical* or *metaphysical* principles, namely "space selects out no special position" or "space-time selects out no special inertial reference frame". Both may be false, and both therefore require critical scrutiny as science develops, in accordance with aim-oriented empiricism, and not dogmatic acceptance or rejection, as required by standard empiricism, with its fixed set of methodological principles (which no one yet has been able to formulate!). Standard empiricism differentiates sharply between the status of *position* and Lorentz invariance – only the former qualifying as a methodological rule of physics, the latter belonging exclusively to the *content* of physics, as a physical theory. But this does violence to Einstein's achievement; it does violence to the new way of doing physics inspired by Einstein, which precisely exploits the fruitful interplay between new theories and new heuristic and methodological principles (along the lines stipulated by aim-oriented empiricism).

3.3 Einstein's discovery of general relativity

Aim-oriented empiricism is even more explicit in Einstein's discovery of general relativity. Einstein exploits the same method of discovery. As before, there are two fundamental conflicting theories: Newton's theory of gravitation (T_1) and special relativity (T_2). These conflict because whereas Newton's theory implies that gravitational influences travel instantaneously, special relativity implies that such influences cannot travel faster than light. As before, Einstein searches for new principles that will guide him to a new unifying theory. His first step is to notice that there is a principle implicit in Newton's theory of gravitation (P_1) which, if generalized (P_1^*), makes it possible to generalize and *improve* the principle of relativity basic to special relativity (P_2). This latter principle seemed unsatisfactory to Einstein because of its restriction to some arbitrarily selected set of inertial reference frames all in uniform motion with respect to each other. Much more satisfactory would be a *general* principle of relativity (P_2^*) which asserts that the laws of nature have the same form in *all* reference frames, however they may be moving or accelerating with respect to each other. But this general principle of relativity seems impossible to implement. It is one thing to say, given a train moving uniformly through a station, that there are two equivalent descriptions: (1) train moving, platform at rest; and (2) train at rest, platform moving (in the opposite direction). It is quite another to say, given that the train crashes into the buffers at the end of the station, that there are two equivalent descriptions: (1) train de-accelerates, platform remains unaccelerated; and (2) train remains unaccelerated, platform de-accelerates. These are not equivalent descriptions: in the first, it is people in the train that suffer from violent de-acceleration, whereas in the second it is people on the platform that suffer. But consider now the following remarkable feature of Newton's law of gravitation (P_1): in a uniform gravitational field all objects accelerate equally, whatever their mass (essentially because inertial and gravitational mass are equal). Generalize this to form the *principle of equivalence* (P_1^*): no local phenomenon distinguishes between (a) uniform acceleration, and (b) being at rest in a uniform gravitational field. Whatever effect a gravitational field has on some phenomenon, it is the same as the effect that the equivalent acceleration would have in the absence of gravitation. This immediately has two consequences. First, it allows us to hold that *all* frames, however accelerating, are equivalent, as long as, in moving from one frame to another accelerating with respect to the first, we can invoke an additional, compensating gravitational field. Thus, in the case of the crashing train we have: (1) train de-accelerates, platform remains stationary; (2) train remains stationary, platform de-accelerates, and a gravitational field exists momentarily to compensate precisely for this de-acceleration. In *both* cases, it is the people in the train who suffer: according to the first description, because of de-acceleration; according to the second, because of the sudden gravitational field (and no compensating de-acceleration, as on the platform). The generalized principle of equivalence (P_1^*) makes it possible, in this way, to hold the generalized principle of relativity (P_2^*). The second consequence of the generalized principle of equivalence (P_1^*) is that, if correct, it enables us to discover the effects that uniform gravitational fields have on phenomena; all we need to do is to consider the effects of uniform acceleration and put these equal to the effects of the corresponding gravitational field in the absence of acceleration. The principle of equivalence (P_1^*) thus has great potential heuristic power for the discovery of the new theory of gravitation, to replace Newtonian theory.[5]

According to special relativity, acceleration affects geometry. Consider a flat, rapidly rotating disc. A rigid rod, of length L at the centre of the disc will, according to special relativity, only have length $L(1 - v^2/c^2)^{1/2}$ at the circumference, given that it is aligned with the motion of rotation which, at the circumference, has the value v. The geometry of the disc, as determined within a reference frame that is not rotating, will be non-Euclidean.[6] Uniform circular motion is accelerated motion. But if acceleration affects geometry, so, too, by the principle of equivalence, must gravitation. We have the possibility that gravitation *is* the (non-Euclidean) curvature of space-time – a possibility which, if true, would bring about a tremendous conceptual unification in the foundations of physics (namely the unification of gravitation and space-time geometry). Postulate therefore that gravitation is indeed the curvature of space-time. The presence of matter curves space-time; and matter moves along geodesics in this curved space-time. Curved space-time can always be reduced to flat Minkowskian space-time in any infinitesimal region by an appropriate choice of coordinate system, in accordance with the principle of equivalence given its final local formulation (P_1^{**}). What remains to be done is to formulate the precise way in which energy–momentum affects the Riemannian curvature of space-time.[7] The field equations of general relativity are the simplest possible solution to this problem. Indeed, granted that the equations involve derivatives no higher than the second, the field equations are determined uniquely to be:

$$R_{ab} - 1/2 g_{ab} R = 8\pi G T_{ab}$$

Here R_{ab} is the Ricci tensor of the metric g_{ab} (the Ricci tensor being derivable from the Riemannian curvature tensor by contraction), R is the Ricci scalar (formed from R_{ab} by contraction), T_{ab} is the energy–momentum tensor, and G is Newton's constant of gravitation.[8] We have arrived at T_3, which reduces to special relativity (T_2) in the absence of gravitation, and which approaches Newtonian theory (T_1) in the limit as gravitational fields become weak and velocities become low in comparison with the velocity of light.

3.4 Did Einstein really employ aim-oriented empiricism?

Does Einstein really put aim-oriented empiricism into practice in developing the special and general theories of relativity, in the way I have just sketched? Is there, here, a genuine method of discovery, given that Einstein failed for over thirty years to develop a satisfactory unified field theory? A few comments are in order.

The full aim-oriented empiricist method of discovery involves the tackling of at least five kinds of problems: (1) conflicts between experimental results and theory, (2) conflicts between two or more well-established fundamental theories, (3) conflicts between such theories and the best available blueprint for physics at level 3 in Figure 2.1 (see Chapter 2),[9] (4) conflicts inherent in the best blueprint itself (or between rival blueprints), and (5) conflicts between established physical theory and the level 4 thesis of physicalism. It could be argued that Einstein only exploits a small part of this method of discovery, in that he is primarily concerned with type (2) problems (and type (1) problems where relevant). But this is, I think, wrong for a number of reasons.

First, the metaphysical thesis that the basic laws of nature have a *unified* structure is an implicit or explicit assumption in all of Einstein's deliberations, which means type (4) problems are, for Einstein, fundamental.

Second, in developing special and general relativity it is precisely the pre-existing metaphysical blueprints of classical physics that Einstein is led to transform: basic assumptions about the nature of space, time, energy, mass, force. In developing new *principles* – such as the principle of Lorentz invariance or the principle of equivalence – Einstein is, at one and the same time, modifying pre-existing blueprint ideas (Newtonian space-time being transformed into Minkowskian space-time, which is in turn transformed into the Riemannian space-time of general relativity).

Third, lurking behind the type (2) problems which concern Einstein (involving clashes between *theories*) there are type (4) *blueprint* problems. Consider the type (2) problem that led to special relativity – the clash between Newtonian mechanics and Maxwellian electrodynamics or, more specifically, the clash between Galilean invariance and the thesis that the constancy of the velocity of light is a law of nature. Around 1900, as we have seen, there was an obvious solution to this problem: interpret electrodynamics in terms of the aether, regard the constancy of the velocity of light as being relative to the aether, and expect Galilean invariance to break down for high velocities with respect to the aether. This amounts, of course, to adopting a blueprint for physics – the aether blueprint. In formulating the problem in the way in which he did, Einstein is in effect *rejecting* this aether blueprint; he is adopting Faraday's view that the field is fundamental, and does not require an underlying aether to make it comprehensible. As I have argued elsewhere (see note 4), there are good reasons for preferring what may be termed the Faraday blueprint to the aether blueprint. The important point, however, is that in formulating his type (2) problem in the way in which he did (crucial for the development of special relativity), Einstein is in effect interpreting Newton's and Maxwell's theories to be two equally fundamental, rival theories, each with its rival blueprint, namely, the Newtonian (or Boscovichean) blueprint of point particles surrounded by spherically symmetrical, rigid fields, and the Faraday field blueprint with variations in the field being transmitted at some *finite* velocity. There is, in short, a type (4) blueprint problem inherent in the type (2) problem that led Einstein to special relativity. This type (4) problem may be formulated, not as a problem about how to reconcile, or choose between, two rival blueprints, but rather as the problem of how to resolve the clash that results from attempting to unify the two blueprints in such a way as to accommodate charged point particles *and* a field.

Fourth, there are grounds for holding that Einstein's fundamental problem soon after 1900 was the type (4) *blueprint* problem I have just indicated – the problem of understanding how charged point particles can interact with the field, or the problem of unifying point particle and field. It is a striking fact that Einstein's three great papers of 1905 can all be interpreted as exploring aspects of this fundamental problem. We have just seen that this is true of the paper introducing special relativity. It is also true of the Brownian motion paper, concerned to establish the existence of atoms – the existence of the particle-like aspect of reality. And it is true above all of the paper which put forward the idea that light has a particle-like aspect in accordance with $E = nh\nu$ (where E is the energy and ν the frequency of the light, h is Planck's constant and n is some integer, the number of light quanta present), this "heuristic" hypothesis of light quanta then being used to explain the photoelectric effect. Here the classical particle/field problem is intensified to an extraordinary extent in that the field itself is revealed to have a particle-like aspect.[10]

As it happens, Einstein himself makes clear in his "Autobiographical Notes" (Einstein, 1969) that he held the classical particle/field problem to be of fundamental importance.[11] Having explained that theories are to be critically assessed from the two distinct standpoints of empirical success and "inner perfection" (unity or comprehensibility) – which in itself commits Einstein to aim-oriented empiricism (see below) – he goes on to assess critically Newtonian mechanics and Maxwellian electrodynamics from the standpoint of inner perfection. We have here, incidentally, an adjunct to, and refinement of, Einstein's method of discovery: *one* theory is here taken at a time, and is assessed from the standpoint of "inner perfection" – from the standpoint, that is, of the capacity of the theory to provide a "perfect" blueprint for all of physics as far as the *form* of the theory is concerned. (In indicating the "inner perfection" *defects* of a theory, one in effect indicates, at least in general terms, what would constitute a "perfect" theory: one indicates, that is, a blueprint.) Einstein discusses six "inner perfection" defects in Newtonian mechanics, namely: (1) arbitrariness in the determination of inertial reference frames from an infinity of alternatives, and inadequacy of introducing absolute space (with respect to which all bodies have absolute acceleration as a solution to this problem); (2) two distinct basic laws (and not *one*), namely: (a) the law of motion ($F = ma$), and (b) the expression for force or potential energy ($F = Gm_1m_2/d^2$); (3) arbitrariness of (b) given (a), there being endlessly many equally good possibilities for (b) given (a); (4) the possibility of the force law being determined by the structure of space (the form of the force law being suggestively simple when viewed in geometrical terms), and yet the *failure* to exploit this possibility; (5) the *ad hoc* character of the equality of inertial and gravitational mass; and (6) unnaturalness of energy being split into two forms, kinetic and potential (see Einstein, 1969, pp. 27–31). As far as electrodynamics is concerned, Einstein discerns one basic defect associated with interpreting the field equations as applying to matter and, in the case of the vacuum, to the aether. Einstein argues (perhaps not altogether accurately) that this defect was overcome by Lorentz in reinterpreting the field equations to hold essentially only for the vacuum, with matter, in the form of charged particles, being the source of the field. Einstein then remarks: "If one views this phase of the development of theory critically, one is struck by the dualism which lies in the fact that the material point in Newton's sense and the field as continuum are used as elementary concepts side by side" (Einstein, 1969, p. 37). Einstein explains why attempts to overcome this basic defect by eliminating the point particle do not succeed; and he concludes: "Accordingly, the revolution begun by the introduction of the field was by no means finished. Then it happened that, around the turn of the century ... a *second fundamental crisis* set in" (ibid., italics mine) – namely the crisis engendered by the first step towards quantum theory, Planck's quantum explanation of his empirical radiation law. If this is the *second* fundamental crisis, then the first is particle/field dualism of classical physics. As it happens, the two crises are intimately interrelated, since Planck's law and quantum theory deal with the *interaction* of field and matter.

There are good grounds, then, for holding that Einstein was concerned with problems from type (2) to type (5), as defined above, and type (1) problems where relevant, a type (4) problem of special concern to Einstein being the problem of how to unify point particles and field.

But did Einstein really invent an authentic method of discovery in view of his failure, during the last thirty years of his life, to discover the unified field theory he so ardently sought?

One reply can be made immediately: the method of discovery, indicated above, though rational, is also non-mechanical and fallible. The failure of the method to lead to a good fundamental new theory over a period of thirty years – even in the hands of Einstein – does not prove that the method is inauthentic.

But there is a much more important reply to be made. Einstein did not use his method of discovery in seeking to formulate his unified field theory. Or rather Einstein *misapplied* this method, in a quite elementary way.

After around 1930, the two fundamental theories that stand in most glaring contradiction with each other are general relativity and quantum theory. In order to implement Einstein's rational method of discovery from about 1930, the first step to take is to extract basic principles, P_1 and P_2, from general relativity and quantum theory respectively, which contradict each other – this even perhaps being the *nub* of the contradiction between the two theories. The task then is to modify P_1 and P_2 (or something else) to form P_3, a new principle which guides us to a new unified theory, T_3, unifying general relativity and quantum theory.[12]

Einstein did none of this. Instead he took as his two theories general relativity and classical electrodynamics and sought to unify these two theories, to form a theory which applied to all phenomena, including quantum phenomena. One may well have doubts as to whether these two theories really do fundamentally contradict each other – even though the theories are clearly *two* distinct theories and not *one* unified theory. They are at least both field theories; they both incorporate Lorentz invariance, at least locally; and they are both classical and deterministic. What is dramatically apparent is that the fundamental contradiction of theoretical physics after 1930 concerns, not the clash between classical general relativity and classical electrodynamics, but rather the clash between general relativity and quantum theory. (One can add that it is perverse to continue to take the unification of gravitation and electromagnetism as *the* unification to strive for, sufficient to create the comprehensive unified field theory, after the discovery of the strong and weak forces in addition to the forces of gravitation and electromagnetism.)

Why did Einstein so crudely and wilfully misapply his rational method of discovery? The answer is straightforward: because of his abhorrence of quantum theory given its orthodox interpretation (OQT).[13] Einstein was absolutely correct to find OQT fundamentally defective from the crucial standpoint of "inner perfection". As I shall argue in a moment, Einstein's attitude towards OQT exemplifies yet again his (sound) commitment in scientific practice to aim-oriented empiricism and scientific realism. Where Einstein went *wrong* was to conclude that quantum theory was therefore entirely devoid of heuristic value – that it "offers no useful point of departure for future development" (Einstein, 1969, p. 87).

What is striking about this is that it is actually a vital feature of Einstein's method of discovery that one deals with theories that are intrinsically *defective*. The defects are clues as to how the theory may be fruitfully modified. As we have seen above, Einstein indicates a number of fundamental defects inherent in Newtonian mechanics and Maxwellian electrodynamics. Einstein even knew, by 1901, as a result of Planck's work, that both theories are fundamentally incorrect. This did not stop him taking these theories as "points of departure". Indeed, it is the *defects* in the theories, as perceived by Einstein, which make his method of discovery so successful: for it is these defects which indicate how the theories are to be modified to overcome the contradictions between them. For Einstein to argue, after 1930, that the *defective* character of quantum theory ensures that the theory cannot form a proper point of departure does violence to the very heart of Einstein's own earlier method of discovery, used in the discovery of special and general relativity with such striking success.

Why did Einstein fail to recognize the fairly obvious point just made? In essence, because his abhorrence of OQT was so intense, so profound, that it was emotionally impossible for him to work seriously with the theory. He did not want to contribute to what he interpreted as a sickness which had entered physics, and which he regarded as symptomatic of the basic sickness of our times. In a sense, Einstein turned his back on quantum theory, and devoted himself to the task of unifying general relativity and classical electromagnetism as a kind of moral protest against the tenor of our times.

In order to substantiate this point I must now break off my discussion of Einstein's successes and failures in implementing aim-oriented empiricism so that I can consider in a little more detail the question of Einstein's attitude to OQT.

3.5 Einstein and quantum theory

His mature attitude can be summarized like this. From the standpoint of empirical criteria, OQT must be judged to be an immense success. From the equally important standpoint of criteria having to do with "inner perfection", with unification, OQT must be judged to be a disaster. This is because the theory cannot be interpreted to be about some hypothetical reality. It was not so much the lack of *determinism* that came to worry Einstein as the lack of *realism*. In his "Autobiographical Notes" he puts it like this:

> Physics is an attempt conceptually to grasp reality as it is thought independently of its being observed. In this sense one speaks of "physical reality". In pre-quantum physics there was no doubt as

> to how this was to be understood. In Newton's theory reality was determined by a material point in space and time; in Maxwell's theory, by the field in space and time. (Einstein, 1969, pp. 82–3)

Einstein goes on to point out that as far as OQT is concerned, there is no quantum equivalent to the classical material point or field. OQT makes probabilistic predictions about the results of performing measurements on an ensemble of similarly prepared systems, but cannot be interpreted as specifying the physical state of the individual system as it evolves in space and time independent of measurement. As Einstein puts it in volume 2 of the same book, in his "Reply to Criticisms":

> What does not satisfy me … [about OQT], from the standpoint of principle, is its attitude towards that which appears to me to be the programmatic aim of all physics: the complete description of any (individual) real situation (as it supposedly exists irrespective of any act of observation or substantiation). (Einstein, 1969, p. 667)

In a letter to Schrödinger in 1950, Einstein expresses himself even more emphatically:

> You are the only contemporary physicist, besides Laue, who sees that one cannot get around the assumption of reality – if only one is honest. Most of them simply do not see what sort of risky game they are playing with reality – reality as something independent of what is experimentally established. They somehow believe that the quantum theory provides a description of reality, and even a *complete* description; this interpretation is, however, refuted, most elegantly by your system of radioactive atom + Geiger counter + amplifier + charge of gun powder + cat in a box, in which the ψ-function of the system contains the cat both alive and blown to bits. Is the state of the cat to be created only when a physicist investigates the situation at some definite time? Nobody really doubts that the presence or absence of the cat is something independent of the act of observation. But then the description by means of the ψ-function is certainly incomplete, and there must be a more complete description. If one wants to consider the quantum theory as final (in principle), then one must believe that a more complete description would be useless because there would be no laws for it. If that were so then physics could only claim the interest of shopkeepers and engineers; the whole thing would be a wretched bungle. (Przibram, 1967, p. 39)

Einstein's opposition to OQT – arising from the lack of realism of the theory – was implacable, even vehement. It was this, after all, which had led to the great rupture between mainstream theoretical physics and Einstein's own work. From 1905 to 1926 Einstein was at the centre of developments in theoretical physics. But from 1926 onwards the ways parted, essentially because Einstein was not able to bring himself to contribute to the development of OQT (confining himself to critical analysis of it). Robert Shankland, who met Einstein a number of times during the years 1950–4, has remarked on the uncharacteristic vehemence of Einstein's opposition to OQT:

> His well-known scepticism on this subject [of quantum mechanics] was clearly evident and his comments on both the subject itself and its leading proponents were often highly critical and even emotional, in contrast to his restrained and quiet explanations of relativity. (In French, 1979, p. 39)

Something of the strength of Einstein's opposition to OQT also emerges from a correspondence which he had with Born on the subject. Einstein makes it quite clear that he finds OQT unacceptable because of its lack of realism. Born persists in a stance of somewhat patronizing incomprehension, Einstein rather sharply writes that he does not wish to continue the discussion, and Pauli is obliged to step in and tick Born off for misunderstanding Einstein, even though he agrees with Born that Einstein's position amounts to asking "how many angels are able to sit on the point of a needle" (see Born, 1971, pp. 199–229).

The strongest statement of Einstein against OQT that I have come across is quoted by Fine (1986, p. 1): "This theory [the present quantum theory] reminds me a little of the system of delusions of an exceedingly intelligent paranoiac, concocted of incoherent elements of thoughts."

These quotations establish beyond all possible doubt that Einstein was committed to full-blooded scientific realism, at least as far as the basic aim of physics is concerned.[14]

Einstein is absolutely correct to hold that OQT cannot be interpreted realistically. As he points out in his letter to Schrödinger, if one attempts to interpret the ψ-function of OQT as providing a complete description of reality, one is led to the (apparently) absurd conclusion that Schrödinger's cat persists as a superposition of being alive and being dead until we open the box and look. And similarly, we would have to conclude that the outcome of any quantum measurement is not some definite state of the apparatus but rather a superposition of macroscopically distinct states – the superposition only collapsing miraculously when we look. The simplest way to demonstrate the impossibility of interpreting OQT realistically, however, arises from the following consideration. If we interpret the ψ-function as describing quantum reality directly, and exclude *measurement* from the basic postulates of the theory, we are left with a theory that is fully deterministic, since quantum states, corresponding to ψ-functions, evolve deterministically in accordance with Schrödinger's equation. Such a version of OQT fails to make contact with the most basic feature of the quantum world – its probabilistic character. In short, just as Einstein declares, OQT must be regarded as a theory which makes probabilistic predictions about the results of performing measurements on systems, but which does not specify the actual physical state of the individual system in the absence of measurement.[15]

How did this extraordinary state of affairs arise? Essentially because, as quantum theory (QT) developed with the work of Bohr, Heisenberg, Schrödinger, Born and others, no solution was found to the quantum wave/particle problem. As we have seen, this problem was first discovered by Einstein with his invention of light quanta – or "photons" as they subsequently came to be called. The problem was further intensified in 1923 when de Broglie proposed that electrons, up till then believed to be particles, have a wave-like aspect associated with them, as was subsequently confirmed experimentally by Davisson and Germer. In order to develop QT as a realistic theory, it would have been necessary to solve the quantum wave/particle problem in such a way that it is possible to specify, consistently and precisely, what sort of physical entities photons and electrons are as they evolve in space and time independently of measurement. This did not happen. Instead, Heisenberg invented matrix mechanics in 1925, intending, from the outset, that the theory should predict the outcome of measurements but should remain silent about what exists physically in the absence of measurement. Schrödinger invented wave mechanics in 1926 with the hope that the wave aspect of quantum entities would turn out to be fundamental. This hope was dashed when it became clear that the ψ-function could not be regarded as describing quantum reality directly, but had to be interpreted as containing probabilistic information about the results of performing *measurements* on an ensemble of similarly prepared systems – as Born was the first to point out.

We can begin to see some of the reasons for Einstein's vehement rejection of OQT as a satisfactory theory (despite its immense empirical success). It was Einstein after all who, in a sense, invented quantum theory. Planck introduced the idea that the energy E of an oscillator of frequency ν is quantized in accordance with $E = nh\nu$ as a calculational device, not as a new hypothesis incompatible with classical physics. Planck's aim was to deduce his empirical law of black-body radiation from the basic postulates of classical physics. He was dismayed to discover that the quantization of energy *contradicted* classical physics, and he spent the next fifteen years or so trying to remove this defect from his derivation. It was Einstein, and initially Einstein alone, who appreciated that Planck's work spelt the downfall of classical physics, a new beginning being required. In this sense, Einstein initiated quantum theory with his paradoxical "heuristic" hypothesis that light consists of discrete quanta with energy $E = h\nu$, even though light also undeniably has a continuous wave-like character. For Einstein around 1905, the fundamental task of the new theory, needed to replace classical physics, would be to solve the riddle of the nature of quantum reality in view of its ostensibly contradictory particle and field aspects. No wonder Einstein was dismayed when the new theory was developed deliberately to *evade* and not to *solve* this basic quantum riddle.

But there is more than this to Einstein's opposition to OQT. As I have stressed above, the failure to solve the quantum wave/particle problem ensures that OQT cannot be interpreted realistically, which in turn ensures that OQT must be interpreted as making (probabilistic) predictions about the results of performing measurements. But this in turn has a variety of disastrous – though rarely noticed – consequences. For it means that OQT only issues in actual physical predictions if some part of *classical physics* (CP) is adjoined to OQT for a treatment of measurement. OQT alone can only issue in *conditional* predictions of the type: *if* a measurement of observable A is made, the outcome will be one or other of the values ($a_1 \ldots a_n$) with probabilities ($p_1 \ldots p_n$), with:

$$\sum_{r=1}^{n} p_r = 1$$

And even this goes too far: strictly speaking, according to OQT, a quantum mechanical state ψ can only be attributed to a system in so far as the system has been subjected to some preparation procedure, which must be specified by means of CP. Thus OQT, devoid of CP, has no physical content whatsoever. It is only OQT + CP which has physical content. But OQT + CP, considered as a fundamental theory of physics, is a disaster. It is (i) grossly *ad hoc* or aberrant, in that it consists of two conceptually incoherent parts, OQT and CP. It is (ii) *imprecise*, because the circumstances in which CP is to be applied are only specified in terms of measurement, and the notion of measurement cannot be made precise (Maxwell, 1972b). It is (iii) *ambiguous* because the theory does not decide unambiguously between probabilism and determinism. It is (iv) *non-explanatory*, not only because of the *ad hoc* character of the theory, but also because the theory is obliged to presuppose some part of what it is intended to explain, namely CP. The theory is (v) *severely restricted in scope* in that it cannot be applied to conditions which exclude the possibility of measurement, such as early states of the universe. It (vi) *excludes the possibility of quantum gravity and quantum cosmology*, since these would require measuring instruments, described in terms of CP, to exist outside space-time and beyond the cosmos, and clearly this is not possible. (These are points I have developed over a number of years: see Maxwell, 1972b, 1973a, 1973b, 1975, 1976b, 1982, and especially 1988, pp. 1–8.)[16]

These six gross defects, especially (i) to (iv), ensure that OQT + CP is unacceptable as a fundamental physical theory. OQT + CP cannot justifiably be held to be part of theoretical scientific *knowledge*. (OQT + CP encompasses a great deal of *empirical* knowledge, but cannot be said to be an acceptable theory, constituting *theoretical* knowledge.) OQT + CP is as unacceptable as the absurd, empirically successful but grossly *aberrant* theories considered in previous chapters. In practice this point is beyond dispute. The vast majority of physicists, from soon after 1926 down to the present day (or down to 1993, when a version of this chapter was first published), have regarded OQT as an entirely acceptable part of scientific knowledge: they have been able to do this because they have been able to pretend that OQT + CP is really just OQT. In almost all the textbooks and physical journals, quantum theory is treated *as if* its postulates are purely quantum mechanical ones. As a result, OQT appears to be thoroughly non-*ad hoc*, precise and explanatory, as conceptually coherent and unified as any classical theory. But all this is an illusion. It is the outcome of pretending that the *physical* theory – the theory that has *physical content* – is OQT rather than OQT + CP. No such thing is possible. OQT, devoid of CP, has no physical content whatsoever. Only an all-pervasive intellectual dishonesty makes it possible to pretend that OQT alone has physical content (or that OQT + CP is really, somehow, just OQT).

All this demonstrates just how sound Einstein's instincts were when he judged OQT to be an unacceptable theory. How unerringly correct Einstein was to declare that Bohr and company "do not see what sort of risky game they are playing with reality"; and how sound his comparison is between OQT and "the system of delusions of an exceedingly intelligent paranoiac, concocted of incoherent elements of thoughts", namely QT *and* CP!

It is important to appreciate that the above six defects of OQT, even though consequences of the impossibility of interpreting OQT realistically, are not defects that only realists will recognize. Any physicist, whether realist or instrumentalist, aim-oriented empiricist or standard empiricist ought, in practice, to regard the above defects as sufficient grounds for finding OQT unacceptable.[17] We have here, in effect, an additional general argument against instrumentalism and for realism. Any fundamental physical theory, and not just OQT, which is interpreted instrumentalistically as predicting only the (observable) outcomes of measurements will be, in the same way, unacceptably (i) *ad hoc*, (ii) imprecise, and (iv) non-explanatory. In other words, theoretical unity implies realism; anti-realism, built into a physical theory (as it is built into OQT) must inevitably, at some point, lead to unacceptable *ad hocness* or aberrance (see Maxwell, 1993b, where this argument is spelled out in more detail).

Even though it is not essential to be an aim-oriented empiricist in order to find OQT unacceptable, it helps. For aim-oriented empiricism provides a clear and cogent *raison d'être* for finding OQT unacceptable even though the theory has met with such outstanding empirical success. Standard empiricism, on the other hand, can provide no such *raison d'être*. If scientific theories ought in the end to be judged solely on the basis of empirical success and failure, then there can be no rational grounds for rejecting OQT, given its immense, its unprecedented, empirical success.

It is in just this way that most of Einstein's contemporaries tended to view his rejection of OQT: as the outcome of unscientific, metaphysical prejudice, or even as an indication of "senility" (as Einstein himself put it). Even Abraham Pais, so knowledgeable about, and so sympathetic towards Einstein, nevertheless regards Einstein's objections to OQT as "unfounded" (Pais, 1982, p. 464).

Einstein's attitude towards OQT, so strikingly at odds with most of his contemporaries, provides further evidence in support of my contention that aim-oriented empiricism is implicit in Einstein's scientific work. If Einstein had assessed OQT in purely standard empiricist terms, he could have had no rational grounds for rejecting OQT – no grounds even for rejecting OQT as a "point of departure" (since this is an *epistemological* judgement, to the effect that OQT is fundamentally *false*).

From the standpoint of standard empiricism, Einstein's implacable opposition to OQT is just plain irrational prejudice. From the standpoint of aim-oriented empiricism, however, Einstein's rejection of OQT emerges as entirely well founded, scientific, rational and objective. OQT is entirely acceptable from the standpoint of empirical considerations, but unacceptable from the equally important standpoint of theoretical unity, comprehensibility. The scandal is that the majority of contemporary physicists do not see this obvious point – or did not, until quite recently.

What is irrational, in other words, is not Einstein's rejection of OQT, but the majority *acceptance* of OQT, the general *blindness* to its gross defects. Einstein, I believe, held this to be the result of the fact that too many physicists put fame before understanding the universe. Einstein felt that, given a choice between winning a Nobel Prize and improving our understanding of the universe, too many physicists would choose the former over the latter. This, for Einstein, amounted to a betrayal of the soul of theoretical physics, the pursuit of a corrupt goal, fame (not for Einstein so very different from the pursuit of power), in preference to the pursuit of the noble goal of improving understanding. And this in turn was, for Einstein, I believe, characteristic of a general sickness of our age: the pursuit of shallow or corrupt goals in life in preference to goals of genuine value.[18] Here is the source of Einstein's inability to contribute to OQT after 1926. It is in this sense that Einstein's pursuit of his unified field theory is a kind of moral protest; this was the clearest way in which he could express his conviction as to what physics ought to be, at its best.

Einstein did not get everything right about OQT. He assumed that the ostensibly highly non-local features of OQT, which seem to contradict special relativity, do not correspond to reality. Here he was wrong.

If two particles, 1 and 2, interact at time t_1 and then separate widely then, in certain circumstances, a measurement performed on 1 at time t_2 enables one to predict with certainty what the result would be of measuring 2. A measurement of the momentum of 1 enables one to predict the precise momentum of 2; or, alternatively, a measurement of the position of 1 enables one to predict the position of 2. It is possible that 2 only acquires a precise momentum or position at time t_2, when one or other kind of measurement is performed on 1. This possibility requires that an influence of some kind travels instantaneously from 1 to 2 to inform 2 as to whether it should acquire a precise momentum or position. If we reject the existence of such instantaneous influences, then in order to explain the correlations between measurements on 1 and 2 we are obliged to hold, it seems, that these correlations are the outcome of correlations established at time t_1, when 1 and 2 interact. But this has the consequence that at time t_2 particle 2 must simultaneously have a precise position *and* momentum (since 2, by hypothesis, cannot "know" instantaneously, at time t_2, whether particle 1, far away, is subjected to a position or momentum measurement). But, according to OQT, no system can be in a state which corresponds to having simultaneously a precise position *and* momentum. Thus OQT implies that correlations cannot be established at t_1 when 1 and 2 interact; they must be established instantaneously, at t_2, when one or other measurement is performed on 1.

That OQT does have this highly non-local character was discovered by Einstein, and was expounded in a famous paper by Einstein, Podolsky and Rosen (1935). Because of the evident clash with special relativity, Einstein concluded that this kind of non-local prediction of OQT is *false*. Particle 2 *does* have a precise position and momentum at time t_2 irrespective of whether measurements are performed on 1 or not, and QT must be interpreted as a purely statistical theory which gives only an *incomplete* description of the evolution of the individual system.

Einstein held that the only reasonable option available was to interpret QT in this way, as an inherently incomplete, statistical theory of "particles". There can be no doubt that this reinforced his conviction that QT did not constitute a proper starting point for future developments, which in turn reinforced Einstein's search for a unified field theory.

Subsequent developments, due to Bohm (Bohm and Aharonov, 1957), Bell (1964), Aspect et al. (1982) and others, have shown that Einstein was *wrong* to dismiss the non-local predictions of OQT as not corresponding to reality: these predictions have now been experimentally confirmed!

This concludes my case for saying Einstein invented and applied aim-oriented empiricism in scientific practice in developing the special and general theories of relativity, and in critically examining quantum theory.

But, it may be asked, did Einstein explicitly *advocate* aim-oriented empiricism? I turn now to a discussion of this question.

3.6 Did Einstein advocate aim-oriented empiricism?

There can be, to begin with, no doubt that Einstein devoted his life to the goal of discovering the unified structure of the universe and that, for him, this constituted an entirely proper aim for science, indeed, the noblest motive for pursuing scientific inquiry. Something of what the desire to understand meant to Einstein emerges from the following passage:

> The most beautiful experience we can have is the mysterious. It is the fundamental emotion which stands at the cradle of true art and true science. Whoever does not know it and can no longer wonder, no longer marvel, is as good as dead, and his eyes are dimmed. It was the experience of mystery, even if mixed with fear, that engendered religion. A knowledge of the existence of something we cannot penetrate, our perceptions of the profoundest reason and the most radiant beauty, which only in their most primitive forms are accessible to our minds – it is this knowledge and this emotion that constitute true religiosity; in this sense, and in this alone, I am a deeply religious man. I cannot conceive of a God who rewards and punishes his creatures, or has a will of the kind that we experience in ourselves. Neither can I nor would I want to conceive of an individual that survives his physical death; let feeble souls, from fear or absurd egoism, cherish such thoughts. I am satisfied with the mystery of the eternity of life and with the awareness and a glimpse of the marvelous structure of the existing world, together with the devoted striving to comprehend a portion, be it ever so tiny, of the Reason that manifests itself in nature. (Einstein, 1973, p. 11)

On one occasion in 1925 he expressed himself to the novelist Esther Salaman in the following terms:

> I want to know how God created this world. I'm not interested in this-or-that phenomenon, in the spectrum of this-or-that element. I want to know His thoughts, the rest are details. (Salaman, 1979, p. 22)

That a basic aim of science is to unify all phenomena is affirmed in numerous passages, such as this, from 1936:

> The aim of science is, on the one hand, a comprehension, as *complete* as possible, of the connection between the sense experiences in their totality, and, on the other hand the accomplishment of this aim *by use of a minimum of primary concepts and relations*. (Seeking, as far as possible, logical unity in the world picture, i.e. paucity in logical elements.) (see Einstein, 1973, p. 293)

Einstein also makes it clear that science at its best assumes that this goal of unification is realizable:

> Certain it is that a conviction, akin to religious feeling, of the rationality or intelligibility of the world lies behind all scientific work of a higher order. This firm belief, a belief bound up with deep feeling in a superior mind that reveals itself in the world of experience, represents my conception of God. In common parlance this may be described as "pantheistic" (Spinoza). (First published 1929; see Einstein, 1973, p. 262)

And on another occasion:

> From the very beginning there has always been present the attempt to find a unifying theoretical basis for all [the] single sciences, consisting of a minimum of concepts and fundamental relationships, from which all the concepts and relationships of the single disciplines might be derived by logical process. This is what we mean by the search for a foundation of the whole of physics. The confident belief that this ultimate goal may be reached is the chief source of the passionate devotion which has always animated the researcher. (First published 1940; see Einstein, 1973, p. 324)

As for scientific realism, Einstein expresses himself with his usual clarity and brevity:

> The belief in an external world independent of the perceiving subject is the basis of all natural science. Since, however, sense perception only gives information of this external world or of "physical reality" indirectly, we can only grasp the latter by speculative means. (First published 1931; see Einstein, 1973, p. 266)

And, on another occasion, as we have already seen:

> Physics is an attempt to grasp reality as it is thought independently of its being observed. (Einstein, 1969, p. 82)

It is all summed up succinctly in a letter to Cornelius Lanczos in 1942:

> You are the only person I know who has the same attitude toward physics as I have: belief in the comprehension of reality through

something basically simple and unified. (Dukas and Hoffmann, 1979, p. 68.)

All this might seem more than enough to demolish decisively the views of those, like Fine and Popper, who hold that Einstein upheld some version of standard empiricism. Unfortunately it is not. In all the above quotations, Einstein can be interpreted as asserting no more than that he, and science, seek to discover, and presuppose the existence of, a unified structure to the universe, *in the context of discovery*. According to this interpretation, Einstein would hold that, *in the context of justification*, nothing must be permanently assumed about the nature of the universe, the sole aim being empirical adequacy, empirical considerations *alone* in the end deciding what is to constitute theoretical scientific knowledge.

On this issue – the crucial issue which divides off standard from aim-oriented empiricism – Einstein seems to have wavered. Consider the following passage:

> Can we hope to be guided safely by experience at all when there exist theories (such as classical mechanics) which to a large extent do justice to experience, without getting to the root of the matter? I answer without hesitation that there is, in my opinion, a right way, and that we are capable of finding it. Our experience hitherto justifies us in believing that nature is the realization of the simplest conceivable mathematical ideas. I am convinced that we can discover by means of purely mathematical constructions the concepts and the laws connecting them with each other, which furnish the key to the understanding of natural phenomena. Experience may suggest the appropriate mathematical concepts, but they most certainly cannot be deduced from it. *Experience remains, of course, the sole criterion of the physical utility of a mathematical construction*. But the creative principle resides in mathematics. In a certain sense, therefore, I hold it true that pure thought can grasp reality, as the ancients dreamed. (First published 1933; see Einstein, 1973, p. 274, italics mine)

This comes tantalizingly close to aim-oriented empiricism. A central tenet of aim-oriented empiricism is that we are rationally entitled to assume that the universe is knowable – there being some fallible, non-mechanical but rational method of discovery available to us – the knowability of the universe implying its comprehensibility. It is just this key element of aim-oriented empiricism which Einstein asserts here, his epistemological and methodological instincts as usual getting almost everything right. (The point is also brilliantly made in one of Einstein's most famous sayings: *"Raffiniert ist der Herrgott, aber boshaft ist er nicht"*– "God is sublime but not malicious".) Unfortunately, in the quotation given above, the italicized sentence provides Popper or Fine with the perfect excuse for interpreting the passage as a defence of standard empiricism. One can argue, of course, that the whole passage only really makes sense if interpreted as asserting: experience remains the *sole* criterion of physical utility *granted that we restrict our attention to simple, unified theories*. This would of course *violate* standard empiricism. But the text, as it stands, is sufficiently ambiguous to leave the matter undecided. Consider next the following passage:

> The very fact that the totality of our sense experiences is such that by means of thinking (operations with concepts, and the creation and use of definite functional relations between them, and the coordination of sense experiences to these concepts) it can be put in order, this fact is one which leaves us in awe, but which we shall never understand. One may say "the eternal mystery of the world is its comprehensibility". It is one of the great realizations of Immanuel Kant that the postulation of a real external world would be senseless without this comprehensibility.
>
> In speaking here of "comprehensibility", the expression is used in its most modest sense. It implies: the production of some sort of order among sense impressions, this order being produced by the creation of general concepts, relations between these concepts, and by definite relations of some kind between the concepts and sense experience. It is in this sense that the world of our sense experiences is comprehensible. The fact that it is comprehensible is a miracle.
>
> In my opinion, nothing can be said *a priori* concerning the manner in which the concepts are to be formed and connected, and how we are to coordinate them to sense experiences. In guiding us in the creation of such an order of sense experiences, success alone is the determining factor. All that is necessary is to fix a set of rules, since without such rules the acquisition of knowledge in the desired sense would be impossible. One may compare these rules with the rules of a game in which, while the rules themselves are arbitrary, it is their rigidity alone which makes the game possible. However, the fixation will never be final. It will have validity only for a special

field of application (i.e., there are no final categories in the sense of Kant). (First published 1936; see Einstein, 1973, p. 292)

This, once again, is tantalizingly close to aim-oriented empiricism. Einstein recognizes clearly that only in a very special kind of universe – a comprehensible universe – is scientific explanation and understanding possible. He recognizes that the particular way the universe is assumed to be comprehensible at any stage in the development of science will lead to rules or principles – such as Galilean or Lorentz invariance, the principle of equivalence, conservation of momentum and energy – without which physics would be impossible. And he points out that these rules are not final: it is to be expected that they will change as science advances. All this accords beautifully with aim-oriented empiricism. What violates aim-oriented empiricism is the suggestion that there is no Kantian synthetic *a priori* proposition built into scientific knowledge. According to aim-oriented empiricism, there are two such propositions, namely, *partial knowability* and *meta-knowability* at levels 7 and 6 of Figure 2.1 (see Chapter 2).[19] We cannot of course know for certain that these propositions are true. They must remain forever *conjectures* – all our knowledge being conjectural in character. Nevertheless, we can be confident that we will always accept these propositions as a part of scientific knowledge. We are rationally entitled to accept *partial knowability* as a part of scientific knowledge because, if it is false, we cannot acquire knowledge of our local circumstances whatever we assume. There can be no circumstances in which it would be in the interests of the pursuit of knowledge of truth to reject partial knowability. And as for meta-knowability, accepting this proposition can only help, and cannot hinder, the pursuit of knowledge of truth, whatever the universe may be like. Both are synthetic *a priori* conjectures, accepted permanently because of the crucial role that they play in the pursuit of knowledge.[20] In short, these two propositions are synthetic *a priori* statements not in the full-blooded Kantian sense that they can be known to be true of all possible experience with absolute certainty, but in the radically qualified Kantian sense that they are conjectures about reality – about the noumenal world – which must remain permanently an integral part of conjectural human knowledge, and which are adopted as knowledge on non-empirical grounds. This crucial tenet of aim-oriented empiricism is, it seems, explicitly rejected by Einstein in the above passage – even though the whole point of the passage, ironically enough, is to affirm it, affirm, that is, that science *cannot* proceed without the assumption that the universe is knowable, even comprehensible.

Einstein's ambivalent attitude to the crucial issue which separates off standard from aim-oriented empiricism gains explicit expression in the following quotation:

> [The aim of science is to arrive] at a system of the greatest conceivable unity, and of the greatest poverty of concepts of the logical foundations, which is still compatible with the observations made by our senses. We do not know whether or not this ambition will ever result in a definite system. If one is asked for his opinion, he is inclined to answer no. While wrestling with the problems, however, one will never give up the hope that this greatest of all aims can really be attained to a very high degree. (First published 1936; see Einstein, 1973, p. 294)

We might interpret this to mean that when Einstein is thinking primarily as a theoretical physicist, he unthinkingly takes the ultimate comprehensibility of the universe for granted – the key component of aim-oriented empiricism. When he comes to reflect philosophically about the aims and methods of his work, however, his (misconceived) philosophical conscience gets the better of him, and he lapses into standard empiricism. Einstein's scientific instincts, in short, are more enlightened than his philosophical reflections – an important point, implicit in my claim that aim-oriented empiricism arose, for Einstein, out of scientific practice, adopted in response to a severe scientific problem.

Are we to conclude, then, that Einstein did not in the end manage to free himself explicitly from the trap of standard empiricism? One point to remember is that throughout his scientific life Einstein's views on the philosophy of science evolved from something close to Machian positivism at the outset (an extreme version of standard empiricism) to a view that comes to resemble aim-oriented empiricism more and more closely towards the end of his life. Einstein himself put the matter like this, in a letter to Lanczos in 1938:

> Coming from sceptical empiricism of somewhat the kind of Mach's, I was made, by the problem of gravitation, into a believing rationalist, that is, one who seeks the only trustworthy source of truth in mathematical simplicity. The logically simple does not, of course, have to be physically true; but the physically true is logically simple, that is, it has unity at the foundation. (see Holton, 1973, p. 241)

This, to begin with, sounds like a clear enough confession of a convinced aim-oriented empiricist. As it stands, it is perhaps something of an oversimplification. In the first place, as we have seen above, elements of aim-oriented empiricist thinking can be found in Einstein's scientific work almost from the outset – from Einstein's first great creative period in 1902–5. Second, Einstein's views concerning the philosophy of science went on developing long after the creation of general relativity, right to the end of his life. Our best hope, then, of finding a clear, unambiguous formulation of aim-oriented empiricism is to look at Einstein's very last writings on philosophy of science. I provide two final quotations. The first comes from Einstein's "Autobiographical Notes", written when he was 67. Einstein is discussing the points of view from which physical theories can be critically assessed, quite generally:

> The first point of view is obvious: the theory must not contradict empirical facts. ... The second point of view is not concerned with the relation to the material of observation but with the premises of the theory itself, with what may briefly but vaguely be characterized as the "naturalness" or "logical simplicity" of the premises (of the basic concepts and of the relations between these which are taken as a basis). This point of view, an exact formulation of which meets with great difficulties, has played an important role in the selection and evaluation of theories since time immemorial. The problem here is not simply one of a kind of enumeration of the logically independent premises (if anything like this were at all unequivocally possible), but that of a kind of reciprocal weighing of incommensurable qualities ... Of the "realm" of theories I need not speak here, inasmuch as we are confining ourselves to such theories whose object is the *totality* of all physical appearances. The second point of view may briefly be characterized as concerning itself with the "inner perfection" of the theory, whereas the first point of view refers to the "external confirmation". The following I reckon as also belonging to the "inner perfection" of a theory: We prize a theory more highly if, from the logical standpoint, it is not the result of an arbitrary choice among theories which, among themselves, are of equal value and analogously constructed. (Einstein, 1969, pp. 21–3)

It is surely clear from this that Einstein came quite explicitly to repudiate all versions of standard empiricism towards the end of his life. There is no suggestion here that the second requirement of "inner perfection" or unity is somehow to be reduced to the first requirement of empirical adequacy: empirical considerations do *not*, for Einstein, *alone* determine choice of theory. Furthermore, Einstein has made it abundantly clear already that, in his view, in choosing only theories which satisfy the requirement of inner perfection, we are in effect assuming that the universe itself is comprehensible – this being a permanent presupposition of scientific knowledge upheld on non-empirical grounds. But in case there is any doubt on this score, here is a passage written in 1950 in which the thesis that there can be no knowledge without the presupposition that the universe is comprehensible is explicitly affirmed:

> It is of the very essence of our striving for understanding that, on the one hand, it attempts to encompass the great and complex variety of man's experience, and that on the other, it looks for simplicity and economy in the basic assumptions. The belief that these two objectives can exist side by side is, in view of the primitive state of our scientific knowledge, a matter of faith. Without such faith I could not have a strong and unshakable conviction about the independent value of knowledge. (Einstein, 1973, p. 357)

I conclude that Einstein came close to articulating aim-oriented empiricism towards the end of his life, even if he did not recognize that this position is required to solve the problem of induction, and did not appreciate that it provides a more rational conception of science than does standard empiricism – and not a less rational conception, as Einstein's references to "faith" and "miracle-creed" tend to suggest.

In the end, however, what really matters is the philosophy of science implicit in Einstein's scientific deeds. Einstein himself held this view. As he put it: "If you want to find out anything from the theoretical physicists about the methods they use, I advise you to stick closely to one principle: don't listen to their words, fix your attention on their deeds" (Einstein, 1973, p. 270). As we have seen above, in order to make rational sense of Einstein's scientific judgements and deeds it is essential to see them from the standpoint of aim-oriented empiricism. More important, Einstein can be said to have invented aim-oriented empiricism in scientific practice during the course of discovering the special and general theories of relativity. His success in discovering these theories owes much to the invention and exploitation of the rational method of discovery of aim-oriented empiricism. This aspect of Einstein's work transformed the whole character of subsequent theoretical physics. Einstein's contributions to theoretical physics are intimately interrelated with his contribution to the philosophy of physics: after Einstein, indeed, physics and philosophy of physics ought to form one integrated discipline – aim-oriented empiricist natural philosophy. The various versions of standard empiricism defended by Popper, van Fraassen and Fine (and most contemporary philosophers of science) all fail to do justice to this vital dimension of Einstein's contribution to science. Indeed, advocacy of standard empiricism after Einstein amounts in itself to a failure to understand an important aspect of Einstein's contribution to science.[21]

4
Non-empirical requirements scientific theories must satisfy: simplicity, unity, explanation, beauty[*]

4.1 The problem

A scientific theory, in order to be accepted as a part of theoretical scientific knowledge, must be sufficiently:

(1) empirically successful;
(2) empirically contentful;
(3) simple, unified, explanatory, beautiful, elegant, harmonious, non-ad hoc, conceptually coherent, invariant, symmetrical, organic, inwardly perfect, non-aberrant (all terms used in this context by scientists and philosophers of science).

It is important to note that this third non-empirical requirement plays a crucial role in science, especially in physics, to the extent, even, of persistently overriding empirical requirements. Given any accepted physical theory, T, however successful empirically, it will always be possible to concoct endlessly many empirically more successful theories, T_1, T_2, etc., if non-empirical requirements can be ignored. T will make endlessly many predictions concerning phenomena not yet observed. Rivals to T can be concocted by modifying T in *ad hoc* ways so that each rival makes a different prediction for some unobserved phenomenon. Then independently testable and corroborated hypotheses can be added to these rivals, the result being a series of theories, T_1, T_2, etc., which have all the empirical success of T, and have excess empirical content over T, this excess content

---

[*] An earlier version of this chapter was posted on the PhilSci Archive in 2004 (see http://philsci-archive.pitt.edu/1759/).

being empirically corroborated. T_1, T_2, etc., are thus empirically more successful than T. Furthermore, almost all accepted physical theories run into empirical difficulties for some phenomena and are, on the face of it, refuted. T_1, T_2, etc., can be further modified in an entirely *ad hoc*, arbitrary fashion, so that these theories predict correctly the phenomena that ostensibly refute T, so that T_1, T_2, etc., are, in addition, empirically successful where T is refuted. In scientific practice, of course, these rivals to T, much more empirically successful than T, are never considered at all because of their failure to satisfy non-empirical requirements. The fact that such empirically more successful theories are persistently ignored because of their unacceptably *ad hoc*, complex, disunified character means that non-empirical considerations are persistently overriding empirical considerations in physics.[1] Non-empirical considerations thus play an irreplaceable and fundamental role in science.

But what is this mysterious non-empirical feature of simplicity, unity, etc., that any acceptable scientific theory must possess? This is the problem I set out to solve in this chapter.

It deserves to be noted that this is an absolutely fundamental problem in the philosophy of science. The solution is required for (a) a specification of scientific method, and (b) the solution to the problem of induction. Both points are demonstrated by the point made above, namely that non-empirical considerations persistently override empirical considerations when it comes to the acceptance of scientific theories.

Non-empirical considerations can have a purely pragmatic role in science: in certain contexts, we choose one formulation over another, or even one theory over another, not because we judge our choice to be more likely to be true, but because it is such that the equations are easier to solve, it is easier to extract useful predictions from the choice we make. Here, I ignore such pragmatic considerations, at least initially, and concentrate exclusively on non-empirical requirements judged to be indicative of truth or knowledge (however fallibly).

The following seven aspects of the problem can be distinguished:

(1) The terminological problem: How can simplicity, unity (etc.) be significant notions, having methodological significance, when the question of whether a theory is simple or complex, unified or disunified, will depend crucially on how the theory in question is formulated? A change of formulation can turn a simple theory into a complex one, and vice versa.

(2) How can degrees of simplicity, unity (etc.) be assessed?

(3) How many *different* features of theories are involved? The plethora of terms used by scientists and philosophers of science in this context does not inspire confidence that people know what they are talking about.

(4) How can one do justice to the fact that conceptions of simplicity or unity evolve with evolving knowledge? Three of Newton's four rules of reasoning concern simplicity (Newton, 1962, pp. 398–400), and yet Newton's notions are different from those of a modern physicist.

(5) How can one do justice to ambiguity of judgements concerning the relative simplicity or unity of theories? Thus Newton's theory of gravitation seems in one way much simpler than Einstein's, but in another way more complex, or at least less unified.

(6) How is persistent preference for simple or unified theories in science, even against the evidence, to be justified? This is, it should be noted, the problem of induction. Solve this, and the problem of induction is solved.

(7) What implications does the solution to these problems have for science itself?

By far the most serious item on this list is (1). I shall concentrate on (1), and at the end will make a few remarks about (2) to (7).

Richard Feynman has provided the following amusing illustration of problem (1) (see (Feynman et al., 1965, 25-10 – 25-11). Consider an appallingly complex universe governed by 10^{10} quite different, distinct laws. Even in such a universe, the true "theory of everything" can be expressed in the dazzlingly simple, unified form: $A = 0$. Suppose the 10^{10} distinct laws of the universe are:

(1) $F = ma$; (2) $F = Gm_1m_2/d^2$; etc.

Let $A_1 = (F - ma)^2$, $A_2 = (F - Gm_1m_2/d^2)^2$, etc., for all 10^{10} distinct laws. Let

$$A = \sum_{r=1}^{10^{10}} A_r$$

The true "theory of everything" of this universe can now be formulated as: $A = 0$. (This is true if and only if each $A_r = 0$, for $r = 1, 2, \ldots 10^{10}$.)

Most scientists and philosophers of science recognize that non-empirical considerations of simplicity, etc., play an important role in science, but no one has been able so far to solve the terminological problem – problem (1). Weyl (1963, p. 155) remarked correctly that "The problem of simplicity is of central importance for the epistemology of the natural sciences". Einstein (1982, p. 23) recognized the problem but confessed that he was not "without more ado, and perhaps not at all" able to solve it. Jeffreys and Wrinch (1921) suggested that simplicity could be identified with paucity of adjustable constants in equations, but unfortunately number of constants can be changed by changes of formulation. Popper (1959, ch. 8) proposed that simplicity is falsifiability, but unfortunately falsifiability can always be increased by adding on independently testable hypotheses which, in general, will drastically decrease simplicity. (Popper's adjunct proposal, in terms of dimension, does not work either, and is in any case subservient to falsifiability.) More recently, Friedman (1974), Kitcher (1981) and Watkins (1984) have sought to identify simplicity or unity with structural, formal or axiomatic features, but these attempts fail (see Salmon, 1989; Maxwell, 1998, pp. 65–8). Even more recently, Weber (1999), Schurz (1999) and Bartelborth (2002) have tackled the problem without success – but see the excellent paper by Maudlin (1996) on unification of theoretical physics.[2] McAllister (1996) constitutes an interesting attempt to solve the problem, and this will be discussed in Chapter 6.

4.2 The proposed solution

Previous attempts at solving the problem have failed because of mistakes concerning two crucial preliminary points.[3]

The first mistake is to formulate the problem, in the first instance, too generally as a problem about *scientific* theories. It is vital, in the first instance, to restrict the problem to fundamental, dynamical *physical* theories. Branches of the natural sciences are not independent of one another; they are interconnected. Biology presupposes chemistry, and even physics; chemistry presupposes physics; geology and astronomy presuppose physics; and phenomenological physics presupposes fundamental physics. All branches of natural science besides theoretical physics, in other words, are constrained by results from some more fundamental science that, in the end, can be traced back to theoretical physics. This is neither a pro- nor anti-reductionist thesis; it is just the simple observation that theories in non-physical branches of natural science are, in the end, in general, exceptions aside, constrained by physics. Only in fundamental theoretical physics does the question of the nature of non-empirical constraints on theories arise in something like a naked, pure form. We must, in the first instance, restrict the problem to that of fundamental, dynamical *physical* theory.

The second mistake is to suppose that simplicity, unity, etc., is a feature of the *theory* itself, its axiomatic structure, its simplicity of formulation, its number of postulates, its characteristic pattern of derivations, its number of adjustable constants. But all this involves looking at entirely the wrong thing. What one needs to look at is not the theory itself, but at the world, or rather at *what the theory says about the world*, the *content* of the theory, in other words. At a stroke, the worst aspect of the problem of what unity *is* vanishes. No longer does one face the *terminological* problem of unity – the problem of the formulation-dependent nature of unity. Suppose we have a given theory, T, which is formulated in N different ways, some formulations exhibiting T as beautifully unified, others as horribly complex and disunified, but all formulations being interpreted in precisely the same way, so as to make precisely the same assertion about the world. If unity has to do exclusively with *content*, then *all these diverse formulations of T, having the same content, have precisely the same degree of unity*. The variability of apparent unity with varying formulations of one and the same theory, T, (given some specific interpretation), which poses such an insurmountable problem for traditional approaches to the problem, poses no problem whatsoever for the thesis that unity has to do with content. Variability of formulation of a theory which leaves its content unaffected is wholly irrelevant: the unity of the theory is unaffected.

But now we have a new problem: How is the unity of the content of a theory to be assessed? What exactly does it mean to assert that a dynamical physical theory has a unified content?

What it means is that the theory has *the same* content throughout the range of possible phenomena to which the theory applies. Unity, in other words, means that there is just *one* content throughout the range of possible phenomena to which the theory applies. If the theory postulates *different* contents, *different* laws, for different ranges of possible phenomena, then the theory is *disunified*, and the more such different contents there are, so the more disunified the theory is. Thus "unity" means "one", and "disunity" means "more than one", the disunity becoming worse and worse as the number of different contents goes up, from two to three to four, and so on. Not only does this enable us to distinguish between "unified" and "disunified" theories, it enables us to assign "degrees of unity" to theories, or to partially order theories with respect to their degree of unity.[4]

To give an elementary example, Newton's theory of gravitation, $F = GM_1M_2/d^2$ is unified in that what the theory asserts is *the same* throughout all possible phenomena to which it applies (all bodies of all possible masses, whatever their constitution, shape, relative velocity or distance apart, at all times and places). An aberrant version of this theory, which asserts that $F = GM_1M_2/d^2$ for times $t \leq t_0$, where t_0 is some definite time, and $F = GM_1M_2/d^3$ for times $t > t_0$, is disunified, because what the theory asserts is *not* the same throughout the range of possible phenomena to which the theory applies.

Note that special terminology could be introduced to make Newtonian theory look disunified, and the aberrant version of Newtonian theory look unified. All we need do is interpret d^N to mean "d^N if $t \leq t_0$ and d^{N+1} if $t > t_0$". In terms of this (admittedly somewhat bizarre) terminology, the aberrant theory has the form "$F = GM_1M_2/d^2$", and Newtonian theory has the "aberrant" form "$F = GM_1M_2/d^2$ for times $t \leq t_0$ and $F = GM_1M_2/d$ for times $t > t_0$". But this mere terminological reversal of aberrance or disunity does not affect the *content* of the two theories: the content of Newtonian theory remains unified, and the content of the aberrant version (which looks unified) remains disunified.

This almost suffices to solve the problem. A little more needs to be added, however, because in practice in physics, assessments of degrees of unity are somewhat more complex than I have indicated so far, owing to the following consideration. In assessing the extent to which a theory is disunified we may need to consider *how* different, or *in what way* different, one from another, the different contents of a theory are. A theory that postulates different laws at different times and places is disunified in a much more serious way than a theory that postulates the same laws at all times and places, but also postulates that distinct kinds of physical particle exist, with different dynamical properties, such as charge or mass. This second theory still postulates *different* laws for different ranges of possible phenomena: laws of one kind for possible physical systems consisting of one type of particle, and slightly different laws for possible physical systems consisting of another type of particle. But this second kind of difference in content is much less serious than the first kind (which involves different laws at different times and places).

What this means is that there are different *kinds* of disunity, different *dimensions* of disunity, as one might say, some more serious than others, but all facets of the same basic idea. We can, I suggest, distinguish at least eight different facets of disunity, as follows.

Any dynamical physical theory, T, can be regarded as specifying an abstract space, S, of possible physical states to which the theory applies, a distinct physical state corresponding to each distinct point in S. (S might be a set of such spaces.) For unity, we require that T asserts that the same dynamical laws apply throughout S, governing the evolution of the physical state immediately before and after the instant in question. If T postulates N distinct dynamical laws in N distinct regions of S, then T has disunity of degree N. For unity, we require $N = 1$. The eight different kinds of disunity can be characterized like this.

(1) T divides space-time up into N distinct regions, $R_1 \ldots R_N$, and asserts that the laws governing the evolution of phenomena are the same for all space-time regions within each R-region, but are different within different R-regions.

Example: the aberrant version of Newtonian theory (NT) indicated above, for which the degree of disunity $N = 2$ in a type (1) way.

(2) T postulates that, for distinct ranges of physical variables (other than position and time), such as mass or relative velocity, in distinct regions, $R_1 \ldots R_N$ of the space of all possible phenomena, distinct dynamical laws obtain.

Example: T asserts that everything occurs as NT asserts, except for the case of any two solid gold spheres, each having a mass of between one and two thousand tons, moving in otherwise empty space up to a mile apart, in which case the spheres attract each other by means of an inverse cube law of gravitation. Here, $N = 2$ in a type (2) way.

(3) In addition to postulating non-unique physical entities (such as particles), or entities that are unique but not spatially restricted (such as fields), T postulates, in an arbitrary fashion, $N - 1$ distinct, unique, spatially localized objects, each with its own distinct, unique dynamic properties.

Example: T asserts that everything occurs as NT asserts, except there is one object in the universe, of mass 8 tons, such that, for any matter up to 8 miles from the centre of mass of this object, gravitation is a repulsive rather than attractive force. The object only interacts by means of gravitation. Here, $N = 2$ in a type (3) way.

(4) T postulates physical entities interacting by means of N distinct forces, different forces affecting different entities, and being specified by different force laws. (In this case one would require one force to be universal so that the universe does not fall into distinct parts that do not interact with one another.)

Example: T postulates particles that interact by means of Newtonian gravitation; some of these also interact by means of an electrostatic force $F = Kq_1q_2/d^2$, this force being attractive if q_1 and q_2 are oppositely charged, otherwise being repulsive, the force being much stronger than gravitation. Here, N = 2 in a type (4) way.

(5) T postulates N different kinds of physical entity,[5] differing with respect to some dynamic property, such as value of mass or charge, but otherwise interacting by means of the same force.

Example: T postulates particles that interact by means of Newtonian gravitation, there being three kinds of particles, of mass m, 2m and 3m. Here, N = 3 in a type (5) way.

(6) Consider a theory, T, that postulates N distinct kinds of entity (e.g. particles or fields), but these N entities can be regarded as arising because T exhibits some symmetry (in the way that the electric and magnetic fields of classical electromagnetism can be regarded as arising because of the symmetry of Lorentz invariance, or the eight gluons of chromodynamics can be regarded as arising as a result of the local gauge symmetry of SU(3)). If the symmetry group, G, is not a direct product of subgroups, we can declare that T is fully unified; if G is a direct product of subgroups, T lacks full unity; and if the N entities are such that they cannot be regarded as arising as a result of some symmetry of T, with some group structure G, then T is disunified.[6] Example: Classical electrodynamics postulates two fields: the electric and the magnetic fields. However, the relative strengths of these fields differ when measured in different inertial reference frames travelling with uniform velocity with respect to each other, although the strength of the combined electromagnetic field does not. But, according to special relativity, laws have the same form with respect to all inertial reference frames. We may hold that the electric and magnetic fields are not two distinct entities. They are, rather, two aspects of one unified entity, the electromagnetic field. To this extent, N = 1 as far as classical electrodynamics is concerned.

(7) If (apparent) disunity of there being N distinct kinds of particle or distinct fields has emerged as a result of cosmic spontaneous symmetry-breaking events, there being manifest unity before these occurred, then the relevant theory, T, is unified. If current (apparent) disunity has not emerged from unity in this way, as a result of spontaneous symmetry-breaking, then the relevant theory, T, is disunified.

Example: Weinberg's and Salam's electroweak theory, according to which at very high energies, such as those that existed soon after the big bang, the electroweak force has the form of two forces, one with three associated massless particles, two charged, W^- and W^+, and one neutral, W^o, and the other with one neutral massless particle, V^o. According to the theory, the two neutral particles, W^o and V^o, are intermingled in two different ways, to form two new neutral particles, the photon, γ, and another neutral massless particle, Z^o. As energy decreases, the W^+, W^- and Z^o particles acquire mass, due to the mechanism known as spontaneous symmetry-breaking (involving the Higgs particle), while the photon, γ, retains its zero mass. This theory unifies the weak and electromagnetic forces as a result of exhibiting the symmetry of local gauge invariance; this unification is only partial, however, because the symmetry group is a direct product of two groups, $U(1)$ associated with V^o, and $SU(2)$ associated with W^-, W^+ and W^o.[7]

(8) According to GR, Newton's force of gravitation is merely an aspect of the curvature of space-time. As a result of a change in our ideas about the nature of space-time, so that its geometric properties become dynamic, a physical force disappears, or becomes unified with space-time. This suggests the following requirement for unity: space-time on the one hand, and physical "particles-and-forces" on the other, must be unified into a single self-interacting entity, U. If T postulates space-time and physical particles-and-forces as two fundamentally distinct kinds of entity, then T is not unified in this respect.

Example: one might imagine that the quantization of space-time leads to the appearance of particles and forces as only apparently distinct from empty space-time. Here, $N = 1$ in a type (8) way: there is just the one self-interacting entity, empty space-time.

For unity, in each case, as I have said, we require $N = 1$. As we go from (1) to (5), the requirements for unity are intended to be accumulative: each presupposes that $N = 1$ for previous requirements. As far as (6) and (7) are concerned, if there are N distinct kinds of entity which are not unified by a symmetry, whether broken or not, then the degree of disunity is the same as that for (4) and (5), depending on whether there are N distinct forces, or one force but N distinct kinds of entity between which the force acts.

(8) does not introduce a new kind of unity, but rather a new, more severe way of counting different kinds of entity. (1) to (7) require, for unity, that there is one kind of self-interacting physical entity evolving in a distinct space-time, the way this entity evolves being specified, of course, by a consistent physical theory. According to (1) to (7), even though there are, in a sense, two kinds of entity, matter (or particles-and-forces) on the one hand, and space-time on the other, nevertheless $N = 1$. According to (8), this would yield $N = 2$. For $N = 1$, (8) requires that matter and space-time are no more than aspects of one basic entity (unified by means of a spontaneously broken symmetry, perhaps).

As we go from (1) to (8), then, requirements for unity become increasingly demanding, with (6) and (7) being at least as demanding as (4) and (5), as explained above.

(1) to (8) may seem very different requirements for unity. In fact they all exemplify the same basic idea: disunity arises when *different* dynamical laws govern the evolution of physical states in different regions of the space, S, of all possible physical states. For example, if a theory postulates more than one force, or kind of particle, not unified by symmetry, then in different regions of S different force laws will operate. If (8) is not satisfied, there is a region of S where only empty space exists, the laws being merely those which specify the nature of empty space or space-time. The eight distinct facets of unity, (1) to (8), arise, as I have said, because of the eight *different* ways in which content can vary from one region of S to another.[8]

4.3 Objections

It may be objected that we never encounter the naked content of a theory, formulation free; we only encounter theories given some formulation. How, then, can we judge whether the *content* does or does not vary through the space S? The answer is that theories are not natural objects we stumble across; *we* formulate theories, and it is for us to ensure, granted we want our theories to be unified, that the content does not change as we move through S. We can arrange, however, that formulation matches content by ensuring that the terminology, the concepts, we use to formulate a theory do not surreptitiously change as we move through S. Given invariant concepts, if the form of the theory is also invariant throughout S, its content will be too, but if, for example, we surreptitiously change our units of length as we move through space, then a theory whose content is spatially invariant will change its form with changes of spatial position (a point which will be taken up again below).

It may be objected that, given any theory, however unified, special regularities will always arise in restricted regions of S, which means disunity. Whether or not the theory is unified is, at best, ambiguous. Thus, given NT, in some regions of S there will be solar systems with planets that rotate in the same direction and conform to Bode's law, whereas in other regions of S these regularities or "laws" will be violated. The answer is to distinguish sharply between accidental and law-like regularities; only the latter are relevant for the assessment of unity. But how is this distinction to be made? The answer is to adopt a suggestion made elsewhere (Maxwell, 1968; 1998, pp. 141–55) that physical laws are true if and only if corresponding physical dynamical (or necessitating) properties exist. According to this suggestion, Newton's law of gravitation can be interpreted as attributing the dynamical property of Newtonian gravitational charge to massive objects. Objects that have this property of necessity obey Newton's law of gravitation. (The empirical content of NT, on this interpretation, is concentrated in the factual assertion: all massive objects possess Newtonian gravitational charge equal to their mass.) If no such property corresponds to a true regularity, then it is merely a true accidental regularity, and not a true law. For unity we require that dynamical properties remain the same throughout S; the regularities of (some) solar systems, mentioned above, are not relevant because these regularities are not law-like, and no dynamical property exists corresponding to them.[9]

It may be objected that physical systems that possess symmetries, which are also symmetries of the theory that determines their evolution, will evolve in accordance with a simplified version of the theory. Thus systems consisting of two spheres equal in every way rotating in a fixed circle about their centre of mass obey a simplified version of the dynamical laws of NT. This means there are regions of S where the dynamical laws are especially simple, and thus different from other regions. Does this mean the theory is correspondingly disunified? The answer is no. We need, again, to consider dynamical properties corresponding to dynamical laws. In the example just considered, if NT is true (interpreted essentialistically) then the spheres in question possess gravitational charge just like all other massive objects. It is just that, in the case of the systems possessing some rotational symmetry, the full, rich implications of the dynamical property of gravitational charge is not made manifest.

It may be objected that we may not know whether two formulations of a theory are just that, two formulations with the same physical content, or two distinct theories with distinct contents. Heisenberg's and Schrödinger's distinct formulations of quantum theory might be an example. This is correct but beside the point. The terminological problem arises when we reformulate a given theory, T, in a variety of ways, some simple and unified, some horribly complex and disunified, but we do this in such a way as to ensure quite specifically that the different formulations have precisely the same content, make precisely the same assertions about the world, this being something that we can always do. The solution to the problem proposed above is not in any way undermined by the fact that it sometimes happens that we do not know whether two formulations of a theory have the same or different contents. Nor is the distinction between form and content undermined: form has to do with what we write down on paper, content with what is being asserted. That we sometimes do not know whether difference of formulation ensures difference of content does not in the least undermine the distinction between form and content. It deserves to be noted, in addition, that one and the same formulation of a theory may be interpreted in more than one way, and thus may have different contents associated with it – a point which, again, does not undermine the theory presented here.

It may be objected that the distinction between dynamical laws which do, and do not, remain the same throughout the space S cannot be maintained. Consider the following two functions: (1) $y = 3x$ for all x, and (2) $y = 3x$ for $x \leq 2$ and $y = 4x$ for $x > 2$. It is tempting to say that (1) remains the same as x changes, but (2) does not, since what (2) asserts changes at $x = 2$. But given the mathematical notion of function as a rule, (2) is just as good a function as (1) and, like (1), "remains the same" as x varies. Functions corresponding to physical theories are somewhat more elaborate than this, but the above point is not affected by that consideration: it seems that the very distinction between "remains the same" and "changes" as one moves through S collapses. Clearly, in order to meet this objection, functions corresponding to physical theories need to be restricted to a narrower notion of function than the above standard mathematical one, if we are to be able to distinguish between functional relationships which do, and which do not, "remain the same" as values of variables change. We need to appeal to what may be called "invariant functions", functions which specify some fixed set of mathematical operations to be performed on "x" (or its equivalent) to obtain "y" (or its equivalent). In the example just given, (1) is invariant, but (2) is not. (2) is made up of two truncated invariant functions, stuck together at x = 2. Functions that appear in theoretical physics are *analytic*; that is, they are repeatedly differentiable. Such functions have the remarkable property that from any small bit of the function, the whole function can be reconstructed uniquely, by a process called "analytic continuation". All analytic functions are thus invariant. The latter notion is, however, a wider one, and theoretical physics might, one day, need to employ this wider notion explicitly, if space and time turn out to be discontinuous, and analytic functions have to be abandoned at a fundamental level.

A similar remark needs to be made about Goodman's (1954) paradox concerning "grue" and "bleen". Modifying the paradox slightly, an object is grue if it is green up to time t, blue after t; it is bleen if it is blue up to time t, green after t. Sometimes it is held that there is perfect symmetry between blue and green, on the one hand, and grue and bleen on the other, especially as "emeralds are green" is equivalent to "emeralds are grue up to t, and bleen afterwards". But this symmetry is merely terminological and, as we have already seen in connection with the aberrant version of Newton's theory of gravitation, discussed in section 4.2, terminological symmetry does not mean there is symmetry of content. That there is not symmetry of content in the grue/bleen case can be demonstrated as follows. If emeralds are grue, a person convinced of this can determine whether t is future or past merely by looking at emeralds. But if emeralds are green, a person convinced of this cannot say whether t is in the future or past by just looking at emeralds. The content, the meaning, of grue and bleen contains an implicit reference to t in a way in which that of green and blue do not. Doubtless symmetry can be created by considering two possible worlds, ours and a Goodmanesque one with special physics and/or physiology of vision so that grue emeralds do not appear to change at t, whereas green ones do. This, however, is to consider conditions quite different from those specified by Goodman. The crucial point to make, in any case, is that dynamical or physical properties, of the kind attributed to physical entities by physical theories (interpreted in a conjecturally essentialistic way), are like blue and green, and unlike grue and bleen, in not containing any implicit reference to specific times or places (or hypersurfaces of S that distinguish one region of S from another). Physical properties must be *invariant* in a sense that corresponds to the *invariance* of allowed functions in physics. The more general notion of property, which includes Goodmanesque properties, is excluded, just as the more general notion of function, which includes (2) above as an "unchanging" function, is excluded.

4.4 Further issues

What of the other aspects of the problem of non-empirical requirements in science mentioned in the introduction? Here are a few brief remarks concerning some of these further issues.

Most of the other terms used to refer to non-empirical requirements can be straightforwardly related to unity. We have seen that this is true of symmetry and invariance. Non-*ad hoc*, organic, inwardly perfect and non-aberrant can be interpreted as appealing to unity, and harmonious, beautiful and conceptually coherent can be interpreted as presupposing unity. A dynamical physical theory can be held to be explanatory in character to the extent that it is (1) empirically contentful, and (2) unified.

Simplicity, however, is quite different. The simplicity of a theory can be interpreted as having to do, not with whether the *same* laws apply throughout the space S, but rather with the *nature* of the laws, granted that they are the same. Some laws are simpler than others. In order to overcome the objection that simplicity is formulation-dependent it is essential, as in the case of unity, to interpret "simplicity" as applying to the *content* of theories, and not to their *formulation*, their *axiomatic structure*, etc. Theories can only, at best, be partially ordered with respect to degrees of simplicity. Even when two theories are amenable to being assessed with respect to relative simplicity, there is always the problem that a change of variables may reverse the assessment. Let the two theories be (1) $y = x$, and (2) $y = x^2$. We judge (1) to be simpler than (2). Let $x^2 = z$. We now have (1) $y = \sqrt{z}$, and (2) $y = z$. Now (2) is simpler than (1). Assessment of relative simplicity of two theories may only be unambiguous when restrictions are placed on the form that physical variables can take, so that only linear transformations of the type $z = Ax + B$ (where A and B are constants) are permitted, for example. It is a further great success of the theory presented here that it succeeds in distinguishing sharply between these two aspects of physical theory, the *unity* and *simplicity* aspects, and succeeds in explicating both.[10]

We can use these two notions to solve the problem of ambiguity of judgement concerning the relative non-empirical merits of Newton's and Einstein's theories of gravitation. Newton's theory is simpler, but Einstein's is more unified in that it eliminates gravitation as a force, and reformulates Newton's first law so that it becomes the assertion that bodies move along geodesics in curved space-time, curvature being caused by mass, or by stress-energy-density more generally. As theoretical physics draws closer to capturing the true theory of everything, it is reasonable to expect that the totality of fundamental physical theory will become increasingly unified and complex.

So far I have stressed that terminological unity and simplicity are irrelevant when it comes to assessing unity and simplicity in a physically significant sense. In scientific practice, however, terminology is chosen so as to reflect physically significant unity and simplicity (Maxwell, 1998, 110–3). Thus if the content of a theory exhibits certain symmetries, terminology is chosen so that it too exhibits these symmetries, so that if the theory is invariant with respect to position or orientation in space, terminology is chosen which reflects this fact. Once a theory is formulated in such "physically appropriate" terminology (as it may be called), two versions of symmetry operations arise as a result: "active" (which make changes to physical systems) and "passive" (which make corresponding changes to the description of unchanged physical systems). Granted that we formulate physical theories exclusively in such "physically appropriate" terminology, then terminological unity and simplicity comes to reflect physical unity and simplicity, and is thus, to that extent, physically significant.

What of the simplicity and unity of theories in sciences other than fundamental physics? Much needs to be said on this topic; the following brief remarks can serve only as pointers to a more adequate treatment. Solutions to the equations of fundamental physical theory, specifying precisely how increasingly complex physical systems evolve in space and time, rapidly become horrendously complex in character. In carrying out derivations, physicists invariably "simplify" results obtained by discarding variable quantities or higher order terms judged to be insignificant in the physical situations under considerations. Just this is done when NT is "derived" from Einstein's theory, or Kepler's and Galileo's laws of motion are "derived" from NT.[11] The outcome is a range of more or less terminologically simple phenomenological laws of only approximate validity. But the simplicity is not, here, merely pragmatic, since such a law has been "approximately derived" from some fundamental physical theory formulated in a "physically appropriate" way, the "approximate derivation" showing what the range of applicability of the law is with what degree of accuracy. Even though such laws are incompatible with the fundamental physical theory from which they have been "approximately derived", nevertheless what the "derivations" reveal is that pragmatic simplicity has been obtained by sacrificing strict derivability and precise empirical accuracy, there being nothing here to counter the underlying unity in nature postulated by fundamental physical theory (in so far as it does postulate this). Laws such as these are prevalent throughout phenomenological physics, astrophysics and parts of physical chemistry. Even where such "approximate derivations" cannot be carried through, for large parts of chemistry, and for biology, nevertheless, as I have already remarked, laws and theories of these sciences are constrained by fundamental physics, and must endeavour to be compatible with fundamental physics, at least in the qualified way just indicated in connection with phenomenological physics. Thus, much of the great explanatory power of Darwinian theory stems from the fact that it postulates mechanisms for evolution – random inheritable variation and natural selection – which are capable of designing living things able to pursue the goals of survival and reproductive success in their given environments, these mechanisms nevertheless being compatible with the purposeless cosmos depicted by physics. Biology must accord with physics in much more specific ways as well, in that the mechanisms of inheritance and development must accord with physics, and so too the multitude of processes that take place in living things.

What implications does the account of non-empirical requirements for theories, given here, have for science? How can justice be done to evolving non-empirical requirements? How is persistent preference for unified theories, even against the evidence, to be justified? I take these three problems together.

At the beginning of this chapter I demonstrated that, in physics, theories that are unified, in senses (1) and (2) at least, are persistently chosen in preference to available, empirically more successful, but disunified theories. To proceed in this way is to make the permanent assumption that the phenomena under consideration are such that all theories of these phenomena that are disunified in senses (1) and (2) are false. If physicists persistently accepted theories that postulate atoms in preference to available, empirically more successful *field* theories, it would be clear that physicists are thereby assuming that all field theories are false. Just the same holds for the persistent rejection of empirically more successful disunified theories.

But rigour demands that assumptions that are substantial, influential, problematic and implicit need to be made explicit, so that they can be critically assessed, so that alternatives can be developed and considered, the hope being that in this way such assumptions can be improved. Thus rigour demands that science makes explicit, and so criticizable and improvable, the substantial, problematic, influential and implicit assumption that the universe, or the phenomena, are such that all disunified theories are false. This assumption, M, can easily be shown to be metaphysical, as follows. Persistent acceptance of theories unified in ways (1) and (2) involves rejecting infinitely many empirically more successful disunified rivals, T_1, T_2, ... T_∞, because they clash with M. In effect, M = $notT_1$ and $notT_2$ ... and $notT_\infty$. In order to verify M we would need to falsify all of T_1, T_2, ... and T_∞, but as there are infinitely many theories, this cannot be done. In order to falsify M we need to verify just one of T_1, T_2, ... or T_∞, but physical theories cannot be verified. Hence M, being neither verifiable nor falsifiable, is metaphysical. It is a permanent metaphysical assumption of science – permanent, at least, as long as all theories disunified in senses (1) and (2) are rejected whatever their empirical success might be.

At once the question arises: how is this assumption M to be critically assessed and, perhaps, improved? In Chapter 2 I argued that once the metaphysical assumption implicit in persistent preference in science for unified theories is acknowledged, it becomes apparent that we need to adopt a new conception of science, which construes science as making a hierarchy of such assumptions, these assumptions asserting less and less as one goes up the hierarchy, and thus becoming more and more likely to be true.[12] These are assumptions about the knowability and comprehensibility of the universe. As we descend the hierarchy, assumptions become more substantial and specific, and much more likely to be false, and in need of revision. Revision is, however, kept as low down in the hierarchy as possible. Those physical theories are accepted which best accord with the evidence and the best available metaphysical assumption, B, say, lowest down in the hierarchy. But B may itself be revised if a rival assumption, B*, is developed which (a) is compatible with the assumption above it in the hierarchy, and (b) supports an empirical research programme that is more successful than the one supported by B. Relatively problematic assumptions high up in the hierarchy thus form a fixed framework within which much more specific, problematic assumptions can be revised in the light of empirical success and failure. As knowledge improves, assumptions and associated methods improve as well; there is something like positive feedback between improving knowledge and improving knowledge-about-how-to-improve-knowledge, the methodological key to the success of modern science. Non-empirical requirements for theory acceptance, corresponding to metaphysical assumptions, improve with improving knowledge. Newton's requirements of simplicity evolve into the symmetry principles of modern physics. For my view as to how acceptance of the hierarchy of metaphysical assumptions is to be justified, see Chapter 7 (see also Maxwell, 1998, ch. 5; 2007, ch. 14; 2017a).

5
Scientific metaphysics

5.1 Introduction

The idea that science cannot proceed without the assumption that the universe is comprehensible in some way – the thesis that it is *physically* comprehensible being a more secure item of knowledge than any accepted physical theory – represents a profound and dramatic revolution in our whole conception of science. It is perhaps not surprising that, even though this idea dates back at least to 1974,[1] and has been expounded and argued for in considerable detail on a number of occasions since,[2] nevertheless, at the time of writing (2016), few scientists or philosophers of science have taken note of, or responded to the arguments for, this revolutionary view.[3] Given the momentous consequences of the idea, and its neglect, it seems to me appropriate to reformulate, in as careful and critical a way as I can, the central argument for aim-oriented empiricism. The argument developed in this chapter improves on the earlier versions, to be found in Chapters 2 and 4.

I argue, first, that persistent acceptance of (more or less) *unified* fundamental dynamic theories in physics, even though endlessly many empirically more successful *disunified* rivals are always available, means that physics makes a persistent untestable (or metaphysical) assumption about the nature of the universe: it is such that some yet-to-be-discovered, more or less unified physical theory is true, and all seriously disunified theories are false. I then invoke the account of what it means to assert of a physical theory that it is unified, developed in the last chapter, to throw light on the question of what physics does, and ought to, assume in assuming that some more or less unified theory is true. This provides us with a way of classifying – indeed, of partially ordering – all metaphysical theses which assert that the universe possesses some kind of comprehensive, more or less disunified dynamic structure – the universe being more or less physically comprehensible, in other words.

At once the fundamental problem arises: how can physics choose between these infinitely many metaphysical theses – versions of *physicalism*, as I call them? Two considerations drive us in opposite directions. On the one hand, we ought to choose that version of physicalism most fruitful for promoting progress in theoretical physics, if true. On the other hand, we ought to choose that minimalist version of physicalism just sufficiently substantial to make physics possible, and thus least likely to be false.

I take these in turn. First, I spell out how the account of unity given in the last chapter provides the means for assessing the relative fruitfulness of rival versions of physicalism; I go on to specify that version of physicalism that is the most fruitful, given the history of physics up till today, and argue that physics should accept this version as its basic metaphysical assumption. Second, I consider the grounds for physics accepting the least substantial version of physicalism that makes physics possible. The problem of choosing between these conflicting considerations is solved by the hierarchical view of aim-oriented empiricism (AOE). This does justice to both apparently conflicting considerations – a strong argument in favour of AOE. In conclusion, I consider two versions of AOE, and indicate, briefly, how the circularity problem can be solved.

The title of this chapter is intended to be provocative. "Scientific metaphysics" sounds like a contradiction in terms in view of Popper's well-known demarcation criterion that rules that metaphysical theses, being unfalsifiable, are not scientific.[4] But of course Popper's falsificationist conception of science, along with others, will be found to be defective precisely because of a failure to acknowledge the role that metaphysical assumptions play in science. Furthermore, as I have indicated, a framework will be developed which makes it possible to appraise (untestable) metaphysical theses empirically, in terms of their "empirical fruitfulness", or fruitfulness for the empirical research programme of theoretical physics. For physics to be rigorous, it will be argued, it is essential that metaphysical theses are acknowledged as key components of theoretical knowledge in physics, and are appraised empirically in terms of their "empirical fruitfulness". Once the conception of physics defended here is accepted, the title entirely loses its air of being self-contradictory.[5]

5.2 Intellectual rigour requires that metaphysical presuppositions be made explicit

Almost all views about science deny that science makes a substantial, persistent, metaphysical (i.e. untestable) assumption about the universe. This is true, for example, of logical positivism, inductivism, logical empiricism, hypothetico-deductivism, conventionalism, constructive empiricism, pragmatism, realism, Bayesianism, induction-to-the-best-explanationism, and the views of Popper, Kuhn and Lakatos. All these views, diverse as they are in other respects, accept a thesis that may be called standard empiricism (SE): in science, theories are accepted on the basis of empirical success and failure, and on the basis of simplicity, unity or explanatoriness, but *no substantial thesis about the world is accepted permanently by science, as a part of scientific knowledge, independently of empirical considerations*. Both Kuhn and Lakatos maintain, it is true, that a "paradigm" or "hard core" may be accepted for a time as a key item of scientific knowledge independently of evidence, even against the evidence; both hold, however, that such a paradigm or hard core will eventually be rejected when an empirically more successful paradigm or hard core emerges. Both Kuhn and Lakatos take SE for granted.[6]

Recently, a new research industry has grown up in the philosophy of science, devoted to "the metaphysics of science".[7] It might be thought that here, SE is repudiated. It is not. SE is implicit in all the works referred to in note 7.

Thus, SE is widely, almost unthinkingly, taken for granted by scientists and non-scientists alike. SE is nevertheless untenable. This is established decisively by the following argument.

Whenever a fundamental physical theory is accepted as a part of theoretical scientific knowledge, there are always endlessly many rival theories which fit the available evidence just as well as the accepted theory. Consider, for example, Newtonian theory (NT). One rival theory asserts: everything occurs as NT asserts up until midnight tonight when, abruptly, an inverse cube law of gravitation comes into operation. A second rival asserts: everything occurs as NT asserts, except for the case of any two solid gold spheres, each having a mass of a thousand tons, moving in otherwise empty space up to a mile apart, in which case the spheres attract each other by means of an inverse cube law of gravitation. There is no limit to the number of rivals to NT that can be concocted in this way, each of which has all the predictive success of NT as far as observed phenomena are concerned but which makes different predictions for some as yet unobserved phenomena.[8] Such theories can even be concocted which are *more* empirically successful than NT, by arbitrarily modifying NT, in just this entirely *ad hoc* fashion, so that the theories yield correct predictions where NT does not, as in the case of the orbit of Mercury, for example (which very slightly conflicts with NT).[9] And quite generally, given any accepted fundamental physical theory, T, there will always be endlessly many *ad hoc* rivals which meet with all the empirical success of T, make untested predictions that differ from T, are empirically successful where T is ostensibly refuted, and successfully predict phenomena about which T is silent (as a result of independently testable and corroborated hypotheses being added on).

As most physicists and philosophers of physics would accept, *two* criteria are employed in physics in deciding what theories to accept and reject: (1) empirical criteria, and (2) criteria that have to do with the simplicity, unity or explanatory character of the theories in question. (2) is absolutely indispensable, to such an extent that there are endlessly many theories empirically more successful than accepted theories, all of which are ignored because of their lack of unity.

Now comes the crucial point. In persistently accepting unifying theories (even though ostensibly refuted), and excluding infinitely many empirically more successful, unrefuted, disunified or aberrant rival theories, science in effect makes a big assumption about the nature of the universe, to the effect that it is such that some yet-to-be-discovered, more or less unified physical theory is true, and no seriously disunified theory is true,[10] however empirically successful it may appear to be for a time. Furthermore, without some such big assumption as this, the empirical method of science would collapse. Science would be drowned in an infinite ocean of empirically successful disunified theories.[11]

If scientists only accepted theories that postulate atoms, and persistently rejected theories that postulate different basic physical entities, such as fields – even though many field theories can easily be, and have been, formulated which are even more empirically successful than the atomic theories – the implications would surely be quite clear. Scientists would in effect be assuming that the world is made up of atoms, all other possibilities being ruled out. The atomic assumption would be built into the way the scientific community accepts and rejects theories – built into the implicit *methods* of the community, methods which would include rejecting all theories that postulate entities other than atoms, whatever their empirical success might be. The scientific community would accept the assumption that the universe is such that no non-atomic theory is true.

Just the same holds for a scientific community which rejects all disunified or aberrant rivals to accepted theories, even though these rivals would be even more empirically successful if they were considered. Such a community in effect makes the assumption that the universe is such that no disunified theory is true. Or rather, more accurately, such a community makes the assumption that "no disunified theory is true *that is not entailed by a true unified theory (plus, possibly, true relevant initial and boundary conditions)*". (A true unified theory entails infinitely many approximate, true, disunified theories.) Let us call this assumption "physicalism".

That physicalism is metaphysical can be shown as follows. Physicalism asserts, "not T_1 and not T_2 ... and not T_∞", where $T_1, T_2, ... T_\infty$ are infinitely many disunified rivals to accepted physical theories. Physicalism cannot be empirically verified, because this would require that all of $T_1, T_2, ... T_\infty$ are falsified, but as there are infinitely many of these theories, each requiring a different falsifying experiment, this cannot be done. Equally, physicalism cannot be falsified, as this requires the verification of at least one of T_1, or T_2, ... or T_∞, which cannot be done, as physical theories cannot be verified empirically. Hence physicalism, being neither verifiable nor falsifiable, is metaphysical.

Thus in persistently rejecting empirically more successful but disunified rivals to accepted physical theories, science makes a persistent metaphysical assumption about the world, namely physicalism. Standard empiricism (SE), and all the above doctrines that include SE as a component, which hold that science makes no persistent metaphysical assumption, are thus untenable.

Let us call the view that science presupposes physicalism "presuppositionism". Presuppositionism is more rigorous than all the above versions of SE *entirely independent of any justification for accepting physicalism as a part of scientific knowledge* (that is in addition to the one given above). In saying this, I am appealing to the following, wholly uncontroversial, requirement for rigour:

The principle of intellectual rigour (PIR): In order to be rigorous, it is necessary that assumptions that are substantial, influential and problematic be made explicit – so that they can be criticized, so that alternatives may be developed and assessed, with the aim of improving the assumptions.[12]

All versions of SE fail to satisfy PIR in just the way in which presuppositionism does satisfy PIR. Presuppositionism makes the assumption of physicalism explicit (and so criticizable and, we may hope, improvable), while all versions of SE deny that science does make any such assumption as physicalism. Thus, quite independent of any claim to solve the problem of induction, presuppositionism is more rigorous, and thus more acceptable, than any of the above versions of SE. And this is the case *even though presuppositionism can provide no justification for accepting physicalism*. It is, indeed, above all when we have no reason whatsoever for supposing physicalism is true that it becomes all the more important to implement PIR, and make the probably false assumption of physicalism explicit, so that it can be critically assessed, so that alternatives can be considered, in the hope that a thesis nearer the truth can be discovered.

Why has this simple argument been ignored by the vast literature on the problem of induction (or underdetermination), referred to in note 8? Three factors are perhaps at work. First, accepting the argument involves acknowledging that science, as it is ordinarily understood (in terms of standard empiricism), lacks rigour. Our understanding of science, and even science itself, need to change if science is to become rigorous and make explicit, and critically assess, implicit, problematic metaphysical presuppositions. Philosophers, tackling the problem of induction, have perhaps been reluctant to take seriously that science, as ordinarily understood, is irrational and needs to be changed. Second, invoking a metaphysical presupposition of unity looks superficially like a well-known and hopelessly invalid approach to the problem of induction: justify scientific theory by an appeal to a metaphysical principle uniformity, and then justify this metaphysical principle by an appeal to the success of science. Aware of the vicious circularity of any such argument, philosophers have instinctively resisted considering the superficially similar, but actually very different, point that rigour requires that substantial, influential, problematic and implicit metaphysical presuppositions need to be made explicit – so that they can be critically assessed and, we may hope, improved. Third, the above elementary argument that physics makes a persistent metaphysical assumption about the nature of the universe has seemed impossible to accept because it has seemed impossible that the metaphysical assumption in question could ever be shown to be *true*, or *probably true*. But this objection profoundly misses the point. As I have already stressed, it is precisely because the metaphysical thesis implicit in persistent acceptance of unified theories when endlessly many empirically more successful disunified rivals are available is nothing more than a pure conjecture, very likely to be false in the specific version of it accepted at any stage in the development of physics, that it is vital that it be made explicit within physics so that it may be critically assessed, so that alternatives may be developed and assessed, in an attempt to improve the specific conjecture that is adopted. The impossibility of providing some kind of *a priori* proof of the truth of the metaphysical conjecture, far from being a good reason for ignoring its existence is, on the contrary, an overwhelming reason to make the conjecture explicit within the context of physics.

Presuppositionism does not, perhaps, entirely solve the problem of induction – the problem of underdetermination – but it does, when further developed, transform that philosophical and scientifically sterile problem into the scientifically fruitful problem of developing and choosing the most fruitful metaphysics for physics, as we shall see below.

5.3 Unity of physical theory

We have seen that physics only accepts theories that are unified, and this commits physics to presupposing physicalism. But what ought physicalism to be interpreted to assert, especially if physics is to comply with PIR? In order to answer this question we first need to solve the problem of what it means to assert of a physical theory that it is *unified*. This problem has long resisted solution.[13] Even Einstein (1969, p. 23) confessed that he did not know how to solve the problem.

However, in Chapter 4, section 4.2, we saw how the problem is to be solved. Above, in the present chapter, I have indicated ways in which theories can be *disunified*. The solution to the problem of what it is for a theory to be unified, spelled out in the previous chapter, in effect extends and develops the above remarks about disunity.

A dynamical physical theory is disunified if its content, what it asserts about the world, is *different*, from one region to another in the space of all possible phenomena to which the theory applies. If the content of the theory differs in N ways, throughout the space of phenomena predicted by the theory, then the theory is disunified to degree N. For unity, we require that $N = 1$.

But, as we saw in the last chapter, there is a refinement. There are *different* ways in which the content of a theory may differ, from one region in the space of possible phenomena to another, some ways being more substantial, more serious as it were, than others. The most serious kind of difference, perhaps, is when the content of a theory differs in different space-time regions. A less serious kind of difference arises when a theory predicts no variation in dynamical laws in space-time, but predicts that there is more than one kind of force, or more than one kind of particle or field (there being one force, or one kind of particle or field in one sub-region of the space of all possible phenomena, and another force, or another kind of particle or field in another sub-region). In all, as we saw, there are *eight* different ways, (1) to (8), in which the content of a theory can differ, in different regions of the space of all possible phenomena that the theory predicts.

As we go from (1) to (8), the requirements for unity become increasingly demanding, with (6) and (7) being at least as demanding as (4) and (5), as explained in section 4.2 of Chapter 4.[14] It is important to appreciate, however, that (1) to (8) are all versions of the same basic idea that T is unified if and only if the content of T is the same throughout the range of possible phenomena to which it applies. When T is disunified, (1) to (8) specify *different* kinds of difference in the content of T in diverse regions of the space, S, of all possible phenomena to which T applies. Or, equivalently, (1) to (8) divide S into sub-regions in *different* ways, T having a different content in each sub-region. For (1), sub-regions contain physical systems in different locations in space-time, the content of T being different in different space-time locations. For (2), sub-regions contain physical systems with different values of physical variables such as mass or relative velocity. For (3), sub-regions contain systems with different dynamically unique objects. For (4), sub-regions contain systems composed of physical entities interacting by means of different forces. For (5), sub-regions contain systems composed of entities interacting by means of the same force, but with different dynamical properties such as values of mass or charge. For (6), sub-regions contain systems composed of different entities that cannot be transformed into each other by means of symmetry operations. For (7), sub-regions contain systems composed of different entities that cannot be construed to differ only because of the product of spontaneous symmetry breaking. For (8), S contains one sub-system consisting of empty space-time, and another consisting of space-time plus some physical entity, and the one cannot be transformed into the other by means of a symmetry operation. We have here eight facets of a single conception of unity.[15]

It needs now to be appreciated that, corresponding to these eight facets of unity, (1) to (8), there are eight different metaphysical theses, eight different versions of physicalism, any one of which might be held to be the best choice of presupposition for physics. If T is the true theory of everything,[16] then we have eight different theses, each of the form "T is unified up to sense (n)" where n = 1, 2, ... 8, corresponding to the eight different kinds of unity. Let us call these eight theses "physicalism(n)", where n = 1, 2, ... 8. It is assumed, here, that in each case, (1) to (8), the *degree* of unity, N, is 1. If we allow N = 1, 2, 3, ..., then there are not eight, but infinitely many different versions of physicalism, depending on the degree of unity, N, that is asserted for any value of n = 1, ... 7. (n = 8 is exceptional in this respect in that, in this case, N can only equal 1 or 2, depending on whether space-time and matter are, or are not, unified.[17])

The different versions of physicalism can be specified to be: physicalism(n, N), with n = 1, ... 7 and N = 1, 2, ... ∞, or with n = 8 and N = 1, 2. A two-dimensional grid is placed over an infinite set of metaphysical theses, distinct versions of physicalism corresponding to distinct appropriate values of the coordinates (n, N), these theses being ordered with respect to degrees of unity.[18]

It deserves to be noted in passing that there is a close connection between the "unity" of a physical theory and its "explanatory power". Explanatory power, one might say, is unity plus empirical content. The explication of "unity" indicated here, and spelled out in more detail in Chapter 4, is thus also an explication of "explanatory power".[19]

5.4 Conflicting desiderata for acceptability of metaphysical theses

Physics must exclude empirically successful disunified theories from consideration if theoretical knowledge in physics is to be possible at all. In persistently excluding such disunified theories, physics thereby makes a persistent metaphysical presupposition, as we saw in section 4.2. But what ought this presupposition to be? Section 4.3 has revealed that there are at least eight candidates, namely physicalism(n) with n = 1, ... 8 (and, potentially, many more if N > 1 for some n). Which of these is the best choice if physics is to comply with PIR?

This question is particularly hard to answer because conflicting desiderata arise when it comes to considering what metaphysical thesis physics should accept.

On the one hand it is reasonable to argue that that thesis should be accepted which can be shown to be the most conducive to progress in theoretical physics so far. In the next section I will demonstrate that this picks out the relatively specific and contentful thesis of physicalism(8) with N = 1.

On the other hand, however, it is reasonable to argue that that thesis should be accepted which has the least content that is just sufficient to exclude the empirically successful disunified theories that current methods of physics do exclude. We have almost no grounds for holding that any version of physicalism is true, or is more likely to be true than some other version. Whatever we choose, we are very likely to choose a thesis that is false. Our best bet, then, is to choose that thesis which has the least possible content which suffices to exclude those disunified theories that are excluded from physics, since the less the content of a thesis – other things being equal – the more likely it is to be true. As we shall see in section 5.6 below, this leads to a choice quite different from physicalism(8).

Which of these conflicting lines of argument should be accepted?

In answering this question, I proceed as follows. In section 5.5, I spell out the argument for holding that physicalism(8) is the most fruitful version of physicalism for physics. In section 5.6, I spell out the argument for accepting the minimal version of physicalism. And in section 5.7, I argue that a new conception of physics resolves the conflict.

5.5 Empirically fruitful metaphysics

Before I plunge into my argument for accepting physicalism(8), there is a preliminary question I must answer: why does the unique fruitfulness of physicalism(8) for physics, supposing it can be established, provide grounds for its acceptance?

The basic idea of PIR, as it applies to physics, is that substantial, problematic, influential and implicit metaphysical assumptions need to be made explicit so that they can be critically assessed, so that alternatives can be developed and considered, in the hope that assumptions more conducive to progress can be developed and accepted. In other words, according to PIR, that assumption ought to be accepted which seems to be the most conducive to progress in theoretical physics.

It is important to appreciate just how profoundly influential over the success or failure of theoretical physics choice of metaphysical thesis, of the kind we are considering, is likely to be. This influence is exercised in two ways. First, the metaphysical presupposition of a community of physicists influences – even determines – the direction in which physicists look in order to develop new physical theories. If physicists are convinced – as many were for much of the nineteenth century – that the universe is made up of point atoms which interact by means of centrally directed, rigid forces, then physicists will persistently seek to develop theories which postulate such entities. Physicists who believe that the basic stuff of the universe is energy may be prompted to develop theories of a rather different type. Second, and even more important, the metaphysical presupposition of physics, implicit in non-empirical methods of physics, influences – or co-determines (with evidence) – what theories are accepted and rejected. The success or failure of physics will be highly dependent on whether the non-empirical methods adopted – and thus the corresponding metaphysical theses presupposed – are, or are not, conducive to the selection of theories capable of meeting with empirical success. Adopting the methodological principle "accept only theories that postulate atoms" amounts to presupposing that the universe is made up of atoms (or at least behaves, to a high degree of approximation, as if it is): if this is correct, this presupposition and associated methodological principle may well lead to empirical success. But if the universe is not made up of atoms, and does not even behave as if it is, adopting this methodological principle and presupposing the associated metaphysical thesis is likely to severely stifle scientific progress.

In short, physics must make some metaphysical presupposition for there to be any theoretical knowledge in physics at all. Since the metaphysical theses in question are about the ultimate nature of the universe, the domain of our ignorance, whatever we assume is almost bound to be false. Accepting a false assumption is likely to severely stifle progress in the theoretical physics. It matters enormously, for the progress of physics, that a good choice of metaphysical thesis is made. Just about the only grounds we have for preferring one thesis to another is that one seems to be more conducive to progress in theoretical physics than another. Thus we ought to prefer that thesis which seems to be the most conducive to progress in physics. This is the choice physics needs to make in order to comply with the requirement of rigour of PIR.

Here, now, are the reasons for holding that physicalism(8) should be the preferred metaphysical thesis for physics, largely because this thesis has proved to be more fruitful for progress in physics than any rival thesis.

First, it deserves to be noted that what needs to be made explicit and accepted, if physics is to comply with PIR, is that thesis which is implicit in the current non-empirical *methods* of physics – methods that determine which theories are to be accepted and rejected on grounds of simplicity, unity, explanatoriness. There can be no doubt that, as far as non-empirical considerations are concerned, the more nearly a new fundamental physical theory satisfies all eight of the above requirements for unity, with $N = 1$, the more acceptable it will be deemed to be.

Furthermore, failure of a theory to satisfy elements of these criteria is taken to be grounds for holding the theory to be false even in the absence of empirical difficulties. For example, high-energy physics in the 1960s kept discovering more and more different hadrons, and was judged to be in a state of crisis as the number rose to over one hundred. Again, even though the standard model (the current quantum field theory of fundamental particles and forces) does not face serious empirical problems, it is nevertheless regarded by most physicists as unlikely to be correct just because of its serious lack of unity. In adopting such non-empirical criteria for acceptability, physicists thereby implicitly assume that the best conjecture as to where the truth lies is in the direction of physicalism(8). PIR requires that this implicit assumption – or conjecture – be made explicit so that it can be critically assessed and, we may hope, improved. Physics with physicalism(8) explicitly acknowledged as a part of conjectural knowledge is more rigorous than physics without this being acknowledged, because physics pursued in the former way is able to subject non-empirical methods to critical appraisal as physicalism(8) is critically appraised, whereas physics pursued in the latter way cannot do this.

The really important point, however, in deciding what metaphysical assumption of unity to accept, is that what needs to be considered is not just current theoretical knowledge, or current methods, but the whole way theoretical physics has developed during the last four hundred, or possibly two thousand years. The crucial question is this: what metaphysical thesis does the best justice to the way theoretical physics has developed during this period in the sense that successive theories increasingly successfully exemplify and give precision to this metaphysical thesis in a way that no rival thesis does? The answer is physicalism(8), as the following considerations indicate.

All advances in theory in physics since the scientific revolution have been advances in unification, in the sense of (1) to (8) above. Thus Newtonian theory (NT) unifies Galileo's laws of terrestrial motion and Kepler's laws of planetary motion (and much else besides): this is unification in senses (1) to (3). Maxwellian classical electrodynamics, (CEM), unifies electricity, magnetism and light (plus radio, infrared, ultraviolet, X-rays and gamma rays): this is unification in sense (4). Special relativity (SR) brings greater unity to CEM, in revealing that the way one divides up the electromagnetic field into the electric and magnetic fields depends on one's reference frame: this is unification in sense (6). SR is also a step towards unifying NT and CEM in that it transforms space and time so as to make CEM satisfy a basic principle fundamental to NT, namely the (restricted) principle of relativity. SR also brings about a unification of matter and energy, via the most famous equation of modern physics, $E = mc^2$, and partially unifies space and time into Minkowskian space-time. General relativity (GR) unifies space-time and gravitation, in that, according to GR, gravitation is no more than an effect of the curvature of space-time: this is a step towards unification in sense (8). Quantum theory (QM) and atomic theory unify a mass of phenomena having to do with the structure and properties of matter, and the way matter interacts with light: this is unification in senses (4) and (5). Quantum electrodynamics unifies QM, CEM and SR. Quantum electroweak theory unifies (partially) electromagnetism and the weak force: this is (partial) unification in sense (7). Quantum chromodynamics brings unity to hadron physics (via quarks) and brings unity to the eight kinds of gluons of the strong force: this is unification in sense (6). The standard model unifies to a considerable extent all known phenomena associated with fundamental particles and the forces between them (apart from gravitation): this is partial unification in senses (4) to (7). The theory unifies to some extent its two component quantum field theories in that both are locally gauge invariant (the symmetry group being $U(1) \times SU(2) \times SU(3)$). All the current programmes to unify the standard model and GR known to me, including string theory or M-theory, seek to unify in senses (4) to (8).[20]

In short, all advances in fundamental theory since Galileo have invariably brought greater unity to theoretical physics in one or other, or all, of senses (1) to (8): all successive theories have increasingly successfully exemplified and given precision to physicalism(8) to an extent which cannot be said of any rival metaphysical thesis, at that level of generality. The whole way theoretical physics has developed points towards physicalism(8), in other words, as the goal towards which physics has developed. Furthermore, what it means to say this is given precision by the account of theoretical unity given in section 5.3 above.

In assessing the relative fruitfulness of two rival metaphysical theses, M_a and M_b, for some phase in the development of theoretical physics that involves the successive acceptance of theories $T_1, T_2, \ldots T_n$, two considerations need to be born in mind. First, how potentially fruitful are M_a and M_b, how specific or precise, and thus how specific in the guidelines offered for the development of new theories? Second, how actually fruitful are M_a and M_b, in the sense of how successful or unsuccessful has the succession of theories, $T_1, T_2, \ldots T_n$, been when regarded as a research programme with M_a or M_b as its key idea? When both considerations are taken into account, physicalism(8) comes out as more fruitful for theoretical physics from Newton to today than any rival thesis (at its level of generality). Physicalism(7) is not as specific as physicalism(8), and thus not as potentially fruitful; it does not do justice to the way GR absorbs the force of gravitation into the nature of space-time, and does not do justice to current research programmes which seek to unify matter and space-time. (All of physicalism(n), n = 1, 2, ... 7, are scientifically fruitful to some extent, but decreasingly so as n goes down from 7 to 6 ... to 1, in view of the decreasing specificity and content of these versions of physicalism.)

The notion of "research programme" appealed to here is similar to, but not the same as, the notion developed by Lakatos (1970). The main differences are as follows. For Lakatos, the "hard core" of a research programme was a testable theory rendered metaphysical by a methodological decision; the main research activity associated with a research programme involved developing successful applications of the theory, guided by the "positive heuristic" stemming from the "hard core". (In all this, Lakatos followed Kuhn's conception of "normal science", giving Lakatosian terms to Kuhnian ideas [see Kuhn, 1970a].) In the text, I have assumed that the metaphysics of a research programme is authentic, inherently untestable metaphysics, the main research task being to develop a succession of theories that progressively capture the metaphysics more and more successfully. The account of "degrees of disunity" given in section 5.3 above provides a precise way of assessing the extent to which successive "totalities of fundamental physical theory" do, or do not, increasingly successfully capture physicalism. Thus, given such a succession, T_1, T_2, ... T_m, with degrees of disunity N_1, N_2, ... N_m, of type (5–7), with $N_1 > N_2 > ... > N_m$, then T_1, T_2, ... T_m do progressively capture physicalism(5–7) more and more successfully. There is nothing like this in the Lakatosian account of research programme, lacking as it does the solution to the problem of unity of theory. Finally, there is a substantial difference in the intended application of the two notions. Whereas I see science as a whole as one gigantic research programme, the hierarchy of versions of physicalism being presupposed as the metaphysical "hard core", it is essential to Lakatos's quasi-Popperian conception of science that science is made up of competing research programmes. This means that, for Lakatos, science cannot be viewed as one gigantic Lakatosian research programme (since, if it were, there could be no competitor). Lakatos does say, it is true, "Even science as a whole can be regarded as a huge research programme with Popper's supreme heuristic rule: 'devise conjectures which have more empirical content than their predecessors'" (Lakatos, 1970, p. 132). But there is here no overall Lakatosian "hard core" or "positive heuristic". This is Popper's conception of science and, for Lakatos, his own conception of research programme is strictly inapplicable to science as a whole. (For a more detailed comparison and critical assessment of the two views, see Chapter 2, section 2.8).[21]

Some philosophers of science hold that the successive revolutions in theoretical physics that have taken place since Galileo or Newton make it quite impossible to construe science as steadily and progressively honing in on some definite view of the natural world (Kuhn, 1970a; Laudan, 1980). If attention is restricted to standard empiricism and physical *theory*, this may be the case. But the moment some form of presuppositionism is accepted, and one considers metaphysical theses implicit in the methods of science, a very different conclusion emerges: *all theoretical revolutions since Galileo exemplify the one idea of unity in nature*. Far from obliterating the idea that there is a persistent thesis about the nature of the universe in physics, as Kuhn and Laudan suppose, all theoretical revolutions, without exception, do exactly the opposite in revealing that theoretical physics draws ever closer to capturing the idea that there is an underlying dynamic unity in nature, as specified by physicalism(8).

There is a further point to be made in favour of physicalism(8). So far, every theoretical advance in physics has revealed that theories accepted earlier are false. Thus Galileo's laws of terrestrial motion and Kepler's laws of planetary motion are contradicted by Newtonian theory, in turn contradicted by special relativity, in turn contradicted by general relativity. The whole of classical physics is contradicted by quantum theory, in turn contradicted by quantum field theory. Science advances from one false theory to another. Viewed from a standard empiricist perspective, this seems discouraging, and has prompted the view that all future theories will be false as well, a view that has been called "the pessimistic induction" (Newton-Smith, 1981, p. 14). Viewed from the perspective of science presupposing physicalism(8), however, this mode of advance is wholly encouraging, since it is required if physicalism(8) is true. Granted physicalism(8), the only way a dynamical theory can be precisely true of any restricted range of phenomena is if it is such as to be straightforwardly generalizable so as to be true of all phenomena. Any physical theory inherently restricted to a limited range of phenomena, even though containing a wealth of true approximate predictions about these phenomena, must nevertheless be strictly false: only a theory of everything can be a candidate for truth!

Not only does the way physics has advanced from one false theory to the next accord with physicalism(8); the conception of unity sketched in section 5.3 successfully accounts for another feature of the way theoretical physics has advanced. Let $T_1, T_2, \ldots T_n$ stand for successive stages in the totality of fundamental theory in physics. Each of $T_1, T_2, \ldots T_n$ contradicts physicalism(8), in that each of T_1 etc. asserts that nature is *disunified*, whereas physicalism(8) asserts that it is *unified*. This might seem to make a nonsense of the idea that $T_1, T_2, \ldots T_n$ is moving steadily and progressively towards some future T_{n+r} which is a precise, testable version of physicalism(8). But what section 5.3 shows is that, even though all of $T_1, T_2, \ldots T_n$ are incompatible with physicalism(8), because they are disunified, nevertheless a precise meaning can be given to the assertion that T_{r+1} is closer to physicalism, or more unified, than T_r. This is the case if T_{r+1} is (a) of greater empirical content than T_r (since these are candidate theories of everything), and (b) of a higher degree of unity than T_r in ways specified in section 5.3. Thus the account of unity given above, involving physicalism(1–8), gives precision to the idea that a succession of false theories, $T_1, \ldots T_n$, all of which contradict physicalism(8), nevertheless can be construed as moving ever closer to the goal of specifying physicalism(8) as a precise, testable, physical theory of everything.

5.6 Metaphysical minimalism

I turn now to the argument designed to show that that version of physicalism should be accepted which is the weakest available which just suffices to exclude theories more disunified than currently accepted physical theories – at present the standard model (SM) and general relativity (GR). We may take this to be the strongest version of physicalism that is compatible with SM+GR. This, it may be argued, leads to physicalism($n = 3$) being accepted – or possibly physicalism($n = 4, N = 4$), where $N = 4$ does justice to the fact that SM+GR includes four distinct forces, electromagnetism, the weak and strong forces, and gravitation (electromagnetism and the weak force being only partially unified in quantum electroweak theory).

But physicalism($n = 3$) is, it may be argued, much too strong. It assumes that SM+GR is consistent. Quantum field theory ordinarily assumes flat Minkowskian space-time, but it can be extended to apply in the curved space-time of GR. Problems arise, however, when attempts are made to treat quantum mechanically described matter or energy as the source of gravitation, and so of the curvature of space-time, according to GR. No version of SM+GR can be applied satisfactorily to the interior of black holes. Insuperable problems arise if one tries to incorporate the quantum measurement of orthodox quantum theory into the framework of GR. SM+GR is, it seems, not a consistent theory. Even GR on its own faces a problem of consistency, in that GR predicts that a singularity forms inside a black hole, which constitutes a breakdown of the continuity of space-time, and thus a kind of inconsistency of the theory. In order to assess the degree of disunity of SM+GR one has to consider a version patched up in an *ad hoc* way so as to create a theory that is at least *consistent*. Such an *ad hoc* patching up will further increase the kind and degree of disunity of the theory.

There is a further point. Orthodox quantum theory (OQT) holds that electrons, atoms, even molecules can be in superpositions of states – at different spatial locations at the same time, for example – whereas macroscopic measuring instruments cannot (measuring instruments detecting *one* outcome and not a *superposition* of outcomes). This means OQT holds that different laws hold at different levels – different masses, levels of complexity or whatever it may be. This in turn means that OQT is disunified in way n = 2. Physicalism(n=2, N=2) is perhaps compatible with OQT, but any more unified version of physicalism is not. This is a very serious level of disunity. What makes the matter even worse is that OQT does not specify what variables are involved – let alone what value of what variables – in the transition from quantum states and superpositions to classical states without such superpositions.[22]

5.7 The hierarchical view

Should physics accept physicalism(8) in line with the argument of section 5.5, and risk committing physics to a highly specific and contentful version of physicalism all too likely to be false, despite its fruitfulness for physics up to the present? Or should physics accept some clumsy version of physicalism indicated in section 5.6, less contentful and thus more likely to be true, but entirely lacking in fruitful guidelines for the development of new physical theories?

No version of presuppositionism which restricts itself to adopting a single (if composite) metaphysical thesis can satisfactorily resolve the conflicting desiderata that are highlighted in these two questions. But this conflict is resolved if we adopt a version of presuppositionism which holds that we need to see physics as adopting a *hierarchy* of theses, from physicalism(8) near the bottom of the hierarchy to physicalism(1) at the top (see Figure 5.1). Physicalism(5–7) are on the same level since they are all but equivalent to one another. As we descend the hierarchy, from level (9) to level (3), theses become increasingly contentful and specific, increasingly potentially fruitful for future progress in theoretical physics but also increasingly likely to be false and in need of revision. As one moves from level (9) to level (3), the corresponding methodological requirements for unity, depicted as sloping dotted lines in Figure 5.1, become increasingly demanding, but also increasingly speculative and uncertain. The totality of physical theory, at any given stage in the development of physics (except when a candidate unified theory of everything has been proposed and accepted) will only satisfy these methodological rules partially; a new theory, in order to be an advance from the standpoint of unity, must lead to a new totality of theory satisfying the methodological rules better than the previous totality.

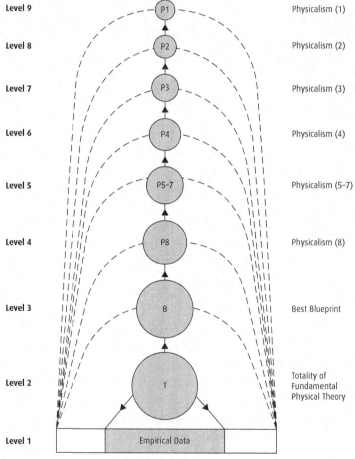

Figure 5.1 Another version of aim-oriented empiricism (Source: author)

This hierarchical view has the following advantages over any version of presuppositionism that restricts itself to a single (possibly composite) thesis. First, the hierarchical view does justice to *both* apparently conflicting desiderata, indicated above, which cannot be done if a single metaphysical assumption is made. The hierarchical view includes *both* the uniquely scientifically fruitful thesis of physicalism(8) and the much less specific and problematic theses of physicalism (1) or (2). Second, the hierarchical view, as a result of making explicit metaphysical theses implicitly presupposed in adopting methods associated with levels (4) to (8), facilitates criticism and revision of these methods, which may well need to be done at some stage (if the corresponding metaphysical theses are false). Such criticism and revision is not facilitated if a single thesis is presupposed. Third, the hierarchical view assists revision of the more contentful and specific versions of physicalism low down in the hierarchy by providing a framework of relatively unproblematic assumptions and methods, at levels (9) to (6), which place restrictions on the way the more specific, problematic versions of physicalism may be revised, should the need to do so arise. If a succession of increasingly empirically successful theories are developed, $T_1, T_2, \ldots$, all of which clash with physicalism(8), but which accord increasingly well with physicalism(7), this might be taken as grounds for rejecting or modifying physicalism(8).

The reasons given above for including the relatively specific, scientifically fruitful metaphysical thesis of physicalism(8) in the hierarchy of accepted theses are reasons also for accepting an even more specific, scientifically fruitful metaphysical thesis, should one be available. A glance at the history of physics reveals that a succession of much more specific metaphysical theses have been accepted, or taken very seriously, for a time, each thesis being an attempt to capture aspects of physicalism. Ideas at this level include the following: the universe is made up of rigid corpuscles that interact by contact; it is made up of point atoms that interact at a distance by means of rigid, spherically symmetrical forces; it is made up of a unified field; it is made up of a unified quantum field; it is made up of quantum strings. These ideas tend to reflect the character of either the current best accepted physical theory, or assumptions made by current efforts to develop a new theory. This is not sufficient to be scientifically fruitful in the way that physicalism(8) is. For this, we require that the thesis in question is such that all accepted fundamental physical theories since Newton can be regarded as moving steadily towards capturing the thesis as a testable physical theory, in the manner indicated in section 5.3. One candidate for such a thesis is Lagrangianism.

Lagrangianism asserts that the universe is such that all phenomena evolve in accordance with Hamilton's principle of least action, formulated in terms of some unified Lagrangian (or Lagrangian density), L. We require, here, that L is not the sum of two or more distinct Lagrangians, with distinct physical interpretations and symmetries, for example, one for the electroweak force, one for the strong force and one for gravitation, as at present; L must have a single physical interpretation, and its symmetries must have an appropriate group structure (the group not being a product of subgroups). We require, in addition, that current quantum field theories and general relativity emerge when appropriate limits are taken.[23]

All accepted fundamental physical theories, from Newton on, can be given a Lagrangian formulation. Furthermore, if we consider the totality of fundamental physical theory since Newton (empirical laws being included if no theory has been developed) then, as in the case of physicalism(8), every new accepted theory has brought the totality of physical theory nearer to capturing Lagrangianism. Thus Lagrangianism is at least as scientifically fruitful as physicalism(8). In fact it is *more* scientifically fruitful since it is very much more specific and contentful. The reasons for accepting physicalism(8) are reasons for accepting Lagrangianism too as the lowest thesis in the hierarchy of metaphysical theses, very much more potentially scientifically fruitful than physicalism(8), but also more speculative, more likely to need revision (see Figure 5.1).

It deserves to be noted that something like the hierarchy of metaphysical theses, constraining acceptance of physical theory from above, is to be found at the empirical level, constraining acceptance of theory from below. There are, at the lowest level, the results of experiments performed at specific times and places. Then, above these, there are low-level experimental laws, asserting that each experimental result is a repeatable effect. Next up, there are empirical laws such as Hooke's law, Ohm's law or the gas laws. Above these there are such physical laws as those of electrostatics or of thermodynamics. And above these there are theories which have been refuted, but which can be "derived", when appropriate limits are taken, from accepted fundamental theory – as Newtonian theory can be "derived" from general relativity. This empirical hierarchy, somewhat informal perhaps, exists in part for precisely the same epistemological and methodological reasons I have given for the hierarchical ordering of metaphysical theses: so that relatively contentless and secure theses (at the bottom of the hierarchy) may be distinguished from more contentful and insecure theses (further up the hierarchy) to facilitate pinpointing what needs to be revised, and how, should the need for revision arise. That such a hierarchy exists at the empirical level provides further support for my claim that we need to adopt such a hierarchy at the metaphysical level.

5.8 Alternative versions of aim-oriented empiricism

The hierarchical view depicted in Figure 5.1 may need to be rejected in its entirety as physics advances. If we exclude from consideration physicalism($n = 1$, $N = \infty$), which permits anything, the hierarchical view assumes that the universe is at least partially physically comprehensible in the sense that phenomena occur in accordance with physical laws which are more or less disunified, the traditional distinction between laws and initial conditions being presupposed. But even though the universe is physically comprehensible, the traditional distinction between laws and initial conditions might not be observed. The true theory of everything might be cosmological in character, and might specify unique initial conditions for the universe.[24] This possibility, and other possibilities of this kind, could no doubt be accommodated within a modified version of the above hierarchical view. But there are other possibilities, of philosophical interest even if of no interest to physics as at present constituted, which cannot be so accommodated. Perhaps God is ultimately responsible for all natural phenomena, or some kind of cosmic purpose or cosmic programme analogous to a computer program (as has been suggested). In these cases the universe would be comprehensible but not physically comprehensible – even though it might mimic a physically comprehensible universe, to some extent.

In order to accommodate these, and other such, possibilities we need to embed the above hierarchical view in the broader view I have called "aim-oriented empiricism" (AOE), depicted in Figure 2.1. However, in order to do this, AOE, as depicted in Figure 2.1, needs to be modified. In this modified figure, we would have in succession, as we go up the hierarchy, empirical phenomena (level 1); accepted fundamental physical theory (level 2); best blueprint (level 3); then, at level 4, the thesis that the universe is perfectly comprehensible physically – physicalism($n=8$); at level 5 the thesis that the universe is all but perfectly comprehensible physically – physicalism($n>4$, $N=1$); at level 6, the thesis that the universe is partially comprehensible physically – physicalism($n>4$, $N<10$); at level 7, the thesis that the universe is partially comprehensible, either physically or in some other way; at level 8, the thesis of meta-knowability: the universe is such that we can discover how to improve our methods of learning; and finally, at level 9, the thesis of partial knowability: the universe is such that we can acquire some knowledge of our local circumstances.

The level 8 thesis, here, asserts that the universe is such that there is some rationally discoverable proposition about its nature (relative to existing knowledge) which, if accepted, makes it possible progressively to improve methods for the improvement of knowledge. "Rationally discoverable" means at least that the thesis is not an arbitrary choice from infinitely many analogous theses. This thesis is to be interpreted as asserting that the universe is not epistemologically malicious, in the sense that apparently improved methods lead to apparent new knowledge which turns out, subsequently, to be illusory, there being no possibility of discovering this before it is revealed. (This is clarified in the next section.) Level 9 asserts that the universe is such that we can continue to acquire knowledge of our local circumstances, sufficient to make life possible.

Such an amalgam of Figures 5.1 and 9.1 (see Chapter 9) is not altogether satisfactory. It is somewhat arbitrary to declare that physicalism($n>4$, $N=9$) represents a partially physically comprehensible universe, but physicalism($n>4$, $N=10$) does not. Some may question that the alternative thesis at level 7 – the thesis that the universe is comprehensible non-physically – has anything to do with physics, or science. One can imagine circumstances, however, in which this thesis might well be preferred to any thesis of partial physical comprehensibility. Suppose God manifests Himself in the sky, and responds to our requests to perform specific miracles by doing just that – converting the orbit of Mercury into a square orbit, for example, or lifting everyone on earth one foot into the air, except where this is impossible or dangerous. How could we not accept, in such circumstances, that God is ultimately in charge of Nature?

Reasons for accepting these theses are similar to those given for accepting the hierarchy of theses of the view depicted in Figures 2.1 and 5.1. As a result of accepting this hierarchy of theses, physics provides itself with a framework of assumptions, and associated methods, high up in the hierarchy, which it will never benefit the pursuit of knowledge to reject; in this way, a framework is created within which very much more substantial and dubious theses, low down in the hierarchy, can be critically scrutinized and, we may hope, improved. The thesis at the top of the hierarchy, at level 9, asserts that the universe is such that we can acquire some knowledge of our local circumstances. If this assumption is false, we will not be able to acquire knowledge whatever we assume. We are justified in accepting this assumption permanently as a part of our knowledge, even if we have no grounds for holding it to be true, since accepting it can only help, and cannot hinder, the acquisition of knowledge whatever the universe is like. The thesis at level 4, physicalism(8), deserves to be accepted because of its extraordinary scientific fruitfulness. All major theoretical developments in physics point towards, and draw closer to, physicalism(8), in that they are invariably major steps in unification, as we saw in section 5.5 above. From the standpoint of scientific fruitfulness, at this level, physicalism(8) has no rival, unless it is the even more substantial thesis of Lagrangianism. Nevertheless, physicalism(8) may be false, and physics may, at some stage, need to adopt a different thesis. Accepting physicalism(8) within the framework of theses just indicated facilitates the development of alternatives, should this be necessary.

The view depicted in Figure 5.1 has, perhaps, a more direct relevance to theoretical physics; AOE as depicted in Figure 2.1 would become relevant if it emerged that the universe differs radically from the way modern science assumes it to be. AOE is more relevant to the philosophy of physics; it is required to solve the problem of induction, as we shall see in Chapter 7 (see also Maxwell, 1998, ch. 5; 2007, ch. 14; 2017a; 2017b.) It is also required to rebut objections of circularity, as we shall now see.

5.9 The circularity problem solved

One feature of the views depicted in Figures 2.1 and 5.1 may be deemed puzzling. They hold that when metaphysical thesis and physical theory clash, physical theory may be revised, but also that metaphysical thesis may be revised. How is such a two-way influence possible? In what follows I consider AOE as depicted in the previous section.

The first point to note is that just such a two-way influence occurs when theory and experiment clash. In general, if a theory clashes with an experiment that has been subjected to expert critical scrutiny and repeated, the theory is rejected. But on occasions, it turns out that it is the experimental result that is wrong, not the theory.[25] In a somewhat similar way, if a new theory increases the conflict between the totality of physical theory and the currently accepted metaphysical thesis, at level 3 of the previous section, the new theory will be rejected (or not even considered or formulated).[26] On occasions, however, a new theory may be developed which increases the conflict between the totality of theory and the current thesis at level 3 but decreases the conflict between the totality of theory and physicalism at level 4. In this case the new theory may legitimately be accepted and the thesis at level 3 may be revised. In principle, as I have already indicated, theses even higher up in the hierarchy may legitimately be revised in this way. A virtue of these hierarchical views is that they make possible and facilitate such two-way revisions.

However, another, potentially more serious, problem faces the two hierarchical views indicated above. Both incorporate what seems to be vicious circularity. Acceptance of theories is influenced by their degree of accord with metaphysical principles, the acceptance of which is in turn, in part, influenced by an appeal the empirical success of physical theories. The claim is that as theoretical knowledge and understanding improves, metaphysical theses and associated methods improve as well. There is something like positive feedback between improving knowledge, and improving knowledge-about-how-to-improve-knowledge. This, it is claimed, is the methodological key to the great success of modern science, namely that it adapts its metaphysical assumptions and methods (its aims and methods) to what it finds out about the nature of the universe.[27] But how can such a circular procedure conceivably be valid?

This is not an objection to the arguments and views put forward so far. No attempt has been made to justify claims to theoretical knowledge. The argument has been modest: granted acceptance of current physical theories and adoption of current methods, then physics is more rigorous (in that it accords better with PIR) if implicit metaphysical assumptions are made explicit, and those assumptions chosen which seem best to promote what we take to be scientific progress. The circularity objection would arise, however, if we were to go beyond the modest aspirations of sections 5.2 to 5.8, and attempt to solve the problem of induction,[28] and justify acceptance of empirically successful unifying theories, within the context of AOE. But this does, in effect, amount to an objection to AOE. For it may be argued that AOE cannot be acceptable because the problem of induction cannot conceivably be solved within its framework: the moment the attempt is made to *justify* acceptance of scientific theories and metaphysical theses as claims to knowledge, vicious circularity sets in. How is this circularity objection to be met?

Here, in a nutshell, is the answer. Permitting metaphysical assumptions to influence what theories are accepted, and at the same time permitting theories to influence what metaphysical assumptions are accepted, may (if carried out properly), *in certain sorts of universe*, lead to genuine progress in knowledge. The level 6 thesis of meta-knowability, of AOE, asserts that *this is just such a universe*. And furthermore, crucially, reasons for accepting meta-knowability make no appeal to the success of science. In this way, meta-knowability legitimises the potentially invalid circularity of AOE depicted in Figure 2.1.

Relative to an existing body of knowledge and methods for the acquisition of new knowledge, possible universes can be divided up, roughly, into three categories: (i) those which are such that the meta-methodology of AOE can meet with no success, not even apparent success, in the sense that new metaphysical ideas and associated methods for the improvement of knowledge cannot be put into practice so that success (or at least apparent success) is achieved; (ii) those which are such that AOE can meet with genuine success; and (iii) those which are such that AOE appears to be successful for a time, but this success is illusory, this being impossible to discover during the period of illusory success. Meta-knowability asserts that this universe is a type (i) or (ii) universe; it rules out universes of type (iii).

Meta-knowability asserts, in short, that the universe is such that AOE can meet with success and will not lead us astray in a way in which we cannot hope to discover by normal methods of scientific inquiry (as would be the case in a type (iii) universe). If we have good grounds for accepting meta-knowability as a part of scientific knowledge – grounds which do not appeal to the success of science – then we have good grounds for adopting and implementing AOE (from levels 8 to 3).

But what grounds are there for accepting the thesis of meta-knowability at level (6)? There are two:[29]

(a) Granted that there is *some* kind of general feature of the universe which makes it possible to acquire knowledge of our local environment (as guaranteed by the thesis at level 8), it is reasonable to suppose that we do not know all that there is to be known about what the *nature* of this general feature is. It is reasonable to suppose, in other words, that we can improve our knowledge about the nature of this general feature, thus improving methods for the improvement of knowledge. Not to suppose this is to assume, arrogantly, that we already know all that there is to be known about how to acquire new knowledge. Granted that learning is possible (as guaranteed by the level 8 thesis), it is reasonable to suppose that, as we learn more about the world, we will learn more about how to learn. Granted the level 8 thesis, in other words, meta-knowability is a reasonable conjecture.
(b) Meta-knowability is too good a possibility, from the standpoint of the growth of knowledge, not to be accepted initially, the idea

only being reluctantly abandoned if all attempts at improving methods for the improvement of knowledge fail.

(a) and (b) are not, perhaps, very strong grounds for accepting meta-knowability; both are open to criticism. But the crucial point, for the present argument, is that these grounds for accepting meta-knowability, (a) and (b), are independent of the success of science. This suffices to avoid circularity.

If AOE lacks meta-knowability, its circular procedure, interpreted as one designed to procure justified knowledge, becomes dramatically invalid, as the following consideration reveals. Corresponding to the succession of accepted fundamental physical theories developed from Newton down to today, there is a succession of aberrant rivals which postulate that gravitation becomes a repulsive force from the beginning of 2050, let us say.[30] Corresponding to these aberrant theories there is a hierarchy of aberrant versions of physicalism, all of which assert that there is an abrupt change in the laws of nature at 2050. The aberrant theories, just as empirically successful as the theories we accept, render the aberrant versions of physicalism just as scientifically fruitful as non-aberrant versions of physicalism are rendered by the non-aberrant theories we actually accept. If we take it as given that we accept non-aberrant theories, the question of what reasons there are for rejecting empirically successful aberrant theories and associated aberrant versions of physicalism does not arise. But the moment we seek to *justify acceptance* of non-aberrant theories and *rejection* of aberrant theories, within the framework of AOE, the question of what reasons there are for rejecting aberrant theories and associated aberrant versions of physicalism arises. If AOE is bereft of meta-knowability, it is not easy to see what these reasons can be. But AOE with meta-knowability included does provide a reason: the aberrant versions of physicalism assert that this is a type (iii) universe, which violates meta-knowability.

Versions of physicalism(n) for which $n = 1$ or 3, and $N > 1$ would seem to violate meta-knowability. But other versions of physicalism with $N > 1$ need not clash with meta-knowability.[31]

5.10 Conclusions

I have argued that if science is to be rigorous it needs to accept explicitly, as a core component of theoretical scientific knowledge, a hierarchy of metaphysical theses (and associated methods) concerning the dynamic unity, the physical comprehensibility, of the universe. I have shown that the notorious problem concerning the unity, the explanatory character, of physical theory can be solved within this hierarchical view of science. This solution, in turn, provides a precise way of assessing the scientific fruitfulness of rival metaphysical theses, from the standpoint of the empirical progressiveness of the research programmes to which they give rise.

These results have dramatic implications for science, for our understanding of science, and for the relationship between science and philosophy. There is a major increase in the (acknowledged) *scope* of scientific knowledge. Whereas standard empiricism implies that science at present provides us with no knowledge about the ultimate nature of the universe (all current fundamental physical theories being false), the hierarchical view holds the opposite. Current science does include knowledge about the ultimate nature of the universe – knowledge which, though theoretical and conjectural, is nevertheless more secure than any accepted theory such as quantum theory or general relativity: physicalism(5–7) is true, even perhaps physicalism(8). Science becomes much more like natural philosophy, in that it incorporates sustained exploration and assessment of metaphysical theses, and associated metholodogical rules, as an integral, vital part of scientific research. Instead of metaphysics and philosophy being banished from science, they become a vital part of science.

Furthermore, as I have argued in detail elsewhere (Maxwell, 1976a, 1984a or 2007, 2000a, 2001a, 2004, 2010a, 2014b, 2016b, 2017a, 2017b), the arguments of this chapter, when extended to take into account, not just the implicit metaphysical assumptions of science, but its implicit value and political assumptions as well, have even more dramatic implications not just for physics or natural science, but for social science, for the humanities, for academic inquiry as a whole.

Perhaps the time has come for philosophers to take note of these arguments that have such revolutionary implications for our intellectual landscape.

6
Comprehensibility rather than beauty

6.1 Beauty or comprehensibility?

Many scientists, and some philosophers of science, have acknowledged that aesthetic considerations play, quite properly, an important role in influencing acceptance and rejection of theories in science, in addition to empirical considerations. A famous example is Dirac, who went as far as to declare: "It is more important to have beauty in one's equations than to have them fit experiment" (quoted in McAllister, 1996, p. 15).

The view that beauty ought to influence choice of theory in science faces, however, a serious problem. Why should beauty be a good indication of truth? Unless the truth is beautiful, and unless we have valid grounds for holding this to be the case, there can be no good reasons, it would seem, for giving preference to beautiful theories in science.

Not only may it seem dubious that we can have grounds for holding that the truth is beautiful; there may well seem to be grounds for holding that it is wildly implausible that the truth should be beautiful, especially in theoretical physics.

Whether we find something beautiful or ugly must depend, to some extent at least, on our personal, subjective, emotional responses to that thing. Aesthetic criteria have their roots deep in the human psyche, and in human culture. But physical reality, that which theoretical physics seeks to grasp, is utterly remote from the human psyche, from human culture. It may well seem utterly implausible that something as anthropomorphic, as personal, as quintessentially human and subjective, as ideas about beauty, should have anything to do with the ultimate nature of the physical universe, utterly impersonal and remote from the circumstances of human life. Beauty may seem to be the last consideration to take into account in assessing the merits of rival fundamental theories in physics.

An extremely interesting and original defence of the thesis that aesthetic considerations do quite properly influence theory choice in science has, however, been put forward recently by James McAllister (1996; see also his 1989, 1990 and 1991). Quite independently, I have, over a number of years, developed a view which resembles McAllister's view in a number of striking ways, but which is also different in important respects, namely aim-oriented empiricism (see Chapters 2, 4 and 5).[1] In this chapter I compare and contrast the two views. My own view, aim-oriented empiricism, has already been expounded and defended. I begin with a sketch McAllister's view. I then discuss how the two views resemble, and differ from, each other. And finally, I discuss the question of which is to be preferred.

6.2 The model of the aesthetic induction

In what follows I call McAllister's account of the role of non-empirical, aesthetic factors in the selection of theories in science "the model of the aesthetic induction" (MAI). Here, in summary, is his view.

According to MAI, the basic aim of science is to develop a body of theory that successfully predicts all observable phenomena. MAI holds that from this aim of "empirical adequacy", we can arrive at the following criteria for assessing theories: success in predicting existing empirical data, success in predicting new phenomena, consistency with other high-level theories, explanatory power and internal consistency.

Many scientists have, however, declared that aesthetic considerations, in addition to the above, play a vital role in both the discovery and acceptance of theories in science. Dirac, Einstein and many others have stressed the importance of aesthetic considerations, such as beauty, elegance, harmony, uniformity amidst variety, simplicity, symmetry. MAI holds that such criteria do indeed have an important role to play in deciding what theories are accepted, to the extent, even, on occasions, of overriding empirical considerations.

But, according to MAI, in so far as such aesthetic considerations exercise a rational influence over choice of theory in science, two crucial points need to be borne in mind. First, theories must be considered to be abstract entities, distinct from this or that linguistic formulation. Second, what matters is not the (subjective) aesthetic judgements themselves, but rather objective, non-aesthetic properties that theories, construed as abstract entities, do actually possess, in virtue of which scientists make their aesthetic judgements.

There are, according to MAI, five classes of properties of theories that are relevant: symmetry, invocation of a model, visualizability/abstractness, metaphysical allegiance and simplicity (related to unity). MAI stresses that many different properties fall under each of these headings. There are different kinds of symmetry; different theories have different kinds of models; some scientists, in some contexts, hold visualizability to be a virtue, while others, in other contexts, prize almost its opposite, namely abstractness; scientists have upheld different metaphysical views at different stages in the development of science, in terms of which they have sought to interpret scientific theories; and there are many different ways of assessing the simplicity of theories, yielding quite different results.

How, then, does the scientific community decide which of these very many different kinds of properties of theories are the relevant or important ones to employ in order to assess the acceptability of theories on non-empirical, or aesthetic, grounds? And what is the justification for so assessing theories, in terms of the preferred properties? How, in particular, can MAI do justice to the fact that aesthetic criteria in science change over time?

The answer is that, at any given stage, a scientific community prefers those new theories that have properties which earlier theories, which have proved to be empirically successful, also possess. If a certain kind of theory, with characteristic aesthetic properties, has met with empirical success in the past then, understandably enough, scientists are influenced to give preference to similar kinds of theories, with similar properties, in the future. This is "the aesthetic induction". At a stroke, the above three questions are answered.

In a little more detail, we can imagine that a scientific community can consider many different aesthetic properties of theories, P, Q, R, ... The community will assign a different weighting, W_P, W_Q, W_R, ... to each of these properties, each weighting determining how influential the corresponding property is in theory choice. The weightings are in turn determined by what kinds of theory, with what properties, have (or have not) met with empirical success in the past. W_P, W_Q, W_R, ... are, in other words, determined by the aesthetic induction.

According to MAI, then, two kinds of criteria are employed in science to choose theories. On the one hand there are criteria, listed above, arrived at by analysis of the basic aim of science of achieving empirical adequacy. And, on the other hand, there are criteria arrived at by the aesthetic induction. The second presupposes the first.

Aesthetic criteria will tend to be conservative, based as they are on empirical performance of theories in the past. New theories, with the potential for great predictive success, may violate existing, conservative aesthetic criteria. When such a theory is developed, there is a rupture in accepted aesthetic criteria. Initially the new theory is judged to be "ugly"; but as its empirical potential becomes manifest, aesthetic criteria are changed to suit the new theory. This is what a scientific revolution amounts to, according to MAI, a conception somewhat different from Kuhn's. In terms of this conception, neither Copernicus's theory, nor Einstein's theory of special relativity, were revolutionary, because neither broke with aesthetic criteria of the past. But Kepler's laws of planetary motion, and quantum theory, were both revolutionary, in that these theories broke dramatically with aesthetic criteria generally accepted at the time.

Finally, though the aesthetic induction might one day favour some particular metaphysical world view, so far this has not happened (McAllister, 1996, 102–4).

6.3 Comparison of the two views

What is rather astonishing about aim-oriented empiricism (AOE) and MAI is that, though arrived at independently, and though giving what are, in some respects, very different pictures of the scientific enterprise, nevertheless the two views have much in common. Both seek to uphold what McAllister calls "the rationalist image of science". Both hold that (some) criteria of theory choice can be justified by an appeal to the aims of science. Both hold that non-empirical criteria of theory choice have an enormously important part to play in science. Both hold that these non-empirical criteria are, in practical applications, quite diverse in character. Both hold that they change over time, as science progresses. And there is considerable agreement as to what these non-empirical criteria are: simplicity, unity, symmetry and compatibility with some metaphysical world view, are all important, for both views.

Both hold that these criteria apply, not to specific formulations of theories, but to what all possible formulations have in common. And both give accounts of scientific revolutions that differ substantially from Kuhn's account.

But there are also dramatic differences. MAI is, for McAllister, "a medium-level model of scientific practice, of a scope intermediate between the loftiest generalization and the historical study" (1996, p. 2). AOE is put forward as a "highest-level model", with implications and applications for all of natural science. (Strictly speaking, it is what I call "generalized AOE" [Maxwell, 1998, pp. 191–2, 185, 191, 208 and 223–4] – embodying the hierarchical structure of AOE, but lacking specific, lower-level theses of AOE – that is a model at the highest level, applying to science throughout history; AOE is restricted to post-Galilean science.)

Again, MAI is a version of standard empiricism (SE), whereas AOE emphatically rejects SE. That MAI is a version of SE is clear from the way the aim of science is characterized as "empirical adequacy". It is also apparent in the way science can, according to MAI, establish a metaphysical world view. This can only happen via the aesthetic induction, and has not as yet come about. According to AOE, by contrast, at levels 7 and 6 there are metaphysical, cosmological theses that are permanently accepted by science, and at levels 5 to 3, there are metaphysical theses which are a part of current scientific knowledge, but which are increasingly likely to require revision with the advance of science, as one descends from level 5 to level 3 (see Figure 2.1 in Chapter 2).

Whereas MAI gives to science just one aim (empirical adequacy), AOE sees science as having a hierarchy of aims, from empirical adequacy, perhaps, at the highest level, down to the aim to turn the best available level 3 blueprint into a precise, true theory of everything, at the lowest level. (And even more specific, and different, aims are assigned to different branches of natural science.)

That AOE postulates this hierarchical structure to the aims of science, whereas MAI does not, leads to different treatments of changing criteria for theory choice. According to MAI, criteria of theory choice are of two kinds: those that are justified by an appeal to the basic aim of science (empirical adequacy), and those that are justified by inductive projection – the aesthetic criteria arrived at by the aesthetic induction. These latter are weaker than the former, and presuppose, for logical reasons, the former (McAllister, 1996, p. 76). According to AOE, by contrast, all criteria of theory choice are arrived at by aim analysis: those that evolve do so because the level 3 aim of science evolves.

Even though MAI and AOE agree that non-empirical criteria of theory choice change with time, they disagree about which criteria change, and what this change involves. According to AOE, something close to physicalism has been implicit in the methods of theoretical physics since Galileo or Newton; the demand for theoretical unity, associated with physicalism, has been more or less unchanging. What has changed is the form that the demand for unity takes, as manifest in dramatically changing level 3 metaphysical blueprints. MAI does not claim that physicalism, and the requirement of unity associated with it, is a part of the unchanging criteria of theory choice (since Galileo, at least). Nor could MAI claim this, as long as it is a version of SE.

According to AOE, the level 4 thesis of physicalism, and the level 3 best metaphysical blueprint, are arrived at by a quasi-Popperian process of conjecture and criticism, the whole direction of progress in theoretical physics since the birth of modern science (or since the Presocratics) being taken into account. The claim is that these theses make explicit what theoretical physics hopes to achieve: they are intended to be the best conjectures as to what the basic aims of theoretical physics should be, at different levels of specificity. These theses are intended to lead to criteria, to methodological principles such as symmetry principles, that will be relevant for future theories, not yet developed. Indeed, according to AOE, the activity of further articulating the best blueprint, and solving problems of unity to which it gives rise, provides science with a rational, if fallible method of discovery.[2] All this contrasts dramatically with criteria arrived at by the aesthetic induction, according to MAI, which are almost bound to be conservative, and more or less inapplicable to revolutionary developments. AOE criteria anticipate and provoke revolution, and judge the existing body of fundamental physical theory as unsatisfactory because of its failure to comply with the demand for unity (the standard model postulates too many particles and forces, and clashes with general relativity); by contrast, MAI criteria are conservative, and are almost bound to be at odds with revolutionary developments (McAllister, 1996, pp. 81–5 and 128–33). AOE criteria are heuristically powerful; MAI criteria are the opposite. Furthermore, AOE criteria, associated with level 3 blueprints, evolve or improve as physics makes progress, and in a way which admits some elements of continuity (see, in particular, Maxwell, 1998, pp. 80–9). MAI sees change, but no overall progress, in non-empirical criteria, and holds, in a quasi-Kuhnian fashion, that revolutions create a rupture in aesthetic criteria, there being no account of the modification and generalization of blueprints, which AOE provides.

MAI and AOE agree that non-empirical criteria apply, not to any specific formulation of a theory, but to what all formulations have in common. But there are somewhat different accounts of what this is. According to MAI, a formulation-independent theory is an abstract entity that exists in its own right, with its own properties distinct from the phenomena the theory postulates (see, for example, McAllister, 1996, pp. 98–100). This leaves obscure what sort of thing such an abstract entity is, and what its relationship is with a linguistic formulation of the theory, and with the phenomena it predicts. According to AOE, the matter is much more straightforward: a formulation-independent theory, T, is the *content* of T, what T predicts, or asserts to be the case. AOE does not appeal to the abstract entities of MAI; it appeals only to possible phenomena, not as actually existing entities, but merely as possibilities. The claim that T exhibits a certain symmetry thus amounts to the claim that the phenomena predicted by T exhibit this symmetry. There is here no mystery about the relationship between a linguistic formulation of T, the abstract entity T, and the phenomena T predicts: the "abstract entity" is just what any linguistic formulation of T asserts to be the case, the *content* of T. This leads to an account of the importance of linguistic-dependent criteria of simplicity (Maxwell, 1998, pp. 110–3), something that MAI does not provide.

A fundamental difference between MAI and AOE, encapsulated in the title of this chapter, is that, whereas MAI holds that aesthetic criteria are important in science, AOE denies this, all non-empirical criteria for theory choice being reducible to the demand that the totality of fundamental physical theory exemplify the level 4 thesis of physicalism or, more specifically, the best available blueprint at level 3. For AOE, what matters is unity or comprehensibility, not beauty.

But this difference is terminological rather than substantial. McAllister defends a projectivist, subjectivist account of the beauty of theories. Beauty is in the eye of the beholder, rather than in the theory itself.[3] Scientists judge certain theories to be beautiful because of non-aesthetic properties that they possess objectively; it is these non-aesthetic properties that are important methodologically and epistemologically, and play the crucial role in the aesthetic induction. One of these is metaphysical allegiance. The demand that the totality of fundamental physical theory should exemplify physicalism, and the best available blueprint, are special cases of metaphysical allegiance. Comprehensibility is just one of McAllister's aesthetic properties. Comprehensibility, one might say, is beautiful. It fits perfectly Hutchinson's characterization of beauty (McAllister, 1996, pp. 17–23) as involving "uniformity amidst variety" (see the discussion of "unity through diversity" in Maxwell, 1998, ch. 3).

A more serious disagreement would seem to be that whereas AOE recognizes only *one* methodologically significant non-empirical property, namely unity or comprehensibility, MAI stresses that there are endlessly many, falling under the five headings of symmetry, invocation of a model, visualizability/abstractness, metaphysical allegiance and simplicity (related to unity).

This disagreement is not quite as big as it might at first appear to be. Here, very briefly, are the similar, but also different, ways in which AOE and MAI treat unity, symmetry, metaphysical allegiance and simplicity.

Unity: AOE and MAI both recognize that the demand for unity takes a number of different forms, but AOE alone holds that these are aspects of just one, single conception of unity. According to AOE, dynamic unity, postulated to exist by physicalism, can be broken in thought in a number of different ways, this creating a number of different kinds of (relative) disunity, and hence a number of different ways in which degrees of unity (or disunity) can be assessed. But these different kinds of disunity all relate to just one conception of unity, namely that which is postulated to exist by physicalism (see Chapter 4, and Maxwell, 1998, pp. 89–93, and p. 280, note 22). MAI, too, stresses that there are different kinds of unification (McAllister, 1996, p. 110) but, unlike AOE, does not relate these to one basic conception of unity.

Symmetry: Here, again, AOE and MAI both recognize that the demand for symmetry takes a number of different forms, but AOE alone holds that these, in so far as they are methodologically legitimate within theoretical physics, all relate to the one basic demand for unity. One of the achievements of AOE is to demonstrate clearly how different kinds of symmetry relate to unity, the demand that theories exhibit symmetries itself being an aspect of the demand for unity (Maxwell, 1998, pp. 89–103, 123–40 and 257–65). MAI recognizes that theories exhibit different kinds of symmetry (McAllister, 1996, 41–4), but fails to recognize that different kinds of symmetry, in theoretical physics at least, are aspects of unity.

Metaphysical allegiance: Once again, both AOE and MAI recognize that an important non-empirical requirement in theoretical physics, upheld by some physicists at least, is that fundamental physical theories should accord with some metaphysical view. Both recognize that metaphysical views associated with physics since Galileo have changed dramatically over time; both recognize that different physicists espouse different metaphysical views at the same time. But AOE and MAI differ here too, in that AOE holds that diverse, evolving level 3 blueprints, in order to be acceptable, need to accord with physicalism, whereas MAI makes no such demand. For AOE, the requirement that a theory exemplifies a metaphysical view, in so far as it is methodologically legitimate, is but an aspect of the basic requirement that the body of fundamental physical theory exemplifies the unity of physicalism (as much as possible). MAI makes no such demand.

Simplicity: Here, yet again, both AOE and MAI recognize that the demand for simplicity takes a number of different forms; both see simplicity as being related to unity, but in somewhat different ways (compare Maxwell, 1998, pp. 111–3 and 157–9, with McAllister, 1996, pp. 109–11). But AOE alone relates the demand for simplicity to the more basic demand for just one kind of unity, dynamic unity postulated by physicalism. According to AOE, unity requires that a theory makes *the same* assertion throughout all possible phenomena to which it applies; simplicity concerns *what* is asserted for any given phenomenon, and thus presupposes unity. Simplicity presupposes unity, but is quite distinct from unity (see Chapter 4, and Maxwell, 1998, pp. 157–9; 2004a, pp. 172–4).

The difference that this reveals in the two views can be summed up like this. AOE postulates just one basic non-empirical requirement, unity, and relates different requirements, of different kinds of unity – symmetry, metaphysical allegiance and simplicity – to this one demand for unity. MAI, by contrast, holds that there just are many different kinds of requirements – unity, symmetry, metaphysical allegiance and simplicity. Unlike AOE, MAI sees no unity in these diverse kinds of requirements – unity, symmetry, metaphysics and simplicity. (In this respect, AOE might be said to give a more unified, and hence more beautiful, account of scientific method than MAI.)

More substantial differences arise in connection with the two remaining kinds of aesthetic properties of theories which MAI holds to be methodologically significant, which I now consider in turn.

Invocation of a model: AOE recognizes that an important consideration in assessing a new physical theory is that it has a form similar to existing empirically successful physical theories. Thus the acceptability of quantum electroweak theory, and chromodynamics is much helped by the fact that these theories are similar in form to the highly empirically successful theory of quantum electrodynamics. All three theories, despite their differences, are locally gauge invariant quantum field theories. According to AOE, this requirement of similarity of form or structure derives, once again, from the requirement of unity (Maxwell, 1998, p. 112). If T_1 and T_2 have some similar structures, then some part of T_1 can be modelled by some part of T_2, and vice versa. According to AOE, having a model is only methodologically significant to this extent, and once again this requirement turns out to be derived from the demand for unity. (Of course, that physical reality is a model of a theory, in the sense that the theory is *true*, is highly significant for AOE; but this is not what

MAI means by a "model".) MAI is, once again, much more open-ended in the kind of models that it is prepared to recognize as methodologically significant, and does not attempt to derive from the demand for unity the requirement that a theory should have some kind of model.

Visualizability/abstractness: According to AOE, neither visualizability nor abstractness are methodologically significant for theoretical physics. What does matter is that a theory can at least be interpreted realistically, as postulating that such and such a physical entity, (or entities), such as a field (or particles) exists, a stepping stone towards the ubiquitous, unified *something* of physicalism. (Actually, AOE demands more. It demands that fundamental physical theories are open to being interpreted in terms of conjectural essentialism (see Maxwell, 1998, pp. 141–55.) If one has acquired an intuitive understanding of a realistic theory, then one may well be able to "visualize" what the theory is about: to this extent, visualizability is methodologically significant, according to AOE, but once again derives from the demand for unity, via the demand for realism. MAI, by contrast, once again, is much more open-ended about visualizability, and makes no attempt to relate it to the demand for unity.

McAllister claims that opposition to orthodox quantum theory (OQT), by Schrödinger, Einstein and others, stemmed from the loss of visualizability and determinism associated with the new theory. But this overlooks the key, entirely legitimate objection to OQT, namely its loss of microrealism, due to the failure to solve the quantum wave/particle problem. Because it failed to specify a consistent quantum ontology, OQT had to be developed as a theory which can, at most, make predictions about the results of performing measurements on quantum systems – measurement being described classically. But this in turn meant that OQT is, quite essentially, made up of two quite different parts stitched together in a grossly *ad hoc* way, namely (1) the quantum part, and (2) some part of classical physics for a treatment of measurement. Despite its immense empirical success, OQT is still today deeply and genuinely problematic, to the point, almost, of being unacceptable, because of its grossly *ad hoc* character, due to its lack of micro-realism (see Chapter 8, and Maxwell, 1972b; 1976b; 1982; 1988; 1998, ch. 7). The mature Einstein was well aware that this is the basic objection to OQT, not lack of visualizability or loss of determinism, as we saw in Chapter 3. Elsewhere I have argued that the grossly *ad hoc* character of OQT, stemming from its lack of micro-realism, provides us with a general argument against instrumentalism and for realism (Maxwell, 1993b). I have also suggested how the quantum wave/particle problem may be solved, and how a fully micro-realistic version of quantum theory may be developed, free of any reference to measurement or classical physics in its basic postulates, able to recover all the successful predictive content of OQT, but also making experimental predictions different from OQT for as yet unperformed experiments (see Chapters 3 and 8, and Maxwell, 1976b; 1982; 1988; 1998, ch. 7; and especially 1994a). (This was done in an attempt to put the rational, but fallible, method of discovery of AOE into scientific practice.) There are, of course, other attempts at developing fully micro-realistic versions of quantum theory (see Bohm, 1952; Ghirardi, Rimini and Weber, 1986; Penrose, 1986; Tumulka, 2006; and Wallace, 2012).[4]

Einstein's mature objection to OQT had to do with the lack of realism of the theory, but he did also, especially earlier, object to its lack of determinism. But here, too, there is a methodologically significant issue at stake, related once again to the demand for unity. A realistic version of quantum theory must be unified, first with special relativity, and then, ultimately, with general relativity. This is a much graver problem, granted probabilistic quantum theory, than it is if quantum theory is deterministic. The demand for unity speaks against probabilistic quantum theory – but not decisively: nature may well be probabilistic, and the task may be to develop probabilistic versions of special and general relativity (Maxwell, 1985a).

As for abstractness, this is, for AOE, without methodological significance, except that, as physical theory draws closer to capturing physicalism, it is almost bound to specify entities increasingly remote from those of ordinary experience. We begin with corpuscles, minute billiard balls, in the seventeenth century; these then transmute into point particles that interact by means of forces; these, in turn, transmute into classical fields, into quantum fields, into curved space-time, into superstrings in ten-dimensional space-time – entities increasingly remote from the familiar billiard ball.

We have seen, so far, that AOE recognizes, ultimately, just one non-empirical criterion, unity or compatibility with physicalism,[5] whereas MAI recognizes many, and makes no attempt to show that these all devolve from just one basic criterion. But I come now to a non-empirical criterion which AOE holds to be absolutely central, but which MAI does not even recognize as an aesthetic criterion at all: explanatory power.

Explanatory power is an ambiguous concept. We may hold that T_1 has more explanatory power than T_2 if (1) T_1 has greater empirical content than T_2, or if (2) T_1 has greater unity than T_2 even though it has the same empirical content. Let us call these type (1) and type (2) explanatory power respectively.

We need also to recognize that criteria legitimately employed in science to assess theories can be put into three categories: (a) empirical, (b) empirical-dependent, and (c) non-empirical. By (a) I mean simply the predictive success of the theory in question; by (b) I mean properties of theories that have to do with how amenable they are to being assessed empirically, such as testability and empirical content; and by (c) I mean properties of theories that have nothing directly to do with empirical success but which are deemed to be indicative of truth, or of potential empirical success.

Type (1) explanatory power is a typical type (b) property of theories. But, according to AOE, type (2) explanatory power is the key type (c) non-empirical property of theories from which, as we have seen, all others, such as symmetry, simplicity or metaphysical allegiance, arise. In seeking to acquire knowledge about the world, we actively hunt for clues as to the kind of universe we are in, and hence the kind of theories we need to develop. *The* big clue that we have (apparently) discovered, is that the universe is more or less comprehensible in some way or other, it being possible to discover explanations for phenomena; this is enshrined in theses of comprehensibility and physical comprehensibility, at levels 5 and 4 of Figure 2.1. The level 5 thesis of comprehensibility is the bold conjecture that the universe is perfectly comprehensible in some way or other – the universe being such that there is some one kind of explanation for all phenomena, couched in terms of God, a cosmic purpose (which everything is designed to fulfil), a cosmic programme, a unified physical entity, or something else. From Galileo on, science has, in effect, made the even bolder conjecture that the universe is *physically* comprehensible, at level 4, and comprehensible in terms of the best available blueprint, at level 3. Type (2) explanatory power, to repeat, is the *key* type (c) non-empirical criterion of theory choice, from which all other type (c) criteria arise. If any property of theories cries out to be *the* aesthetic property of beauty, which scientists quite properly take note of as being methodologically significant, it is type (2) explanatory power.

And yet, astonishingly, MAI does not even include explanatory power in its list of aesthetic properties of theories, despite its open-ended, all-inclusive approach to listing such properties (in such sharp contrast to AOE).

MAI does hold that the requirement of type (2) explanatory power is a permanent criterion of science, one that can be arrived at by aim-analysis, taking the aim of science to be empirical adequacy (McAllister, 1996, p. 11). It is clear that type (2), and not merely type (1), explanatory power is intended here, for McAllister writes that a successful explanatory theory is deemed to have "identified a pattern or mechanism underlying the data" (1996, p. 11). But such an analysis could only, at most, justify adopting the requirement of type (1) explanatory power; it does not justify adopting type (2) explanatory power as a requirement – not unless the truth, the universe that is, is permanently presumed to have a more or less unified dynamic structure (a presumption which contradicts SE). McAllister provides no argument in support of the contention that favouring theories with type (2) explanatory power can be justified by an appeal to the aim of empirical adequacy. He does refer to an approach to the problem of induction, espoused by Braithwaite and Mellor, according to which we are justified in proceeding as if regularities or patterns exist in nature because this gives us the best hope of acquiring knowledge whatever the universe may be like (McAllister, 1996, pp. 100–1). It is this argument, perhaps, which McAllister assumes justifies taking type (2) explanatory power as a permanent criterion for theory choice, arrived at by aim-analysis, taking the aim of science to be empirical adequacy.

But there are three things wrong with this.

First, the Braithwaite–Mellor justification of induction does not work, as I shall show in the next section.

Second, many different kinds of explanation are possible; the universe may be comprehensible (phenomena being explainable) in many different ways, and to many different degrees, as the different theses from levels 3 to 7 of AOE attest. Here, above all, science needs to be flexible and responsive, constantly modifying the kind of explanations to be sought in the light of empirical success and failure, in the kind of way in which the hierarchical methodological structure of AOE is designed to facilitate. If ever there was a role for the aesthetic induction, it would surely be here, in connection with explanatory power. But in excluding type (2) explanatory power from the list of aesthetic properties, and in making it a fixed, unchanging requirement of theory choice, MAI fails to exploit this vital need for science constantly to modify and improve the kind of explanations that it seeks. It is just this, by contrast, that is the key idea behind AOE.

Third, if McAllister's argument were successful, so that giving preference to theories that exhibit type (2) explanatory power could be justified by an appeal to the aim of empirical adequacy, then this would be a disaster for MAI, for it would obviate entirely the need for science to consider aesthetic properties of theories, and to employ the aesthetic induction. As I have argued above, all aesthetic properties of theories that have any methodological significance can be derived from the demand for unity – that is, the demand for type (2) explanatory power. Once type (2) explanatory power is acknowledged to be methodologically significant, no other aesthetic properties of theories are required by science.

I conclude this section by mentioning three further differences between AOE and MAI.

First, reasons given in defence of MAI for holding that aesthetic considerations are methodologically significant in science arise from the fact that scientists themselves have stressed their importance, and they do indeed seem to influence what theories are chosen in science. Reasons given in defence of AOE for holding that type (c) non-empirical considerations are methodologically important are much stronger: science becomes impossible if such considerations are not deployed to rule out endlessly many empirically successful but grossly *ad hoc* theories.

Second, MAI, despite being a contribution to the rationalist conception of science, does not provide a basis for systematically correcting scientific practice. But AOE does. As I have already remarked, if a view genuinely increases our understanding of science, it would be surprising if it did not have implications for scientific practice. AOE passes this test, in emphasizing the need for explicit articulation of metaphysical theses at levels 3 and 4, and explicit tackling of the problems thereby generated.

Third, MAI and AOE conceive of the relationship between science and the philosophy of science differently. MAI takes the conventional view for granted: philosophy of science is a meta-discipline which seeks to spell out and justify methods implicit in successful scientific practice, but which is quite distinct from science itself. AOE upholds the unorthodox view that the philosophy of science is an integral part of science itself, influenced by and seeking to influence science, articulating and critically assessing actual and possible aims and methods for science, at various levels, the fundamental aim being to contribute to scientific progress. A new level 3 aim for physics, i.e. a new blueprint, plus associated new methods, might constitute a major contribution to theoretical physics, as well as being a contribution to the philosophy of physics. Einstein's special theory of relativity is an example. It puts forward both a modified blueprint (Newtonian space-time becoming Minkowskian space-time), and modified methods (Galilean invariance becoming Lorentz invariance): it is thus a major contribution to physics itself, which is also a contribution to the aims and methods of physics – that is, to the philosophy of physics.

6.4 Assessment

Which is to be preferred, AOE or MAI? The two views need not, of course, be regarded as rivals. AOE is a highest-level model, whereas MAI is a medium-level model; one could consider accommodating MAI within AOE. This would require, however, that MAI be modified quite extensively, as the previous section has shown.

Interpreting AOE and MAI as rival rationalist accounts of science, my chief criticism of MAI is that it is a version of SE, and thus suffers from the defects that all versions of SE suffer from. Given any empirically successful theory, T, there will always be endlessly many *ad hoc* rivals to T, even more empirically successful than T, which will never even be considered within science, let alone considered and rejected. In persistently rejecting such *ad hoc* rivals, even more empirically successful than T, science makes a persistent assumption about the nature of the universe. This contradicts SE – and contradicts MAI.[6]

McAllister might seek to evade this conclusion by arguing, as he does in his book, that non-empirical, aesthetic criteria that rule out acceptance of empirically successful, *ad hoc* rival theories, are too diverse in character, too changeable over time, to amount to the implicit acceptance of any persistent assumption. But such an argument collapses the moment one takes into account radically *ad hoc* theories of the kind considered in this book, in Chapters 2 and 5, and in Maxwell (1998, pp. 47–54). Rejection (or rather complete neglect) of such radically *ad hoc* theories persists throughout revolutions and all changes in aesthetic fashions in science. The persistent rejection of such theories unquestionably commits science to making a substantial metaphysical assumption about the nature of the universe.

McAllister might, at this point, appeal to the pragmatic justification of induction of Braithwaite and Mellor, already referred to above (McAllister, 1996, pp. 100–1). According to this argument, science proceeds, and is justified in proceeding, as if it assumes there are regularities to be discovered, but does not actually assume that regularities exist. But even if this argument is valid, it does not in any way invalidate my point above, that in persistently rejecting empirically successful *ad hoc* theories, science implicitly makes a persistent metaphysical assumption about the world. It should be noted that kinds of *ad hoc* theories can be formulated that specify regularities, in that these theories are invariant with respect to position and time (no specific places or times being specified by the theories). These theories might be said to specify *ad hoc* regularities.

But, in any case, the Braithwaite–Mellor attempt at solving the problem of induction does not succeed. Restricting science to the search for regularities is both too narrow, and not narrow enough. Too narrow, because it is conceivable that we can live and acquire knowledge but not by searching for regularities in phenomena. God might get in touch with us, explain His purposes, keep us informed about what is going to happen. Getting in touch with God by means of prayer and meditation, and not by searching for regularities, might be the way to acquire knowledge; and various other science fiction possibilities can be imagined (see Maxwell, 1998, p. 185). Such possibilities are excluded by the search for knowledge as characterized by Braithwaite and Mellor; this means these possibilities are just dogmatically assumed to be false. But the Braithwaite–Mellor approach is also not narrow enough, because, as I have indicated above, if science is to be possible, *ad hoc* regularities must be persistently excluded from consideration. And, as we have seen, there is no sharp distinction between the *ad hoc* and the non-*ad hoc*. In Chapter 4 I listed eight kinds of disunity – in effect, eight different ways in which regularities might be *ad hoc*, which range from the severely *ad hoc* (distinct regularities in different space-time regions) to the scarcely *ad hoc* at all (space-time and matter not being unified). What does the policy of "inductive projection" (McAllister, 1996, p. 101) amount to? Does it involve merely excluding permanently all theories that are type (1) *ad hoc* (distinct regularities in different space-time regions)? Again, this is both too narrow, and not narrow enough. Exactly the same objection arises wherever the line is drawn, between regularities that are too *ad hoc* to be considered by science, and those that are sufficiently non-*ad hoc* to be open to scientific consideration. We cannot, at this point, simply invoke the aesthetic induction, and declare that we discover, by induction, where the line is to be drawn between the acceptably and unacceptably *ad hoc*, because, as McAllister himself has so clearly shown, for logical reasons, the aesthetic induction can only proceed once methods have been arrived at by aim-analysis (McAllister, 1996, p. 76).

Another approach might be to favour permanently in science theories that are as non-*ad hoc* as possible, in all eight senses, but not to draw a rigid line between the acceptably and unacceptably *ad hoc*. This would allow something like the aesthetic induction to proceed in science (although not in quite the open-ended way in which McAllister envisages). But even this attempt at solving the problem is both too narrow and not narrow enough. Endlessly many universes are possible in which we may live and acquire knowledge, and yet this inductive policy would not be appropriate for acquiring knowledge. It biases the search for knowledge in the direction of physicalism. But physicalism may be false; the universe may be comprehensible in some other way, or not comprehensible at all.

My claim is that the best possible way in which we can go about seeking knowledge is to do so employing the hierarchical aims-and-methods structure of (generalized) AOE. We must make some kind of guess as to what kind of universe we are in, in order to proceed at all. At the top of the hierarchy we need to put those relatively contentless guesses which are such that their truth is required for acquisition of knowledge to be possible at all. These are justifiably permanent items of scientific knowledge. As we descend the hierarchy, we need to put increasingly contentful guesses, chosen because these seem to be the most fruitful from the standpoint of engendering methods that seem to offer the best help with acquiring empirical knowledge. As we proceed, we revise these guesses in the light of the relative empirical success and failure of rival research programmes, based on rival low-level metaphysical guesses. We try to keep such revisions as low as possible in the hierarchy when we seem to be achieving overall success, and only allow revisions to ascend higher up in the hierarchy when success is not being achieved, and higher-level revisions seem to be required.

This hierarchical conception of scientific method enables science to respond sensitively to what it seems to discover about the nature of the universe, lower-level aims and methods being adjusted in the light of apparent empirical success and failure, and within a framework of fixed, relatively unproblematic, higher-level aims and methods. All attempts at justifying induction pragmatically that are known to me, along lines advocated by Braithwaite and Mellor, fail because they fail to take note of the resources of (generalized) AOE. They all attempt to justify methods that are demonstrably not as efficient as those of AOE in enabling us to acquire knowledge of nature. They fail to encapsulate the responsiveness, the flexibility, the open-endedness and precision, of AOE.

And this is true of MAI as well. Indeed, as we saw in section 6.4 above, the aesthetic induction has conservatism built into it, and cannot help engender revolutionary new ideas for revolutionary new theories, whereas AOE is designed to do just that. It embodies a rational, if fallible, method of discovery for theoretical physics.

7
A mug's game? Solving the problem of induction with metaphysical presuppositions

> I think that I have solved a major philosophical problem: the problem of induction. This solution has been extremely fruitful, and it has enabled me to solve a good number of other philosophical problems. However, few philosophers would support the thesis that I have solved the problem of induction. Few philosophers have taken the trouble to study – or even to criticize – my views on this problem, or have taken notice of the fact that I have done some work on it. (Popper, 1972, p. 1)

This is how Karl Popper opens his book *Objective Knowledge*. There are at least two oddities about what Popper says here. First, Popper is wrong; he did not solve the problem of induction. Second, even by 1971, when this passage was first published, Popper's work on the problem of induction had received a great deal of attention.[1]

Popper's words might, however, be uttered by me with far greater justice. For I really have solved the problem of induction. The solution has been extraordinarily fruitful, and has enabled me to solve a number of other philosophical problems.[2] But few philosophers – if any – would agree that I have solved the problem. Few, indeed, have taken the trouble to study, or criticize, my work, or are even aware that I have done some work on the problem.[3]

I think I know why this is the case. First, it is no doubt the fate of most of us seeking to contribute to philosophy: our work sinks without trace, without comment. Second, the problem of induction has been around for a very long time; anyone claiming to solve the problem is almost bound to be wrong. Third, there is a kind of "negative judgement through persistent neglect" effect. The first version of my proposed solution was published over forty years ago in 1974: if there was anything in it, surely someone would have noticed and taken up the idea by now. Fourth, as Popper points out elsewhere (1963, ch. 2), "analytic" philosophy has tended to be more interested in analysis of concepts than in proposed solutions to fundamental philosophical problems. Fifth, my solution amounts to a radical improvement of Popper's attempted solution. Popper was hostile to this, and Popperians today are hostile to it, precisely because I have the temerity to claim that I have radically improved Popper's ideas. Anti-Popperians are indifferent because they know Popper has failed to solve the problem, and they assume my approach inherits Popper's failure. Finally, and perhaps most damagingly, my proposed solution involves recognizing that science makes a persistent metaphysical assumption of "uniformity" or "unity". Philosophers at once know that any attempt to solve the problem of induction along these lines is hopeless. As Bas van Fraassen once put it: "From Gravesande's axiom of the uniformity of nature in 1717 to Russell's postulates of human knowledge in 1948, this has been a mug's game" (van Fraassen, 1985, pp. 259–60). There is no need to study or criticize my proposed solution to the problem of induction: I am playing a well-known mug's game.

There is not much that I can do about the first five reasons for ignoring my work on the problem of induction; I can, however, at least set out to demolish the sixth reason. This is what I propose to do in what follows. I first give a brief sketch of my proposed solution to the problem of induction (spelled out in much greater detail elsewhere); I then demolish the thesis that it amounts to van Fraassen's "mug's game".[4] My hope is that this may provoke one or two readers to take note of what I have done elsewhere.[5]

7.1 Aim-oriented empiricism and the problem of induction

In order to solve the problem of induction, it is both necessary and sufficient to construe science from the standpoint of aim-oriented empiricism[6] (already encountered in Chapters 2, 3, 5 and 6).[7] Profound, far-reaching consequences come from rejecting all versions of the current orthodoxy of standard empiricism, and accepting aim-oriented empiricism instead. The reasons for rejecting standard empiricism and accepting aim-oriented empiricism deserve to be subjected to especially fierce scrutiny. I therefore, in this section, carefully spell out these reasons once more. Any reader already convinced may skip this section. In subsequent sections, I argue that aim-oriented empiricism solves the "methodological" and "theoretical" problems of induction, demolish the thesis that aim-oriented empiricism represents a mug's game, and conclude by showing how the view solves what may be called the "practical" problem of induction.

The fundamental line of thought behind aim-oriented empiricism (AOE) can be indicated like this. Theoretical physics, and therefore all of natural science (since theoretical physics is fundamental for natural science), persistently selects fundamental physical theories that help to unify the whole of theoretical physics. Thus Newtonian theory (NT) unifies Galileo's laws of terrestrial motion and Kepler's laws of planetary motion (and much else besides). Maxwellian classical electrodynamics, (CEM), unifies electricity, magnetism and light (plus radio, infrared, ultraviolet, X-rays and gamma rays). Special relativity (SR) brings greater unity to CEM (in revealing that the way one divides up the electromagnetic field into the electric and magnetic fields depends on one's reference frame). SR is also a step towards unifying NT and CEM in that it transforms space and time so as to make CEM satisfy a basic principle fundamental to NT, namely the (restricted) principle of relativity. SR also brings about a unification of matter and energy, via the most famous equation of modern physics, $E = mc^2$, and partially unifies space and time into Minkowskian space-time. General relativity (GR) unifies space-time and gravitation, in that, according to GR, gravitation is no more than an effect of the curvature of space-time. Quantum theory (QM) and atomic theory unify a mass of phenomena having to do with the structure and properties of matter, and the way matter interacts with light. Quantum electrodynamics unifies QM, CEM and SR. Quantum electroweak theory unifies (partially) electromagnetism and the weak force. Quantum chromodynamics brings unity to hadron physics (via quarks) and brings unity to the eight kinds of gluons of the strong force. The standard model unifies to a considerable extent all known phenomena associated with fundamental particles and the forces between them (apart from gravitation). The theory unifies to some extent its two component quantum field theories in that both are locally gauge invariant (the symmetry group being $U(1) \times SU(2) \times SU(3)$). String theory, or M-theory, holds out the hope of unifying all phenomena.[8]

It might be thought that, during the last four hundred years or so, science has been pursued in a thoroughly open-minded, unbiased fashion, theories being selected impartially on the basis of empirical success alone, the emergence of increasing theoretical unity being a surprising and purely empirical discovery – unifying theories just being much more empirically successful than disunified rivals. Nothing could be further from the truth. In fact, in connection with every accepted unifying theory – NT, CEM and the rest – there have always been endlessly many, easily formulated, disunified rival theories very much more empirically successful than the theories that have been accepted.[9]

Thus, given NT, for example, one rival theory might assert: everything occurs as NT asserts up till midnight tonight when, abruptly, an inverse cube law of gravitation comes into operation. A second rival theory might assert: everything occurs as NT asserts, except for the case of any two solid gold spheres, each having a mass of 1,000 tons, moving in otherwise empty space up to a mile apart, in which case the spheres attract each other by means of an inverse cube law of gravitation. A third rival asserts that everything occurs as NT asserts until 3 kilograms of gold dust and 3 kilograms of diamond dust are heated in a platinum flask to a temperature of 450°C, in which case gravitation will instantly become a repulsive force everywhere. And so on. There is no limit to the number of rivals to NT that can be concocted in this way, each of which has all the predictive success of NT as far as observed phenomena are concerned but makes different predictions for some as yet unobserved phenomena.[10] Such theories can even be concocted which are *more* empirically successful than NT, by arbitrarily modifying NT, in just this entirely *ad hoc* fashion, so that the theories yield correct predictions where NT does not, as in the case of the orbit of Mercury, for example (which very slightly conflicts with NT).[11]

This last point may be made more generally, as follows. Most accepted physical theories, for most of the time that they exist, are confronted by various empirical difficulties. Let T be any one of the above unifying accepted theories – NT, CEM, or whatever. Typically, T is confronted by the following empirical conditions. There is a domain A of phenomena for which the predictions of T are wholly successful; there is a domain B for which T fails to predict the phenomena because the equations of the theory cannot be solved; there is a domain C where T is ostensibly refuted (because the predictions of T clash with the phenomena of C, but this may be due, not to T yielding false predictions, but to experimental error, or relevant physical conditions not being taken into account); and finally there is a domain D of phenomena which T fails to predict because they lie outside the scope of T. (Here the phenomena, in A to D, are to be understood as consisting of low-level empirical or experimental laws.) It is now easy to concoct rivals to T that are much more empirically successful than T, as follows. One such rival asserts: as far as phenomena in A are concerned, everything occurs as T asserts; as far as phenomena in B are concerned, the phenomena occur in accordance with established empirical laws; and the same for C, and for D. This rival to T, T* let us call it, reproduces all the empirical success of T (in A), successfully predicts phenomena that T is not able to predict (in B), successfully predicts phenomena that refute T (in C), and successfully predicts new phenomena that lie beyond the predictive scope of T (in D). It might be demanded that T* should predict new phenomena; but this demand can be met too, since "phenomena", here, are laws with content in excess of actual experiments that have been performed. T* satisfies every imaginable requirement for being an empirically more successful theory than T.[12]

And this has been the situation for all the accepted fundamental physical theories indicated above, for most of the time that they have been in existence: endlessly many rival, disunified theories have been available, far more successful empirically than the accepted, unifying theories, and these empirically more successful, grossly disunified or, as I have called them, "aberrant" theories (see Maxwell, 1974, p. 128) are all ignored.

As most physicists and philosophers of physics would accept, *two criteria are employed in physics in deciding what theories to accept and reject*: (1) empirical criteria, and (2) criteria that have to do with the simplicity and unifying capacity of the theories in question. (2) is absolutely indispensable, to such an extent that there are endlessly many theories empirically more successful than accepted theories, which lack unity, and are not even considered as a result.

Now comes the crucial point. In persistently accepting unifying theories, and excluding infinitely many empirically more successful, disunified or aberrant rival theories, science in effect makes a big assumption about the nature of the universe, to the effect that it is such that no disunified theory is true, however empirically successful it may appear to be for a time. Furthermore, without some such big assumption as this, the empirical method of science collapses. Science would be drowned in an infinite ocean of empirically successful disunified theories.

If scientists only accepted theories that postulate atoms, and persistently rejected theories that postulate different basic physical entities, such as fields – even though many field theories can easily be, and have been, formulated which are even more empirically successful than the atomic theories – the implications would surely be quite clear. Scientists would in effect be assuming that the world is made up of atoms, all other possibilities being ruled out. The atomic assumption would be built into the way the scientific community accepts and rejects theories – built into the implicit *methods* of the community, methods which include rejecting all theories that postulate entities other than atoms, whatever their empirical success might be. The scientific community would accept the assumption that the universe is such that no non-atomic theory is true.

Just the same holds for a scientific community which rejects all disunified or aberrant rivals to accepted theories, even though these rivals would be even more empirically successful if they were considered. Such a community in effect makes the assumption: the universe is such that no disunified theory is true.

Thus the idea that science has the aim of improving knowledge of factual truth, *nothing being presupposed about the nature of the universe independently of evidence*, is untenable. Science makes one big, persistent assumption about the universe, namely that it is such that no disunified or aberrant theory is true.[13] It assumes that the universe is such that there are no pockets of peculiarity, at specific times and places, or when specific conditions arise (gold spheres, gold and diamond dust, etc.), that lead to an abrupt change in laws that prevail elsewhere. Science assumes, in other words, that there is a kind of uniformity of physical laws throughout all phenomena, actual and possible. Furthermore, science *must* make this assumption (or some analogous assumption) if the empirical method of science is not to break down completely. The empirical method of science of assessing theories in the light of evidence can only work if those infinitely many empirically successful but disunified theories are permanently excluded from science independently of, or rather in opposition to, empirical considerations; to do this is just to make the big, permanent assumption about the nature of the universe.[14]

Let us call this assumption of unity U; and let us call the view, just outlined, that in persistently only accepting unifying theories science presupposes U, "presuppositionism".

Most current views about science deny that science makes a substantial, persistent assumption about the universe. This is true, for example, of logical positivism, inductivism, logical empiricism, hypothetico-deductivism, conventionalism, constructive empiricism, pragmatism, realism, induction-to-the-best-explanationism, and the views of Popper, Kuhn and Lakatos.[15] It is true, too, of more recent work on the methods and metaphysics of science.[16] All these views, diverse as they are in other respects, accept a thesis which may be called standard empiricism (SE): in science, theories are accepted on the basis of empirical success and failure, and on the basis of simplicity, unity or explanatoriness, but *no substantial thesis about the world is accepted permanently by science, as a part of scientific knowledge, independently of empirical considerations*. It deserves to be noted that even Feyerabend, and even social constructivist and relativist sociologists and historians of science, uphold SE as the best available ideal of scientific rationality. *If* science can be exhibited as rational, they hold (in effect), then this must be done in a way that is compatible with SE. The failure of science to live up to the rational ideal of SE is taken by them to demonstrate that science is not rational. That it is so taken demonstrates convincingly that they hold SE to be the only possible rational ideal for science (an ideal which cannot, it so happens, in their view, be met).

Presuppositionism is of course incompatible with SE, and thus incompatible will all the above doctrines. One crucial point needs to be noted about the argument so far: presuppositionism is more rigorous than all the above versions of SE *entirely independent of any justification of U, or justification for accepting U as a part of scientific knowledge* (that is in addition to the one given above). In saying this I am appealing to the following wholly uncontroversial requirement for rigour.

> (R) In order to be rigorous, it is necessary that assumptions that are substantial, influential and problematic be made explicit – so that they can be criticized, so that alternatives may be developed and assessed (see Maxwell, 1984a, p. 224; 1998, p. 21).

All versions of SE fail to satisfy (R) in just the way in which presuppositionism does satisfy (R). Presuppositionism makes the assumption U explicit (and so criticizable and, we may hope, improvable), while all versions of SE deny that science does make any such assumption as U. Thus presuppositonism is more rigorous than all versions of SE *even in the absence of any kind of justification of U*. Indeed, it is precisely because the version of U that is implicitly accepted by physics at any stage in its development is a pure conjecture, bereft of justification, almost bound to be false, that it needs to be acknowledged, made explicit within physics, so that it can be critically assessed and, we may hope, improved. In short, quite independent of any claim to solve the problem of induction, presuppositionism is more rigorous, and thus more acceptable, than any of the above versions of SE. This has a major implication for all attempts at solving the problem of induction: no such attempt can succeed if any version of SE is presupposed, since these all lack rigour. Attempts at solving the problem of induction must at least begin with presuppositionism, unless a better view of science emerges.

Far from presupposing the uniformity or unity of nature being a mug's game, it is the other way round: attempting to construe science in such a way that science does *not* presuppose the uniformity or unity of nature is the mug's game, since all such views of science fail to satisfy elementary requirements for rigour, namely (R), and thus cannot provide a basis for solving the problem of induction that can hope to succeed. Presuppositionism is the only non-mug's game in town unless, as I have said, something better turns up.

Presuppositionism is, however, as it stands, untenable. This is because it is not at all clear what the assumption U is, or ought to be. It is vital to appreciate that there are endlessly many different assumptions of unity which science may be construed to make, almost all of which are false (since they contradict each other). Even more urgent than any problem of justification, there is the following problem: How can the assumption of unity being made by science at present, which is implicit in current scientific views as to what counts as theoretical unity, and almost bound to be false, be *improved*?

What is at issue is not the traditional philosophical problem of *justification* (which presupposes that U is true), but rather the scientific (and quasi-Popperian) one of *improving* what is almost bound to be false.[17]

In surveying the different possible ways in which the universe may be unified, one important point to appreciate is that there is no single, sharp distinction between unity and disunity or aberrance. By "unity" we could mean merely that physical laws are the same throughout space and time. Or we could mean, in addition, that physical laws remain the same as other variables change, such as velocity, temperature, or mass (so that, for example, Newton's inverse square law of gravitation does not abruptly become an inverse cube law as masses of 1,000 tons are reached). Or, more restrictively still, we could mean (in addition) that there is only *one* force in nature, and not three or four distinct forces (such as gravitation, the electromagnetic force, and the weak and strong forces of nuclear physics). More restrictively still, we could mean that there is just one kind of particle in existence, or one kind of physical entity, a self-interacting field spread throughout space and time. Finally, and even more restrictively, we could mean that space, time, matter and force are all unified into one, unified, self-interacting entity.

Even more restrictive assumptions can be made, which specify the kind of entity or entities out of which everything is composed. And at the other end of the spectrum, much looser, less restrictive assumptions could be made which, if true, would still make science possible. Thus science could assume that the universe is such that local observable phenomena occur, most of the time, to a high degree of approximation, in accordance with some yet-to-be-discovered physical theory that is not too seriously disunified.

It is always possible, of course, that the universe only appears to be physically unified (to some extent). Perhaps, as theoretical physics advances, everything will become increasingly complex (as even some physicists believe[18]). Perhaps a malicious God is in charge, who has been controlling the universe up to now in such a way that it is as if physics prevails everywhere, but who, shortly, will startle us all by causing a series of dramatic, large-scale miracles to occur which violate all known laws. Perhaps as we probe deeper into physical reality we will discover that the universe exemplifies, not physical laws, but something that is closer to a computer program (as some people have suggested). The universe may be comprehensible, but not *physically* comprehensible. That is, it may be that *something* exists – God, a society of gods, an overall cosmic purpose, a cosmic "computer" program – which controls or determines the way events occur, and in terms of which, in principle, everything can be explained and understood: but this *something* may not be a unified physical entity, a unified pattern of physical law, and thus the universe, though comprehensible, is not *physically* comprehensible. Finally, the universe may not be comprehensible at all, and yet it might still be possible for us to live, and to acquire some knowledge of our local circumstances.

How do we choose between these endless possibilities? Science must make some kind of choice. It is all-important that science makes the correct choice, or at least a good choice, since this choice will determine what (non-empirical) methods are employed by science to assess theories. If science chooses a cosmological thesis that is radically false, then science will only consider false theories, and will exclude from consideration all theories that might take us towards the truth. Science will come to a dead end. The more restrictive the chosen cosmological assumption is, so the more potentially helpful it will be in selecting theories, but also the more likely the assumption is to be radically false, thus imposing a block on scientific progress. On the other hand, the looser, the more unrestrictive the assumption is, so the more likely it is to be true, but the less helpful it will be in excluding empirically successful "disunified" theories. (Other things being equal, the less one says, the more likely it is that what one says is true. "The universe is not a chicken" is almost certainly true about ultimate reality, just because it says so little, there being an awful lot of ways in which the universe can not be a chicken.)

It is all-important that science makes the right assumption about the ultimate nature of reality; and yet it is just here, concerning the ultimate nature of reality, that we are most ignorant, and are almost bound to get things wrong. How on earth are we to proceed?

The solution to this dilemma – the fundamental epistemological and methodological dilemma of science – is to make, not *one* cosmological assumption, but a *hierarchy* of assumptions, the assumptions becoming less and less restrictive, asserting less and less, as one goes up the hierarchy (see Figures 2.1 and 5.1), and associated text spelling out aspects of this hierarchical view.

The figures make things look complicated, but the basic idea is extremely simple. By displaying assumptions and associated methods – aims and methods – in this hierarchical fashion, we create a framework of high-level, relatively unspecific, unproblematic, fixed assumptions and methods within which low-level, much more specific, problematic assumptions and methods may be revised as science proceeds, in the light of the relative empirical success and failure of rival scientific research programmes to which rival assumptions lead. If currently adopted cosmological assumptions, and associated methods, fail to support the growth of empirical knowledge, or fail to do so as apparently successfully as rival assumptions and methods, then assumptions and associated methods are changed, at whatever level appears to be required.[19] Every effort is made, however, to confine such revisions to cosmological theses as low down in the hierarchy of theses as possible. Only persistent, long-term, dramatic failure (at levels 1 and 2) would lead us to revise ideas above level 3, let alone above level 4; only an earthquake in our understanding of the universe would lead us to revise ideas above level 5. In this way we give ourselves the best hope of making progress, of acquiring authentic knowledge, while at the same time minimizing the chances of being taken up the garden path, or being stuck in a cul de sac. The hope is that as we increase our knowledge about the world, we improve the cosmological assumptions implicit in our methods, and thus in turn improve our methods. As a result of improving our knowledge, we improve our knowledge about how to improve knowledge. Science adapts its own nature to what it learns about the nature of the universe, thus increasing its capacity to make progress in knowledge about the world – the methodological key to the astonishing, accelerating, explosive growth of scientific knowledge.

It is this conception of science, postulating more or less specific, problematic, evolving aims and methods for science within a framework of more general, relatively unproblematic, more or less fixed aims and methods, that I call *aim-oriented empiricism* (AOE).[20] (For further details see previous chapters, and Maxwell, 1998, chs. 1 and 3–6; 2004a; 2007, especially ch. 14; 2017a; 2017b.) The basic idea, let me re-emphasize, is that the fundamental aim of science of discovering how, and to what extent, the universe is comprehensible is deeply problematic; it is essential that we try to improve the aim, and associated methods, as we proceed, in the light of apparent success and failure. In order to do this in the best possible way, we need to represent our aim at a number of levels, from the specific and problematic to the highly unspecific and unproblematic, thus creating a framework of fixed aims and meta-methods within which the (more or less specific, problematic) aims and methods of science may be progressively improved in the light of apparent empirical success and failure.[21]

This hierarchical view of AOE is put forward to solve the fundamental problem confronting presuppositionism, indicated above. It is put forward to solve the problem of *improving* the basic metaphysical assumption of science, implicit in persistent scientific preference for unifying theories even against the evidence, granted that some such assumption must be made, and that it is almost bound to be false. The claim is that the hierarchical framework of AOE provides the best possible means for discovering metaphysical assumptions which best aid the task of improving knowledge of truth; AOE provides the best possible means for *improving* choice of metaphysical assumption. There is no attempt to justify the *truth* of metaphysical assumptions. At most, there is a justification for choosing metaphysical thesis A over B granted that the aim is to make that choice which gives the best promise of aiding the search for knowledge of truth. Justification is involved only in the quasi-Popperian sense that the best possible justification of metaphysical assumptions that we can have is to expose these assumptions to the most searching criticism possible, to criticism best designed to promote progress in knowledge.

Something like AOE has always been implicit in scientific practice (otherwise science would have come to an end). AOE becomes all but scientifically explicit with the work of Einstein in discovering special and general relativity (as we saw in Chapter 3). Aspects of this work that are characteristic of AOE are the fundamental role played by the search for theoretical unity, and the vital role played by symmetry principles (such as Lorentz invariance and the principle of equivalence). These latter are fallible and revisable, and have the dual role of being both physical and methodological principles, all of which is integral to AOE.

AOE, as indicated above, is intended to depict the metaphysical components of scientific knowledge given science as it exists today. AOE takes the specific form that it does in part because of what we have learned from Galileo onwards (or from the Presocratics onwards). History, in other words, is built into AOE. In the future, when we have learned more, AOE will be somewhat different. But however dramatic future revolutions in knowledge may be, we still ought to represent our knowledge in the same hierarchical form, with the same thesis, at level 7, at the top. Let us call this view "generalized AOE". When it comes to considering whether AOE succeeds in solving the problem of induction in a non-circular way, we need to consider various possible versions of generalized AOE which differ from AOE. The crucial question is, "Can sufficiently good grounds be given for preferring AOE to all other rival versions of generalized AOE that one can think of?" That is the proper way to formulate the problem of induction. (One striking feature of the problem of induction, as it is usually formulated, is its scientific sterility: work on the problem of induction has made no contribution to science, with the possible exception of Popper's work. But when the problem is formulated in the way just indicated, it is clear that it is potentially a highly fruitful problem for science: a version of generalized AOE that is genuinely an improvement over AOE is likely to be a major contribution to science itself.)

7.2 How aim-oriented empiricism solves the problem of induction

At this point, the basic objection to the whole approach being advocated here may be reiterated. Either AOE solves the problem of induction, or it does not. If it does not, no more needs to be said. If it does, then an element of justification must enter in. This in turn means that AOE must commit van Fraassen's mug's game. Choice of theory, at level 2, is justified in part by being compatible with choice of metaphysical thesis at level 3 or 4; this latter choice is in turn justified in terms of the success of science. We have here the vicious circularity of the mug's game. And it is inescapable. Interpreting AOE as a framework for detecting error, for criticism, does not help; even given this interpretation, there must be some justification for regarding metaphysical thesis U_2 as a better choice, an improvement over, more likely to be true than, thesis U_1: here, unavoidably, justification is present, which introduces the vicious circularity of the mug's game.

The first thing that needs to be said in response to this is that, as I have already emphasized, there is no question of justifying the *truth* (to some degree of certainty or probability) of any of the theses at levels 3 to 7. These theses remain, throughout, pure conjectures. I concur with Popper's thesis that all our knowledge is ultimately conjectural. (Whether such a view can claim to be the solution to the problem of induction is an issue I will take up below.)

At most, then, there is a justification for *accepting* such and such a thesis as a part of (conjectural) scientific knowledge, or *preferring* thesis A to thesis B. Second, the top thesis is accepted on grounds which have nothing to do with the success of science at all. It is accepted because its truth is a necessary precondition for the acquisition of knowledge to be possible at all.

The thesis at level 7 asserts that the universe is such that it is possible for us to acquire some knowledge of our local circumstances (sufficient for it to be possible for us to continue to live). We are justified in accepting this thesis entirely in the absence of any justification for its truth (or probable truth), just because we have nothing to lose; accepting this thesis as a part of our knowledge can only help, and cannot obstruct, the task of acquiring knowledge whatever the universe is like (see Maxwell, 1998, pp. 186–7).

This elementary argument for permanently accepting this level 7 thesis can of course be challenged. What is beyond question, however, is that no circularity is involved here at all. The argument in support of accepting the level 7 thesis makes no appeal to the success of science whatsoever. Science is not even mentioned.

I might add that a part of the point of exhibiting the metaphysical assumptions of science in the form of a hierarchy, from level 3 to 7, is to overcome a fatal objection to one traditional approach to solving the problem of induction, versions of which have been argued for by, for example, Reichenbach (1961, sections 38–41), Braithwaite (1953, pp. 255–92) and Mellor (1991). This argues that we are rationally entitled to assume that there are sufficient regularities in nature for the inductive methods of science to meet with success because, if such regularities do not exist, no method will procure knowledge. But this argument tries to establish too much; it is not valid. Counterexamples can be imagined. The world might be such that "the inductive methods of science" meet with no success at all, and yet we can still acquire sufficient knowledge to live. Natural phenomena might be governed by gods: in order to get nature to do what we want it to do, we might need to make sacrifices, or pray. (For further suggestions along these lines, see Maxwell, 1998, p. 185.) The thesis of AOE, at level 7, might be called a "principle of uniformity", but it is very much weaker than the assertion that there are regularities such that "the inductive methods of science" meet with success. The fatal objection to the Reichenbach–Braithwaite–Mellor (RBM) approach is that (1) *either* it seeks to justify acceptance of a "principle of regularity" which, if accepted, suffices to justify science, but the argument is invalid; *or* (2) it is valid, but the "principle of regularity or uniformity" whose acceptance is justified is much too weak to justify science. AOE adopts (2), and recognizes that the acceptance of other, more restrictive "principles of uniformity" needs to be justified on other grounds; RBM, not acknowledging the hierarchy of principles, are doomed to opt for (1). There is another, related objection to RBM: "the inductive methods of science", at least as conceived of by RBM, are *not* the best available. They do not have the flexibility of the methods of AOE, which allow for the possibility of methods (associated with theses low down in the hierarchy) being *improved* in the light of improving knowledge, feedback being facilitated by the hierarchical structure of AOE between improving knowledge and improving knowledge-about-how-to-improve-knowledge (i.e. improving aims and methods). The traditional "inductive methods of science", as a result of their inflexibility, are both too restrictive, and not restrictive enough. Like most other traditional attempts at solving the problem of induction, RBM try to justify the unrigorous, and thus the unjustifiable. The status quo needs to be changed, improved, not justified.

What about the thesis of "meta-knowability" at level 6?

Here are two arguments for accepting meta-knowability which make no appeal whatsoever to the success of AOE science.

> (i) Granted that there is *some* kind of general feature of the universe which makes it possible to acquire knowledge of our local environment (as guaranteed by the thesis at level 7), it is reasonable to suppose that we do not know all that there is to be known about what the *nature* of this general feature is. It is reasonable to suppose, in other words, that we can improve our knowledge about the nature of this general feature, thus improving methods for the improvement of knowledge. Not to suppose this is to assume, arrogantly, that we already know all that there is to be known about how to acquire new knowledge. Granted that learning is possible (as guaranteed by the level 7 thesis), it is reasonable to suppose that, as we learn more about the world, we will learn more about how to learn. Granted the level 7 thesis, in other words, meta-knowability is a reasonable conjecture.

(ii) Meta-knowability is too good a possibility, from the standpoint of the growth of knowledge, not to be accepted initially, the idea only being reluctantly abandoned if all attempts at improving methods for the improvement of knowledge fail.

(i) and (ii) are not, perhaps, very strong grounds for accepting meta-knowability; both are open to criticism. But the crucial point, for the present argument, is that these grounds for accepting meta-knowability, (i) and (ii), are independent of the success of science. This suffices to avoid circularity.[22]

But what about reasons for accepting theses at levels 5, 4 and 3? Are not these inevitably viciously circular? The thesis that the universe is comprehensible, at level 5, is accepted because no other idea, compatible with meta-knowability, has been so fruitful in generating empirically progressive research programmes; the thesis that the universe is physically comprehensible, at level 4, is accepted because no other thesis, compatible with the level 5 thesis, has been so fruitful in generating empirically progressive research programmes;[23] and likewise for the thesis at level 3. In short, theories at level 2 are accepted because of empirical success and compatibility with level 3, 4 or 5 theses; and these theses are accepted because of their empirical fruitfulness. This would seem to be viciously circular in the most blatant fashion imaginable.

I have three arguments in refutation of this charge.

First, there is no question of the truth of theories being justified by an appeal to metaphysical theses, the truth of which is in turn justified by the success of science, for the simple reason that AOE is thoroughly conjectural, and to that extent Popperian, in character, there being no attempt to justify the truth of either theories or metaphysical theses.

Second, physicalism is *incompatible* with accepted fundamental physical theories, so there could be no question of the truth of one being justified by an appeal to the truth of the other. Physicalism is deployed to *criticize*, and to try to *improve*, accepted fundamental theories, not to *justify* their truth.

Third, and decisively, in so far as *acceptance* of physical theories is in part justified by an appeal to physicalism, whose acceptance is in turn justified by an appeal to the (apparent) success of science, which does involve a kind of circularity, this is licensed and legitimized by the level 6 thesis of meta-knowability. This asserts that the universe is such that there is some rationally discoverable thesis which, if accepted, makes possible the progressive improvement of more specific assumptions and methods in the light of the empirical success and failure of the research programmes to which they give rise. *If* meta-knowability is true, then progressively improving more specific metaphysical assumptions in the light of which seem to lead to the greatest empirical success, while at the same time choosing those empirically successful theories which best accord with these metaphysical assumptions, is just what needs to be done to make scientific progress. Meta-knowability, if true, justifies the element of circularity that is involved.

The gross invalidity of the genuinely viciously circular argument can be highlighted as follows. The argument seeks to justify acceptance of theory T by an appeal to metaphysical thesis M, and then justify acceptance of M by an appeal to the empirical success of T. But this argument works just as well (or ill) if we choose some empirically successful but horribly *ad hoc* rival to T, say T*, and a suitably *ad hoc* variant of M, say M*. We can now argue, with equal validity (i.e. none) that we justify acceptance of T* by appealing to M*, and justify acceptance of M* by appealing to the empirical success of T*. We have here a way of testing whether or not a putative solution to the problem of induction is, or is not, viciously circular: it must provide some valid way of ruling out arguments that appeal to *ad hoc* theories and theses like T* and M*.

AOE, granted the level 6 thesis of meta-knowability, does provide this. Given that M accords with meta-knowability in being rationally discoverable, all *ad hoc* rivals of M (i.e. M*) are ruled out because these are not "rationally discoverable": M* is just one of infinitely many equally viable theses. Thus, *if* meta-knowability is accepted, AOE is not viciously circular – not circular in any invalid sense. Meta-knowability in effect asserts that the universe is such that no *ad hoc* or aberrant version of argumentation which appeals to T and M – a version which appeals to some T* and M* – can meet with success, because all M*-type metaphysical theses are false.

It is of course absolutely vital that arguments for accepting meta-knowability do not themselves appeal to the success of science (for this would simply reintroduce vicious circularity at a higher level). The argument given above for accepting meta-knowability is weak, but it does not appeal, in any way whatsoever, to the success of science. Thus AOE is free of vicious circularity.[24]

7.3 Two versions of critical rationalism

Even if AOE does not play van Fraassen's mug's game, nevertheless how can it conceivably solve the *practical* problem of induction given its quasi-Popperian character?

Let me say at once that two versions of Popperianism deserve to be distinguished. On the one hand there is Popper's own view, which I shall call, with ironic intent, "dogmatic critical rationalism". This stresses merely the vital role that criticism has for rationality. Criticism is deployed, one might say, in an uncritical or almost dogmatic fashion. In contrast to this there is the version of critical rationalism which I wish to defend, which might be called "critical critical rationalism". This takes seriously the implications of a point emphasized, but not adequately followed up, by Popper, namely, that the whole point of *rational* criticism is to promote progress – and in connection with science, to promote progress in knowledge (and understanding). This means that theses which are demonstrably such that not accepting them can only harm and cannot help progress in knowledge whatever the universe is like, do not require (rational) criticism. They deserve to be permanently accepted. The cosmological thesis at level 7 of the hierarchy of AOE is accepted on these grounds – in sharp contrast to anything found in Popper's work. Furthermore, it is all important, according to critical critical rationalism, to highlight that part of our knowledge which, we conjecture, it is most fruitful to criticize, from the standpoint of achieving progress in knowledge. Mere criticism is not good enough; we need to be critically critical, critical of criticism itself, directing criticism to that which we conjecture it is most fruitful to criticize from the standpoint of achieving progress. A basic idea behind the hierarchy of AOE is just to display the metaphysical presuppositions of science in such a way that that which, we conjecture, it is most fruitful to criticize be brought to the fore, fruitful criticism being especially facilitated. Criticism needs to be directed, above all, at theses at levels 1, 2 and 3 – theses from 4 to 6 becoming increasingly unfruitful to criticize as we ascend the hierarchy, due to their increasing lack of factual content and increasingly indispensable role in the search for knowledge.

7.4 The practical problem of induction

But how does any of this help – the reader may ask with rising impatience – with solving the practical problem of induction? I now address this question head on.

It is important to appreciate that there are three parts to the problem of induction. There is the *methodological* problem: to specify the precise methods involved in the choice of theory in science. There is the *theoretical* problem: to show that we are justified in accepting the scientific theories we do accept, granted the aim is to improve theoretical knowledge and understanding of the universe. And, perhaps hardest of all, there is the *practical* problem: to show that we are justified in accepting those results of science that we do accept, granted our aim is to use these results as a basis for action, such as constructing bridges and producing drugs, lives potentially being lost if the predictions of science turn out to be false. The vast literature on the problem of induction is almost entirely devoted to the practical problem, but this is like trying to fly before you can crawl. The proper place to begin is with the methodological problem: if this has not been solved, to the extent of specifying *rigorous* methods for science, all efforts to solve the other two problems will be squandered on trying to justify the unjustifiable. This, in essence, is the reason for the long-standing failure of attempts to solve the problem(s) of induction.

Popper's falsificationism, like all other versions of SE, fails at the first hurdle (as I have, in effect, already pointed out above). Methods actually employed in physics involve persistently choosing unifying theories in preference to more empirically successful disunified rival theories. This in turn involves making a big, persistent metaphysical assumption, to the effect that all disunified theories are false. Rigour demands that this (implicit) metaphysical assumption be made explicit, within science, so that it can be criticized and, we may hope, improved. But falsificationism cannot do this, because its criterion of demarcation declares metaphysics to be non-scientific. And no version of SE can do this either, because the metaphysical assumption, implicit in persistent scientific preference for unifying theories against the evidence, is repudiated, denied, by all versions of SE. All SE views about scientific method lack rigour.

And related to the lack of rigour, there is a lack of precise characterization. Presuppositionism, as we have seen, leads to AOE, and to the view that metaphysical assumptions and associated methods *evolve* with evolving knowledge. No version of SE (including falsificationism) can do adequate justice to this evolving, positive feedback aspect of scientific method.[25]

Finally, in order merely to specify precisely what the methods of physics are, it is necessary to specify what unification in theoretical physics *is*. We have seen in Chapter 4 what this amounts to: the more the *content* of the totality of fundamental theory in physics is the same throughout all the possible phenomena to which this totality of theory applies, so the more *unified* it is. What matters is what the totality of theory asserts about the world, not the character or structure of the theory itself. As far as non-empirical requirements for acceptability are concerned, in order to be ultimately acceptable, the totality of fundamental theory must be unified in the strongest sense.

This conception of unity of theory, and the associated non-empirical methodological requirement for theory acceptance, fall naturally out of AOE, and are unproblematic when viewed from that standpoint. But granted SE, this conception of unity of theory, and associated methodological requirements, are unacceptable. For, in demanding that theoretical physics accepts this conception of unity, and implements the associated methods, one thereby commits physics to presupposing physicalism. That contradicts SE. Hence, those who take some version of SE for granted, cannot adopt this conception of theoretical unity.

Physicists and philosophers of physics have long been baffled by what theory unity *is*. Even Einstein was baffled (see Einstein, 1969, pp. 20–5, and Maxwell, 1998, pp. 105–6). In section 4.1, Chapter 4, I indicated some of the failed attempts that have been made to say what unity of theory really is.[26] But all attempts to solve the problem within the framework of SE are doomed to fail. In order to solve the problem, one needs to abandon SE, accept that persistent preference for unified theories, in the relevant sense of "unified", means that physics makes a persistent metaphysical assumption about the nature of the universe, and thus follow the line of argument, spelled out above, which leads to AOE.

In short, the problem of induction cannot be solved, granted SE, partly because SE lacks rigour (to the extent of being inconsistent), partly because it cannot solve even the easiest part of the problem of induction – namely the *methodological* part. It cannot do this because it cannot specify those important, non-empirical methods that have to do with theoretical *unity*. AOE, by contrast, solves the purely methodological problem with ease.

Having failed to solve the methodological problem, it follows at once, of course, that falsificationism and other versions of SE all fail to solve the theoretical and practical problems of induction as well.

This has a bearing on the question of whether the problem of induction is solvable at all. Most contemporary philosophers, historians and sociologists of science seem to have concluded that the practical or justificational problem is unsolvable, just because repeated efforts to solve the problem seem to have got nowhere. But what the above point shows is that there is an entirely straightforward reason for this failure. All these failed attempts failed even to formulate the problem correctly! To do that one needs first to have identified the correct methods of science.

There is another way in which the problem may be misformulated so as to render it insoluble. The formulation may make epistemological demands that are so high that they are quite impossible to fulfil. One such formulation would be, "How can our confidence that empirically successful, accepted scientific theories are true be justified?" This makes impossibly high epistemological demands. All fundamental dynamical physical theories so far put forward, whatever empirical success they may have achieved, are false![27] Formulated in this way, the problem is insoluble. A slightly less epistemologically ambitious formulation would be, "How can our confidence that the empirical predictions of empirically successful, accepted theories are true, be justified?" But this also asks for too much. All physical theories so far proposed, however empirically successful, yield false empirical predictions. A still less epistemologically ambitious formulation would be, "How can our confidence that empirically successful, accepted theories yield true empirical predictions, within the standard range of phenomena (and accuracy) for which they have already been shown to be successful, be justified?" But even this may ask for the impossible. Perhaps our customary confidence in science is misplaced. Perhaps just this is revealed by the correct solution to the practical or justificational problem.

In short, in order to avoid struggling to achieve the impossible, we need to formulate the problem in a somewhat more open-ended way than any of the above. The following stands a better chance of being solvable: How can our confidence that empirically successful, accepted theories yield true empirical predictions, within the standard range of phenomena (and accuracy) for which they have already been shown to be successful, be justified *in so far as such confidence is justified*?

We cannot just assume, from the outset, that the solution to the problem must justify our "pre-Humean" confidence in common sense and scientific knowledge – the confidence we had, that is, before learning of Hume's devastating arguments. For this may, again, be asking for the impossible. Perhaps Hume demonstrated, decisively, that such "pre-Humean" confidence is misplaced and unjustifiable. The correct solution to the justificational problem would, in this case, demonstrate just this to be the case. This, of course, is the view defended here.

We have seen that theoretical scientific knowledge makes assumptions about the cosmos. But there is more to it than that. Even very modest common-sense knowledge, such as, "I can walk across the room," or, "This room will endure for the next ten seconds," makes assumptions about the entire cosmos, about the nature of ultimate reality. Such very modest common-sense theses imply (a) that the entire cosmos is such that no cosmic explosion is occurring anywhere which will spread with nearly infinite speed to engulf and destroy the earth in the next few seconds.

We only possess knowledge of such modest common-sense items in so far as we possess knowledge of the cosmological thesis (a). Science and common sense both make cosmological assumptions.

If these cosmological assumptions can be established to be true in such a way that we are justified in being confident of their truth, then there is the hope that we can be justifiably confident of the truth of both some empirical predictions of empirically successful theories, and modest items of common sense. But if the relevant cosmological theses are such that they cannot conceivably be established to be true in such a way that we are justified in being confident of their truth, then it follows straight away, from elementary logic, that we cannot conceivably be confident of the truth of either the predictions of scientific theories or the items of common sense. If our most modest, immediate, apparently secure items of common sense have implications for the nature of the entire cosmos that are irredeemably speculative and conjectural, then our modest items of common sense must themselves be irredeemably speculative and conjectural as well.

The relevant cosmological theses are indeed irredeemably speculative and conjectural, as everyone would surely admit. The conclusion is thus inescapable: scientific knowledge, and modest common-sense knowledge, are both irredeemably speculative and conjectural as well. If we demand of the solution to the justificational problem of induction that it restores our pre-Humean confidence in scientific and common-sense knowledge, then we demand the impossible. The attempt to restore such pre-Humean confidence can only undermine the rationality of science, in so far as it lulls us into a false sense of security, and leads us to believe that parts of our knowledge do not need critical scrutiny.

One demand that can be made of the correct solution to the justificational problem of induction is that it puts empirical data and scientific theories onto an equal epistemological footing. Ordinarily we assume we can be confident of the truth of factual statements about our immediate, observed surroundings: this is a table, that is a book, this is a Bunsen burner, and so on. Before encountering Hume (and his twentieth-century descendents, such as Einstein, Popper and Kuhn), we may feel equally confident of the truth (or approximate truth) of empirically successful, accepted scientific theories. But Hume's argument, reformulated a little, has the effect of opening up a gulf between evidence and theory. Any theory has infinitely many empirical consequences – and consequences for distant times and places. However many consequences we verify empirically, we will always be infinitely far away from verifying the theory. Even if particular items of evidence are known with absolute certainty, theories must, it seems, remain pure conjectures. They can be falsified, perhaps, but remain permanently unverifiable, to any degree whatsoever. The solution to the justificational problem must, it may be felt, remove this gulf in epistemological status between empirical data and theory.

AOE does just that, by making clear that empirical data, just like theories, contain permanently conjectural cosmological implications. All knowledge, theoretical, empirical and common sense, is irredeemably conjectural because of these conjectural cosmological implications.

But if this is the case, what grounds are there for holding that AOE solves the justificational problem of induction? How does AOE do any better than Popper's conjecturalist position?

AOE, as we have seen, solves the methodological problem, whereas Popper's falsificationism fails to do this. As a result, AOE is able to solve a very important part of the justificational problem which falsificationism, notoriously, fails to solve. This is the problem of discriminating decisively between those conjectures about whose truth we are so confident that we are prepared to entrust our lives to their truth, and conjectures about whose truth we have no confidence at all. Every time we fly in an aeroplane, cross a suspension bridge, or imbibe medicine, we entrust our lives to the correctness of relevant items of scientific knowledge, and confidently take for granted that rival conjectures that can be concocted (ostensibly even more empirically successful but horribly *ad hoc*) which predict we will die are all false. Accounting for this difference deserves to be regarded as the nub of the justificational problem. AOE accounts fully for this dramatic difference, whereas falsificationism entirely fails to do this.

John Worrall has dramatized the problem as follows. We are, let us suppose, standing on top of the Eiffel Tower, and we are confronted by two rival conjectures: one says that if we jump we will float gently down to earth without harm; the other says we will fall in the usual way to our death (Worrall, 1989). Only lunatics think the first a viable possibility; the rest of us are absolutely confident in the truth of the second conjecture. How is this confidence to be justified? No version of SE comes up with an adequate answer, especially as aberrant versions of Newtonian theory or general relativity can be concocted which predict that jumping on this occasion will lead to a soft, harmless landing, and which are empirically *more* successful than the non-aberrant versions of these theories. Can AOE justify our confidence that if we jump we will be killed?

If we grant the truth of the theses of AOE, from level 4 to 7, a straightforward answer can be given. Physicalism tells us that a unified pattern of physical law governs all phenomena. By far our best efforts at discovering invariant (or unified) laws governing such things as bodies in free fall near the earth's surface are Newton's theory of gravitation and, better still, Einstein's theory. No rival theory is even remotely as good at complying with the two requirements of (1) empirical success and (2) compatibility with physicalism. Theories that are empirically more successful and predict a gentle landing can be concocted, but these clash horribly with physicalism, and deserve to be rejected for that reason. But Newton's or Einstein's theory (plus additional information about such things as the mass of the earth) predict with stark clarity: jumping leads to rapid acceleration at roughly 32 feet per second squared. Above all, a theory which accords with physicalism as well as Newton's or Einstein's theory, but predicts that jumping will lead to a gentle floating to the ground, is nowhere on the horizon. Thus, given the truth of physicalism, there is absolutely no question, no grounds for serious doubt, whatsoever: jumping is for suicides only.

But we are not given the truth of physicalism. At most, arguments deployed above give grounds for accepting physicalism granted our aim is to improve our conjectural knowledge of truth. There are arguments justifying *acceptance* of theses at levels 3 to 7, but no arguments justifying the *truth* of these theses. And it is the latter we require, it seems, to solve Worrall's problem, and the practical problem of induction more generally.

I have two replies to this objection.

First, even in the absence of any kind of justification of the truth of physicalism AOE succeeds, nevertheless, in distinguishing decisively between conjectures we are confident are true, to the extent even of entrusting our lives to their truth, and conjectures (even empirically more successful conjectures) about whose truth we have no such confidence.

Second, the demand that the truth of physicalism must receive some kind of justification before it becomes rationally acceptable for practical purposes is not just impossible to fulfil; it deserves to be rejected in that it stems from an unrigorous, untenable conception of science, and human knowledge more generally. If, and only if, some version of SE is correct, and science is based on evidence, and on metaphysical theses whose truth has been justified (if there are any), is the demand to justify the truth of physicalism itself justifiable. But all versions of SE are unrigorous and untenable. Hence, the SE demand to justify physicalism is itself unjustifiable, and must be rejected. What has been demonstrated above is that *all* significant factual knowledge, common sense and scientific, implies (and thus presupposes) cosmological theses: rigour requires that these unjustifiable cosmological theses are made explicit, so that they can be critically assessed and, we may hope, improved. To demand that such cosmological theses cannot be accepted unless their truth is justified condemns science to lack of rigour, because it ensures that unjustifiable cosmological theses will not be, and cannot be, accepted as a part of scientific knowledge. The demand deserves to be rejected.

Human knowledge has always had this inescapable cosmological dimension built into it. The illusion that science could dispense with such unjustifiable cosmological conjectures only crept in with the general acceptance of SE, some time after Newton and before the end of the nineteenth century. What needs to be done is not to justify the truth of physicalism, but rather to justify the claim that this cosmological conjecture has played a more fruitful role in the advance of science than any rival at that level. Just this was done above. *Science does not prove its cosmological conjectures; it sets out to improve those it has inherited from the past.* Physicalism is the best available, at that level of generality, and that suffices to solve the Worrall problem, and the justificational problem of induction. We are justified in entrusting our lives to the standard empirical predictions of those theories (a) which have met with sufficient empirical success, and (b) which, together with other such empirically successful theories, are more nearly compatible with our best metaphysical theses concerning the comprehensibility and knowability of the universe. Our best metaphysical theses, in turn, are those which have generated the most empirically progressive scientific research programmes. The circularity that seems to be involved here is legitimized by acceptance of meta-knowability.

7.5 Cosmological conjectures need acknowledgement and improvement

I now spell out in a little more detail the point just made, which is central to the solution to the practical problem of induction.

When viewed from the perspective of SE, it looks as if, in order to solve the practical problem, sufficiently good grounds must be given for the *truth* of physicalism, from scratch as it were. Acknowledging, initially, only knowledge of particular empirical facts, we must somehow provide an argument for the truth of physicalism which is so good that it justifies us in being entirely confident of the truth of physicalism even when our lives are at stake. Let us call this the "SE requirement" for solving the practical problem of induction. Given this requirement, the prospects for solving the practical problem of induction seem hopeless.

But this requirement itself deserves to be rejected. It is only acceptable if what I shall call "the SE prescription" is acceptable. But this latter is an intellectual disaster and deserves to be rejected. Hence the SE requirement must be rejected as well.

By the SE prescription I mean this: in order to develop science in a properly scientific, rigorous way, so that it is capable of delivering authentic, reliable knowledge, science must eschew all metaphysical presuppositions in the context of justification and base acceptance of scientific theories as far as possible solely on empirical considerations without reference to conjectural metaphysics. If this is correct, then it makes perfect sense to demand of any attempt to solve the practical problem of induction that it satisfies the SE requirement. But the SE prescription – eschewing metaphysical presuppositions in order to render science scientific and rigorous – is, we have seen, an intellectual disaster. It has entirely the opposite effect of the one intended. If taken seriously, instead of enhancing the rigour of science, it would destroy science and, indeed, all knowledge. The arguments of this and previous chapters demonstrate that science devoid of metaphysics is not possible. Selecting theories on the basis of empirical success and failure, no kind of assumption being made about what kind of world this is, cannot succeed if rigorously pursued, because science would be overwhelmed by endlessly many empirically successful aberrant theories which would stultify science and render technological application impossible. (Or, if requirements of simplicity and unity are invoked, in addition to empirical requirements, then metaphysical assumptions of unity are being presupposed, but in a surreptitious fashion.) Science has survived and progressed despite, and not because of, acceptance of SE by the scientific community. Science has managed to do this by implementing SE in only a highly unrigorous and hypocritical fashion (implicit metaphysical presuppositions exercising a highly influential role over choice of theory).

The SE prescription is, then, an intellectual disaster. Unfounded metaphysical or cosmological conjectures about the comprehensibility and knowability of the universe are essential for science, and cannot be eliminated without disaster. What needs to be implemented, instead, is the "AOE prescription": roughly, endeavour to *improve* metaphysical assumptions explicit or implicit in current science and knowledge; do this by modifying existing assumptions in that direction which seems to be the most fruitful from the standpoint of acquiring empirical knowledge within a fixed framework of assumptions and methods which are such that these are required for the acquisition of knowledge to be possible at all. In other words, put generalized AOE into practice.

Before the advent of SE, the pursuit of science, or of knowledge more generally, invariably went on within a framework of religious and metaphysical assumptions. Christianity, the corpuscular hypothesis, and Galileo's thesis that the book of nature is written in the language of mathematics, played especially important roles in the sixteenth and seventeenth centuries as far as the birth and development of modern science was concerned. The orthodox SE prescription insists that science must be dissociated from such dubious, unfounded religious and metaphysical doctrines. But this cannot be done; at best, unrigorous and hypocritical science results. The impression that it has been done creates the insoluble problem of induction (as traditionally construed, from the SE perspective).

Instead, we need to see AOE science as explicitly *improving* on antecedently upheld religious and metaphysical theses. There is, in science, a substantial component of faith – but, ideally, it is *rational* faith, openly acknowledged as conjectural in character, subjected to sustained criticism, and undergoing persistent modification in the direction of that which seems to lead to the most empirically progressive research, at levels 1 and 2. Science does not eliminate metaphysics; it implements a method which makes it possible for us to develop and choose those metaphysical ideas most fruitful for progress in empirical knowledge.

But once the SE prescription has been rejected as intellectually disastrous, and the AOE prescription accepted instead, it is clear that the SE requirement for solving the practical problem of induction must itself be rejected as intellectually unreasonable and unacceptable. The only rationale for adopting it arises from the idea that it is entirely proper to put the SE prescription into practice. But putting this into practice makes science impossible (for reasons wholly in addition to the resulting insolubility of the practical problem of induction). The SE requirement presumes a state of knowledge that has resulted from implementing the intellectually destructive SE prescription: this state of knowledge is an intellectual disaster, and must be rejected, and along with it the SE requirement.

Taking the SE requirement seriously is rather like an athlete having both legs amputated and then expecting to win the 100 metres at the Olympics. Render science, and indeed all knowledge, impossible (by throwing away vital metaphysics) and it should occasion no surprise that a situation is created in which the practical problem of induction becomes impossible to solve as well.

For the last four centuries ever since Galileo, or perhaps for the last two and a half thousand years ever since the Presocratics, physicalism has been by far the most fruitful metaphysical thesis available from the standpoint of promoting progress in science, or in knowledge more generally. No rival metaphysical thesis has been remotely as fruitful. AOE correctly depicts our current scientific knowledge. Physicalism, in short, is justifiably a basic tenet of current (conjectural) scientific knowledge, our best attempt, at that vital level, of acquiring knowledge of truth, more secure, indeed, than any fundamental physical theory, such as quantum theory or general relativity. It makes very good sense not to jump off the Eiffel Tower if you want to stay alive, for the reasons given above.

But if science must accept, and not eschew, religious or metaphysical ideas, what grounds are there, it may be asked, for preferring physicalism to a thesis such as that God exists and benevolently arranges for the cosmos to be such as to make science possible? For my reply, see Maxwell (1998, pp. 206–8; 2001a, pp. 6–10; 2002c).

The argument given above is a special case of a more general argument concerning Cartesianism. The proper fundamental problem of epistemology is the problem of how we can best *improve* knowledge (improve what we have inherited from the past). The Cartesian prescription says: throw everything away we have inherited from the past except that which cannot be doubted, or is most secure; taking this as a secure base, build up rigorous, reliable knowledge. This Cartesian prescription has exercised a profound influence on philosophy, on both so-called rationalists and on empiricists. Influencing the former it leads to the search for secure principles founded on reason; influencing the latter it leads to the idea that observational knowledge alone constitutes the only acceptable Cartesian secure base, it being necessary to build the rest of knowledge up from this secure base. This, of course, is the SE prescription; it is a version of the Cartesian prescription. But the Cartesian prescription must be rejected. The proper way to set about *improving* knowledge is to acknowledge the conjectural character of what we have inherited from the past, subject it to critical scrutiny, and put generalized AOE into practice. In direct opposition to Cartesianism, this involves in part taking most seriously ideas that are most *vulnerable* to being found to be false, namely the falsifiable theories of science. Far from giving priority to ideas immune to doubt, we need to give priority to ideas most vulnerable to doubt. In this way, as Popper stresses, we make it possible for us to learn from our mistakes. But this Popperian prescription needs itself to be modified, as we have seen, so that some unfalsifiable, metaphysical ideas, inherited from the past, continue to be accepted and developed as a part of scientific knowledge, those in particular being accepted which either (a) must be true if knowledge is to be possible at all, or (b) are most fruitful in leading to progress in empirical knowledge. The Cartesian prescription deserves to be resoundingly rejected, and along with it the SE prescription and SE requirement.

The problem of induction is not just a philosophical puzzle à la Wittgenstein. Its long-standing insolubility is indicative of a fundamental defect in our understanding of science and its relationship with metaphysics and philosophy – a fundamental defect in our whole culture. Science does not stand opposed to metaphysics and philosophy; it is metaphysics and philosophy carried on employing the improved methods of investigation of empiricism: observation and, above all, experimentation (a point enshrined in the seventeenth-century terms of "experimental" and "natural" philosophy). A basic task for philosophers today is to try to get across to the scientific community just how vital metaphysics and philosophy, properly conducted, are for science, so that scientists and philosophers can begin to collaborate on implementing AOE science, thus recreating natural philosophy.[28] But this is unlikely to happen as long as AOE continues to be dismissed, so unjustly, as a mug's game.

8
Does probabilism solve the great quantum mystery?

8.1 Orthodox quantum theory is the best and worst of theories

What sort of entities are electrons, photons and atoms given their wave-like and particle-like properties? Is nature fundamentally deterministic or probabilistic? Any decent theory of the quantum domain, able to provide us with genuine knowledge and understanding of its nature, ought to provide answers to these childishly elementary questions. Orthodox quantum theory (OQT) evades answering these questions by being a theory, not about quantum systems as such, but rather about the results of performing measurements on such systems.[1]

This state of affairs came about as follows. Bohr, Heisenberg, Dirac and the other creators of OQT did not know how to solve the quantum wave/particle dilemma. This created a grave problem for those seeking to develop quantum theory: How can one develop a consistent theory about entities that seem to be both wave-like and particle-like, as in the two-slit experiment, for example? Around 1925, Heisenberg hit upon the strategy of evading this fundamental dilemma by developing what subsequently became matrix mechanics as a theory exclusively about the results of performing measurements on quantum systems, this version of quantum theory thus not needing to specify the nature of quantum systems when not undergoing measurement. Schrödinger, a little later in 1926, developed wave mechanics in the hope that it would be a precise theory about the nature of quantum systems. This theory, Schrödinger hoped, would show the electron to be wave-like in character. But then Born successfully interpreted the Ψ function of Schrödinger's wave mechanics as specifying the probability of detecting the particle in question. According to Born's crucial interpretative postulate, $|\Psi|^2.dV$ gives the probability of detecting the particle in volume element dV if a position measurement is performed. Schrödinger proved that his theory and Heisenberg's matrix mechanics are equivalent: the outcome, a sort of synthesis of the two theories, is OQT.

OQT is an extraordinarily successful theory empirically, perhaps the most successful in the whole of physics when one takes into account the range, immense diversity, and accuracy of its predictions. But not only does it fail to solve the great quantum mystery of what sort of entities electrons and atoms can be in view of their apparently contradictory particle and wave properties. It also fails to answer the other childishly elementary question: Is the quantum domain deterministic or probabilistic? The basic dynamic equation of OQT, Schrödinger's time-dependent equation, is deterministic in character. It tells us that quantum states, given by Ψ, evolve deterministically in time, as long as no measurements are made. But this does not mean OQT asserts that the quantum domain is deterministic. First, given OQT, Ψ cannot be interpreted as specifying the actual physical state of a quantum system, just because OQT fails to solve the wave/particle dilemma, and thus fails to provide a consistent specification of the physical nature of quantum systems when not being measured. Given OQT, Ψ must be interpreted as containing no more than information about the outcome of performing measurements. Second, OQT in general makes *probabilistic* predictions about the outcome of performing measurements, not (apart from exceptional circumstances) *deterministic* predictions. But one cannot conclude from this that OQT asserts that the quantum domain is fundamentally *probabilistic* in character, some physical states of affairs only determining what occurs subsequently only *probabilistically*. This is because, according to OQT, probabilistic outcomes only occur when we intervene, and make a *measurement*. In the absence of measurement, nothing probabilistic occurs at all, according to OQT. Indeed, if the process of measurement is treated quantum mechanically, then nothing probabilistic occurs at all, precisely because the basic dynamic equation of OQT, Schrödinger's time-dependent equation, is deterministic.

The inability of OQT to answer these two elementary questions is in itself a serious failure of the theory. But there are, as a consequence, a host of further failures and defects. Because OQT is about the results of performing measurements on quantum systems (and not about quantum systems *per se*, due to its failure to solve the wave/particle problem), in order to come up with physical predictions, OQT must consist of two parts, (1) quantum postulates, and (2) some part of classical physics for a treatment of measurement. (2) is indispensable.

(1) alone, precisely because OQT lacks its own quantum ontology, cannot predict anything physical at all – or at least can only make conditional predictions of the form: if such and such a measurement is made, such and such will be the outcome with such and such a probability. Thus OQT = QP + CP, where "QP" stands for the quantum mechanical postulates of the theory, and "CP" stands for the classical postulates, required for measurement.

In what follows, a quantum "measurement" is a process that actually detects quantum systems; a process which prepares a quantum system to be in a certain quantum state, but does not detect the system, is a "preparation" rather than a "measurement".

OQT, construed as QP + CP, as it must be, is a seriously defective theory. (a) OQT is *imprecise*, due to the inherent lack of precision of the notion of "measurement". How complex and macroscopic must a process be before it becomes a measurement? Does the dissociation of one molecule amount to a measurement? Or must a thousand or a million molecules be dissociated before a measurement has been made? Or must a human being observe the result? No precise answer is forthcoming. (b) OQT is ambiguous, in that if the measuring process is treated as a measurement, the outcome is in general probabilistic, but if this process is treated quantum mechanically, the outcome is deterministic. (c) OQT is very seriously *ad hoc*, in that it consists of two incompatible, conceptually clashing parts, QP and CP. OQT only avoids being a straightforward contradiction by specifying, in an arbitrary, *ad hoc* way, that QP applies to the quantum system up to the moment of measurement, and CP applies to the final measurement result. (d) OQT is non-explanatory, in part because it is *ad hoc*, and no *ad hoc* theory is fully explanatory, in part because OQT must presuppose some part of what it should explain, namely classical physics. OQT cannot fully explain how classical phenomena emerge from quantum phenomena because some part of classical physics must be presupposed for measurement. (e) OQT is limited in scope in that it cannot, strictly speaking, be applied to the early universe in conditions which lacked preparation and measurement devices. Strictly speaking, indeed, it can only be applied if physicists are around to make measurements. (f) OQT is limited in scope in that it cannot be applied to the cosmos as a whole, since this would require preparation and measurement devices that are outside the cosmos, which is difficult to arrange. Quantum cosmology, employing OQT, is not possible. (g) For somewhat similar reasons, OQT is such that it resists unification with general relativity. Such a unification would presumably involve attributing some kind of quantum state to space-time itself (general relativity being a theory of space-time). But, granted the basic structure of OQT, this would require that preparation and measurement devices exist outside space-time, again not easy to arrange.

These nine defects, the two basic failures with which we began and the seven consequential defects, (a) to (g), are, taken together, very serious indeed.[2] Despite its immense empirical success, OQT must be declared to be an unacceptably defective theory. It is the best of theories, and the worst of theories.[3]

In opposition to this conclusion, it may be argued that all physical theories, even a classical theory such as Newtonian theory (NT), must call upon additional theory to be tested empirically. In testing predictions of NT concerning the position of a planet at such and such a time, optical theory is required to predict the results of telescopic observations made here on earth. But this objection misses the point. NT is perfectly capable of issuing in *physical* predictions without calling upon additional theory, just because it has its own physical ontology. NT, plus initial and boundary conditions formulated in terms of the theory, can issue in the *physical* prediction that such and such a planet is at such and such a place at such and such a time, whether anyone observes the planet or not, without calling upon optical theory or any other theory. This OQT cannot do. It cannot do this because it lacks its own quantum ontology, having failed to solve the quantum wave/particle problem. In order to deliver an unconditional physical prediction, OQT must call upon some part of classical physics, as a matter of necessity, so that the theory can refer to something physically actual. The case of NT and OQT are quite different, because NT postulates actually existing physical bodies whether observed or not, whereas QP does not; for that, one requires OQT, that is, QP + CP.

It may be objected that even if non-relativistic quantum theory fails to solve the wave/particle problem, relativistic quantum theory, or quantum field theory, does solve the problem in that it declares that what exists is the quantum field, "particles" being discrete excitations of the field. But this objection misses the point as well. Orthodox quantum field theory (OQFT) is just as dependent on measurement, and thus on some part of classical physics, as non-relativistic OQT is. The quantum states of the quantum field of OQFT have to be interpreted as making probabilistic predictions about the results of performing measurements, just as in the case of OQT. A version of quantum field theory which succeeded in specifying the nature of the quantum field in a fully satisfactory way, so that the theory has its own quantum ontology entirely independent of any part of classical physics, would be able to issue in physical predictions about actual physical states of affairs entirely independently of measurement. Such a theory would be able to predict and explain macroscopic, quasi-classical phenomena as arising from the quantum field alone, without calling upon some part of classical physics for a treatment of measurement. This OQFT cannot do.

8.2 Probabilism to the rescue

What needs to be done to cure OQT of its serious defects? The primary task must be to specify precisely and unambiguously the nature of quantum entities so that quantum theory (QT) can be formulated as a testable theory about how these entities evolve and interact, without there being any mention of measurement or observables in the postulates of the theory at all. The key point that needs to be appreciated, I suggest, in order successfully to complete this task, is that the quantum domain is fundamentally probabilistic.[4] It is this that the manifestly probabilistic character of QT is trying to tell us.

The approach to solving the mysteries of the quantum domain that I am suggesting here has been long ignored largely because of the accidents of history. When quantum theory (QT) was being developed and interpreted, during the first three decades of the last century, two opposing camps developed: the Bohr–Heisenberg camp, which argued for the abandonment of micro-realism, and the abandonment of determinism; and the Einstein–Schrödinger camp, which argued for the retention of realism, and the retention of determinism. One result of this polarization of views was that the idea of *retaining* realism but *abandoning* determinism was overlooked. But it is just this overlooked option, I maintain, which gives us our best hope of curing the defects of QT. One might call this option *probabilistic micro-realism*.

Once we acknowledge that the quantum domain is fundamentally probabilistic, so that the basic laws governing the way quantum systems interact with one another are probabilistic, it is clear that *measurement* cannot be a satisfactory necessary and sufficient condition for probabilistic transitions to occur. Probabilistic transitions must be occurring in nature whether or not physicists are around to dub certain processes "measurements". The very notion of measurement is in any case, as we have seen, inherently imprecise. We require a new, precise, necessary and sufficient condition for probabilistic transitions to occur, to be specified in fundamental, quantum mechanical terms.

Furthermore, once the fundamentally probabilistic character of the quantum domain is acknowledged, it immediately becomes clear how the key quantum wave/particle problem is to be solved. If the quantum domain is fundamentally probabilistic, then the physical entities of this domain – electrons, atoms and the rest – cannot possibly be classical, deterministic entities – classical particles, waves or fields. Quite generally, we should hold that there is a one-to-one correspondence between the dynamical laws of a physical theory on the one hand, and the entities and their physical properties postulated by the theory, on the other hand. In speaking of the entities, and the properties of entities, postulated by a physical theory, we are thereby speaking, in other terms, of the dynamical laws of the theory. Hence, change dynamical laws in some basic way, and we thereby change postulated physical entities and their properties. In particular, change dynamical laws dramatically, so that they become probabilistic instead of being deterministic, and the nature of postulated physical entities must change dramatically as well. Quantum entities, interacting with one another probabilistically, must be quite different from all physical entities so far encountered within deterministic classical physics.[5]

The defects of OQT have arisen, in other words, because physicists have sought to interpret probabilistic quantum theory in terms of classical waves and particles, deterministic metaphysical ideas appropriate to earlier classical physics but wholly inappropriate to the new quantum theory. The failure of this entirely misguided attempt then led to despair at the possibility of solving the (misconstrued) wave/particle problem, despair at the possibility of specifying the precise physical nature of quantum entities. This despair in turn led to the development of OQT as a theory about the results of performing measurements – a theory which, it seemed, did not need to specify the precise nature of quantum entities. But the outcome is a theory burdened with the nine serious defects indicated above.

Thus the traditional quantum wave/particle problem is the *wrong* problem to pose. We should ask, not "Are quantum entities waves or particles?", but rather, (1) what kinds of possible, unproblematic, fundamentally probabilistic physical entities are there? And (2) are quantum entities one kind of such unproblematic probabilistic entity? The failure to put right the serious defects of OQT has persisted for so long because physicists have abandoned hope of solving the traditional quantum wave/particle problem, not realizing that this is entirely the *wrong* problem to try to solve in the first place. Once it is appreciated that (1) and (2) are the *right* problems to try to solve, new possibilities, long overlooked, immediately spring to mind.

First, physical entities that interact with one another probabilistically may be dubbed *propensitons*. Two kinds of unproblematic propensiton can immediately be distinguished: *continuous* propensitons, which evolve probabilistically continuously in time, and *intermittent* propensitons, which evolve deterministically except for intermittent moments in time when appropriate physical conditions arise, and the propensitons undergo probabilistic transitions.

There is a second obvious distinction that can be made, between propensitons which spread out spatially in time, increasing the volume of space they occupy with the passage of time, and propensitons which do not spread spatially in this way. Let us call the first *spatially spreading* propensitons, and the second *spatially confined* propensitons.

We are in new territory. In our ordinary experience of the world, and within deterministic physics, we never encounter propensitons. Probabilistic outcomes, obtained when we toss a penny or a die, can always be put down to probabilistic changes in initial conditions. Classical statistical mechanics presupposes that the underlying dynamic laws are deterministic. Having no experience of them, propensitons will, inevitably, when we first encounter them, strike us as mysterious, even unacceptably weird. But these feelings of unfamiliarity ought not to lead us into deciding that theories which postulate such entities are inherently unacceptable. In particular, the four kinds of propensity indicated above should be regarded as equally viable, *a priori*. Whether a theory that postulates one or other type of propensiton is acceptable or not should be decided upon in the usual way, in terms of its empirical success, and the extent to which it is unified, simple, explanatory.

Granted that quantum systems are *some* kind of propensiton, which of the four kinds of unproblematic propensiton just indicated should we take quantum systems to be? There is here a very important consideration to be borne in mind. Despite suffering from the nine defects indicated above, nevertheless OQT is perhaps the most empirically successful physical theory ever formulated. The range, variety and accuracy of its empirical predictions are unprecedented. No other physical theory has been subjected to such sustained severe experimental testing, and has survived without a single refutation. There are good grounds for holding that OQT has got quite a lot right about the nature of the quantum world. Our strategy, then, ought to be, in the first instance at least, to stick as close to OQT as possible, and modify OQT just sufficiently to remove the defects of the theory. The structure of OQT mirrors that of the intermittent, spatially spreading propensiton. On the one hand, quantum states evolve deterministically, in accordance with Schrödinger's time-dependent equation;

on the other hand, there are, on the face of it, probabilistic transitions associated with measurement. Quantum states spread out spatially when evolving deterministically, and tend to become localized when measurements are made. All this mirrors the character of the intermittent, spatially spreading propensiton, the only unsatisfactory feature of OQT being that the theory stipulates that probabilistic transitions occur when measurements are made.

A very elementary kind of spatially spreading intermittent propensiton is the following. It consists of a sphere, which expands at a steady rate (deterministic evolution) until it touches a second sphere, at which moment the sphere becomes instantaneously a minute sphere, of definite radius, somewhere within the space occupied by the large sphere, probabilistically determined. The second sphere undergoes the same instantaneous probabilistic transition. Then both minute spheres again undergo steady, deterministic expansion, until they touch once more, and another probabilistic localization occurs.

A slightly more sophisticated version of this elementary spatially spreading intermittent propensiton is the following. The sphere is made up of variable "position probability density", such that, when the sphere localizes probabilistically, in the way just indicated, it is most probable that it will be localized where the position probability density is most dense. A law specifies how position probability density is distributed throughout the sphere. We might even imagine that the position probability density exhibits a wave-like distribution. Such a propensiton, given appropriate conditions for probabilistic localization, might even exhibit interference phenomena in a two-slit experiment!

Quantum entities, such as electrons, photons and atoms, are, I suggest, spatially spreading intermittent propensitons. Their physical state is specified by the ψ function of QT. The deterministic evolution of these quantum propensitons is specified by Schrödinger's time-dependent equation:

$$i\hbar \frac{\partial \psi(t)}{\partial t} = -\frac{\hbar^2}{2m} \nabla^2 \psi(t) + V\psi(t)$$

The crucial questions that need to be answered to specify precisely the probabilistic properties – or propensities – of quantum systems are these:

(a) What is the precise quantum mechanical condition for a probabilistic transition to occur?

(b) Given the quantum state, ψ, at the instant before the probabilistic transition, how does this determine what the possible outcome states are, $\phi_1, \phi_2, \ldots \phi_N$?
(c) How does ψ determine the probability, p_r, that the outcome of the probabilistic transition will be ϕ_r, for $r = 1, 2, \ldots N$?
(d) How can (a) to (c) be answered so that the resulting fundamentally probabilistic version of quantum theory reproduces all the empirical success of OQT?

A number of different answers can be given to (a) to (d).

One possibility is the proposal of Ghirardi, Rimini and Weber (see Ghirardi and Rimini, 1990), according to which the quantum state of a system such as an electron collapses spontaneously, on average after the passage of millions of years, into a highly localized state. When a measurement is performed on the quantum system, it becomes quantum entangled with millions upon millions of quantum systems that go to make up the measuring apparatus. In a very short time there is a high probability that one of these quantum systems will spontaneously collapse, causing all the other quantum entangled systems, including the electron, to collapse as well. At the micro level, it is almost impossible to detect collapse, but at the macro level, associated with measurement, collapse occurs very rapidly all the time.

Another possibility is the proposal of Penrose (1986; 2004, ch. 30), according to which collapse occurs when the state of a system evolves into a superposition of two or more states, each state having associated with it a sufficiently large mass located at a distinct region of space. The idea is that general relativity imposes a restriction on the extent to which such superpositions can develop, in that it does not permit such superpositions to evolve to such an extent that each state of the superposition has a substantially distinct space-time curvature associated with it.

The possibility that I favour, put forward before either Ghirardi, Rimini and Weber's proposal, or Penrose's proposal, is that probabilistic transitions occur whenever, as a result of inelastic interactions between quantum systems, new "particles", new bound or stationary systems, are created (Maxwell, 1972b, 1976b, 1982, 1988, 1994a).[6] A little more precisely:

> **Postulate 1A**: Whenever, as a result of an inelastic interaction, a system of interacting "particles" creates new "particles", bound or stationary systems, so that the state of the system goes into a superposition of states, each state having associated with it

different particles or bound or stationary systems, then, when the interaction is nearly at an end, spontaneously and probabilistically, entirely in the absence of measurement, the superposition collapses into one or other state.

Two examples of the kind of interactions that are involved here are the following:

$$e^- + H \rightarrow \begin{array}{l} e^- + H \\ e^- + H^* \\ e^- + H + \gamma \\ e^- + e^- + p \end{array}$$

$$e^+ + H \rightarrow \begin{array}{l} e^+ + H \\ e^+ + e^- + p \\ (e^+/e^-) + p \\ p + 2\gamma \end{array}$$

Here e^-, e^+, H, H*, γ, p and (e^+/e^-) stand for electron, positron, hydrogen atom, excited hydrogen atom, photon, proton and bound system of electron and positron, respectively.

What exactly does it mean to say that the "interaction is very nearly at an end" in the above postulate? My suggestion, here, is that it means that forces between the "particles" are very nearly zero, except for forces holding bound systems together. In order to indicate how this can be formulated precisely, consider the toy interaction:

$$a + b + c \rightarrow \begin{array}{ll} a + b + c & (A) \\ a + (bc) & (B) \end{array}$$

Here, a, b and c are spinless particles, and (bc) is the bound system. Let the state of the entire system be $\Psi(t)$, and let the asymptotic states of the two channels (A) and (B) be $\Psi_A(t)$ and $\Psi_B(t)$ respectively. Asymptotic states associated with inelastic interactions are fictional states towards which, according to OQT, the real state of the system evolves as $t \rightarrow +\infty$. Each outcome channel has its associated asymptotic state, which evolves as if forces between particles are zero, except where forces hold bound systems together.

According to OQT, in connection with the toy interaction above, there are states $\phi_A(t)$ and $\phi_B(t)$, such that:

(1) For all t, $\psi(t) = c_A\phi_A(t) + c_B\phi_B(t)$, with $|c_A|^2 + |c_B|^2 = 1$;
(2) as $t \to +\infty$, $\phi_A(t) \to \Psi_A(t)$ and $\phi_B(t) \to \Psi_B(t)$.

The idea is that at the first instant t for which $\phi_A(t)$ is very nearly the same as the asymptotic state $\Psi_A(t)$, or $\phi_B(t)$ is very nearly the same as $\Psi_B(t)$, then the state of the system, $\psi(t)$, collapses spontaneously either into $\phi_A(t)$ with probability $|c_A|^2$, or into $\phi_B(t)$ with probability $|c_B|^2$. Or, more precisely:

> **Postulate 1B**: At the first instant for which
> $|\langle \Psi_A(t)|\phi_A(t)\rangle|^2 > 1-\varepsilon$ or $|\langle \Psi_B(t)|\phi_B(t)\rangle|^2 > 1-\varepsilon$,
> the state of the system collapses spontaneously into $\phi_A(t)$ with probability $|c_A|^2$, or into $\phi_B(t)$ with probability $|c_B|^2$, ε being a universal constant, a positive real number very nearly equal to zero.

The evolutions of the actual state of the system, $\psi(t)$, and the asymptotic states, $\Psi_A(t)$ and $\Psi_B(t)$, are governed by the respective channel Hamiltonians, H, H_A and H_B, where:

$$H = -\left(\frac{\hbar^2}{2m_a}\nabla_a^2 + \frac{\hbar^2}{2m_b}\nabla_b^2 + \frac{\hbar^2}{2m_c}\nabla_c^2\right) + V_{ab} + V_{ac} + V_{ac}$$

$$H_A = -\left(\frac{\hbar^2}{2m_a}\nabla_a^2 + \frac{\hbar^2}{2m_b}\nabla_b^2 + \frac{\hbar^2}{2m_c}\nabla_c^2\right)$$

$$H_B = -\left(\frac{\hbar^2}{2m_a}\nabla_a^2 + \frac{\hbar^2}{2m_b}\nabla_b^2 + \frac{\hbar^2}{2m_c}\nabla_c^2\right) + V_{bc}$$

Here, m_a, m_b and m_c are the masses of "particles" a, b and c respectively, and $\hbar = h/2\pi$ where h is Planck's constant.

The condition for probabilistic collapse, formulated above, can readily be generalized to apply to more complicated and realistic inelastic interactions between "particles".

According to the micro-realistic, fundamentally probabilistic version of quantum theory, indicated above, the state function, $\psi(t)$, describes the actual physical state of the quantum system, from moment to moment. Quantum systems may be called "propensitons". The physical (quantum) state of the propensiton evolves in accordance with Schrödinger's time-dependent equation as long as the condition for a probabilistic transition to occur does not obtain. The moment it does obtain, the state jumps instantaneously and probabilistically, in the manner indicated above, into a new state. (All but one of a superposition of states, each with distinct "particles" associated with them, vanish.) The new state then continues to evolve in accordance with Schrödinger's equation until conditions for a new probabilistic transition arise.

Propensiton quantum theory (PQT), as we may call this micro-realistic, fundamentally probabilistic version of quantum theory, can recover all the experimental success of OQT. This follows from four points. First, OQT and PQT use the same dynamical equation, namely Schrödinger's time-dependent equation. Second, whenever a position measurement is made, and a quantum system is detected, this invariably involves the creation of a new "particle" (bound or stationary system, such as the ionization of an atom or the dissociation of a molecule, usually millions of these). This means that whenever a position measurement is made, the conditions for probabilistic transitions to occur, according to PQT, are satisfied. PQT will reproduce the predictions of OQT (given that PQT is provided with a specification of the quantum state of the measuring apparatus). Third, all other observables of OQT, such as momentum, energy, angular momentum or spin, always involve (i) a preparation procedure which leads to distinct spatial locations being associated with distinct values of the observable to be measured, and (ii) a position measurement in one or other spatial location. This means that PQT can predict the outcome of measurements of all the observables of OQT. Fourth, in so far as the predictions of OQT and PQT differ, the difference is extraordinarily difficult to detect, and will not be detectable in any quantum measurement so far performed.

In principle, however, OQT and PQT yield predictions that differ for experiments that are extraordinarily difficult to perform, and which have not yet, to my knowledge, been performed. Consider the following evolution:

```
                    collision   superposition   reverse collision
                                  a + b + c
    a + b + c    ⟶                              ⟶            a + b + c
                                  a + (bc)
      (1)          (2)              (3)            (4)            (5)
```

Suppose the experimental arrangement is such that, if the superposition at stage (3) persists, then interference effects will be detected at stage (5). Suppose, now, that at stage (3) the condition for the superposition to collapse into one or other state, according to PQT, obtains. In these circumstances, OQT predicts interference at stage (5), whereas PQT predicts no interference at stage (5) (assuming the above evolution is repeated many times). PQT predicts that in each individual case, at stage (3), the superposition collapses probabilistically into one or other state. Hence there can be no interference.

8.3 Further questions

It may be asked how $\psi(t)$ can possibly represent the real physical state of a quantum system given that $\psi(t)$ is a complex function of space and time. The answer is that $\psi(t)$ can always be construed to depict two real functions of space and time.

It may be asked how $\psi(t)$ can possibly represent the real physical state of a quantum system consisting of two (or more) quantum entangled "particles", since in this case $\psi(t)$ is a function of six-dimensional configuration space plus time (or, in general, a function of 3N configuration space plus time, where N is the number of quantum entangled "particles" that go to make up the system in question). In the case of two "particles", we can construe $\psi(r_1, r_2, t)$, where r_1 and r_2 are the spatial coordinates of "particles" 1 and 2 respectively, as depicting the propensity state of the system in real 3-dimensional physical space, as follows. $|\psi(r_1, r_2, t)|^2 dV_1 dV_2$ represents the probability of the system interacting in a localizing (wave-packet-collapsing) way such that "particle" 1 interacts in volume element dV_1 about spatial coordinates r_1, and "particle" 2 interacts in volume element dV_2 about spatial coordinates r_2. The quantum entangled nature of the system means that as r_2 is changed, so the probability of "particle" 1 interacting in dV_1 about r_2 will, in general, change too.

It may be objected that postulate 1(A+B) provides no mechanism for quantum systems to be localized. This is not correct. If a highly localized system, S_1, interacts inelastically with a highly unlocalized system, S_2, in such a way that a probabilistic transition occurs, then S_1 will localize S_2. If an atom or nucleus emits a photon which travels outwards in a spherical shell, and which is subsequently absorbed by a localized third system, the localization of the photon will localize the emitting atom or nucleus with which it was quantum entangled.

Postulate 1(A+B) above has been formulated for rearrangement collisions. But the postulate is intended to apply to inelastic interactions that lead to the creation (or annihilation) of new particles, as in interactions such as $e^- + e^+ \to 2\gamma$. Such interactions require that one employs relativistic QT, which is beyond the scope of the present chapter. It deserves to be noted, however, that the root idea that probabilistic transitions occur when new "particles" are created can be interpreted in a number of different ways:

(1) There is the option considered above. The inelastic interaction must be such that distinct "particle" channels have, associated with them, distinct asymptotic states which evolve in accordance with distinct Hamiltonians. This means at least that distinct "particles" have different masses associated with them (so that an excited state of a bound system is, potentially, a different "particle" from the ground state, since the excited state will be slightly more massive than the ground state).

(2) As above, except that, for two interaction channels to differ it is not sufficient that "particles" associated with the two channels have distinct masses; either there are different numbers of "particles" (counting a bound system as one "particle") associated with different channels, or there is at least one "particle" which has a different charge, or force, associated with it.

(3) For a probabilistic transition to occur, rest mass must be converted into energy of "particles" without rest mass (e.g. photons), or vice versa.

(4) For a probabilistic transition to occur, fermions must be converted into bosons, or vice versa.

Only experiment can decide between these options. The import of this chapter, and of previous papers published by the author,[7] is that a major research effort ought to get underway, both theoretical and experimental, devoted to exploring and testing rival collapse hypotheses. Only in this way will a version of quantum theory be developed that is free of the defects of OQT and which also meets with greater empirical success than OQT. Only in this way will physics succeed in providing some kind of answer to the two childishly elementary, interrelated questions with which we began this chapter.

8.4 Quantum confusions a part of a historical pattern

I conclude with a historical remark. I have argued that the long-standing failure to solve the mysteries of the quantum domain – and so to develop a fully acceptable version of quantum theory – is due to the misguided attempt to understand the probabilistic quantum domain in terms of deterministic metaphysical ideas appropriate to the earlier theories of classical physics. As a result of the failure to solve the wholly misguided traditional wave/particle problem, Heisenberg, Bohr, Born and others developed quantum theory as a theory about the results of performing *measurements*, which seemed successfully to avoid the need to specify precisely the nature of quantum systems, but which unintentionally led to the creation of a theory with severe, if somewhat surreptitious, defects.

This pattern of confusion has occurred on at least two earlier occasions in the history of physics. On these occasions, too, physicists have attempted to interpret a new theory in terms of old, inappropriate metaphysics; the failure of this misguided effort then leads to despair at the possibility of interpreting the new theory realistically. It leads to instrumentalism, in other words, to the view that physical theories have to be interpreted as being about observable phenomena, and not about unobservable physical entities such as particles and fields. Eventually, however, the new theory may be interpreted in terms of new appropriate metaphysics. Physicists, one might say, are brilliant when it comes to equations, but not so brilliant – or at least very conservative – when it comes to metaphysics.

An example is Newton's theory of gravitation, which postulates a force at a distance between bodies with mass. The reigning metaphysical idea in Newton's time was the corpuscular hypothesis, the thesis that nature is made up of tiny corpuscles which interact only by contact. This thesis functioned as a standard of intelligibility: no fundamental physical theory could claim to be intelligible if it could not be interpreted in terms of the corpuscular hypothesis. The impossibility of interpreting Newton's theory of gravitation in terms of the corpuscular hypothesis initially led some of Newton's most eminent contemporaries to reject Newton's theory. Thus Huygens, in a letter to Leibniz, writes:

> Concerning the Cause of the flux given by M. Newton, I am by no means satisfied [by it], nor by all the other Theories that he builds upon his Principle of Attraction, which seems to me absurd ... I have

often wondered how he could have given himself all the trouble of making such a number of investigations and difficult calculations that have no other foundation than this very principle. (Quoted in Koyré, 1965, pp. 117–8)

Newton in a sense agreed, as is indicated by his remark:

That gravity should be innate, inherent and essential to matter, so that one body may act upon another, at a distance through a vacuum, without the mediation of anything else ... is to me so great an absurdity, that I believe no man who has in philosophical matters a competent faculty of thinking can ever fall into it. (Quoted in Burtt, 1932, pp. 265–6)

The impossibility of interpreting the law of gravitation in terms of the corpuscular hypothesis, in terms of action-by-contact, led Newton to interpret the law instrumentalistically, as specifying the way bodies move without providing any kind of explanation for the motion, in terms of unobservable forces. Subsequently, however, Boscovich and others were able to conceive of a metaphysical view more appropriate to Newton's new theory, according to which nature is made up of point particles, with mass, each point particle being surrounded by a rigid, spherically symmetric, centrally directed field of force which varies with distance. Reject the corpuscular hypothesis and adopt, instead, this new Boscovichean metaphysics, and Newton's theory ceases to be incomprehensible, and becomes the very model of comprehensibility.

Another example is provided by James Clerk Maxwell's theory of electrodynamics. Maxwell himself, and most of his contemporaries and immediate successors, sought to interpret the electromagnetic field in terms of a material substratum, the hypothetical aether, itself to be understood in Newtonian terms. A tremendous amount of effort was put into trying to understand Maxwell's field equations in terms of the aether. Faraday, who appreciated that one should take the electromagnetic field as a new kind of physical entity, and explain matter in terms of the field rather than try to explain the field in terms of a kind of hypothetical matter (the aether), was ignored. The unrealistic character, and ultimate failure, of mechanical models of the electromagnetic field led many to hold that the real nature of the field must remain a mystery. The most that one could hope for from Maxwell's equations, it seemed, was the successful prediction of observable phenomena associated with electromagnetism. This instrumentalistic attitude remained even after the advent of Einstein's special theory of relativity in 1905, which might be interpreted as giving credence to the idea that it is the field that is fundamental. Gradually, however, Einstein and others came to adopt the view that one should see the field as a new kind of physical entity, quite distinct from corpuscle and point particle.

There are two lessons to be learned from these episodes, one for quantum theory specifically, the other for theoretical physics in general. In the first place, quantum theory, if fundamentally probabilistic, needs to be formulated as a theory about fundamentally probabilistic physical entities – propensitons – however weird these may seem given our common sense and classical intuitions. We require a fully micro-realistic version of quantum theory which, though testable, says nothing about "observables" or "measurement" in the basic postulates of the theory at all. Second, if theoretical physics is to free itself from the obstructive tendency to interpret new theories in terms of old, inappropriate metaphysics, physicists need to recognize that metaphysical ideas are inevitably an integral part of theoretical physics, and need to be developed and improved in the light of new theoretical developments. In previous chapters I have argued that, in order to construe science as a rational enterprise, we need to see physics as making a hierarchy of metaphysical assumptions concerning the comprehensibility and knowability of the universe, these assumptions becoming increasingly insubstantial, and thus increasingly likely to be true, as we ascend the hierarchy.[8] According to this "aim-oriented empiricist" view, this hierarchy creates a framework of reasonably secure, permanent assumptions (and associated methods) high up in the hierarchy, within which much more specific, substantial and fallible assumptions (and associated methods), low down in the hierarchy, can be revised and improved. If ever the physics community came to accept and put into scientific practice this aim-oriented empiricist methodology, then the best available metaphysical ideas might lead the way to the discovery of new physical theories, instead of obstructing interpretation and understanding of theories that have been discovered (and thus also obstructing the discovery of new theories). In one exceptional case in the history of physics, the new metaphysics came first, led the way, and actually made possible the subsequent discovery of the new theory. This happened when Einstein discovered general relativity. Einstein first hit upon the metaphysical idea that gravitation is due to the curvature of space-time, and then subsequently discovered how to capture this idea precisely in the field equations of general relativity. In stark contrast to the cases of Newtonian theory, Maxwellian classical electrodynamics and quantum theory, general relativity was discovered as a result of the prior development of new appropriate metaphysics, instead of the discovery of the new theory, if anything, being obstructed by current metaphysical ideas, the theory being misunderstood and misinterpreted by such ideas, once discovered. That Einstein's discovery of general relativity should stand out in this way is not, in my view, surprising: in Chapter 3 I argued that Einstein both put into practice, and upheld, a conception of science close to that of aim-oriented empiricism.

9
Science, reason, knowledge and wisdom: a critique of specialism

9.1 Introduction

In this chapter I argue for a kind of intellectual inquiry which has, as its basic aim, to help all of us to resolve rationally the most important problems that we encounter in our lives, problems that arise as we seek to discover and achieve that which is of value in life. Rational problem solving involves articulating our problems, proposing and criticizing possible solutions. It also involves breaking problems up into subordinate problems, creating a tradition of specialized problem solving – specialized scientific, academic inquiry, in other words. It is vital, however, that specialized academic problem solving be subordinated to discussion of our more fundamental problems of living. At present specialized academic inquiry is dissociated from problems of living – the sin of specialism, which I criticize.

I proceed by discussing two rival views about the nature of intellectual inquiry that I call *universalism* and *specialism*.[1] My claim is that at present the whole institutional structure of academic inquiry, by and large, presupposes specialism. Of the two views under consideration it is, however, universalism, and not specialism, which provides us with a rational conception of intellectual inquiry. Failure to put universalism into practice has profoundly damaging consequences for science and scholarship, and indeed for life, for our whole modern world. Ideally intellectual inquiry ought to help us to tackle rationally those problems of living which we encounter in seeking to discover and achieve that which is of value in life. Intellectual inquiry ought, in other words, to devote reason to the enhancement of wisdom (wisdom being defined here as the capacity to discover and achieve that which is of value in life, for oneself and others – wisdom thus including knowledge and understanding, but much else besides). In fact, at present, scientific, academic inquiry gives priority to the achievement of knowledge only, rather than to the achievement of wisdom. It is essentially the general adoption of specialism which is responsible for the persistence of this highly undesirable state of affairs.

9.2 Universalism

According to universalism, a proper, basic task of intellectual inquiry is to help us to improve our answers to four universal – or fundamental – questions, namely:

(1) What kind of world is this?
(2) How do we fit into the world and how did we come to be?
(3) What is of most value in life and how is it to be achieved?
(4) How can we help develop a better human world?[2]

In particular, according to universalism, it is a basic task of intellectual inquiry to help us to tackle these four fundamental problems in *a rational* fashion and, where many people are involved, in a *cooperatively rational* fashion. Rational problem solving is understood here to involve, at the very least, putting into practice the two rules or methods:

(a) Articulate, and seek to improve the articulation of, the problem to be solved.
(b) Propose and critically assess possible solutions.[3]

There is of course more to rational problem solving than this.[4] But these two rules are understood by universalism to constitute the nub of rationality.

Thus, according to universalism, the central and fundamental task of intellectual inquiry is to improve the articulation of the above four problems, and to propose and critically assess possible solutions to them. All other intellectual activity is subservient to this.

It deserves to be noted, in passing, that these four fundamental problems can be regarded as components of just one, even more fundamental problem – *the* fundamental problem of all thought and life: How can our human world, and the world of sentient life more generally, imbued with perceptual qualities, consciousness, free will, meaning and value, exist and best flourish, embedded as it is in the physical universe?[5]

It is important that the four basic problems are coalesced into this one fundamental problem, so that we are brought face to face with the horrendous issues that arise as to how all that we hold precious in life can conceivably exist and flourish if we really are just a part of the physical universe, everything we do and are being precisely determined by physics.

A basic idea of universalism is that ideally it is we ourselves who answer the above four questions, as we live. The proper task of reason, of thought, of intellectual inquiry, is to help us to arrive at answers that we really do wish to give to these questions, answers that do justice to what is true and genuinely of value, rather than to determine the answers for us. Intellectual inquiry is our servant, not our master. It is not in itself any kind of authority or oracle.

There are two further extremely important, elementary rules or methods of rational problem solving:

(c) Break up the basic problem to be solved into subordinate, specialized, easier-to-solve problems.
(d) Interconnect attempts to solve basic and specialized problems, so that the one may influence and be influenced by the other.[6]

According to universalism, an immense amount of intellectual activity arises, quite properly, as a result of putting these two heuristic rules into practice. That is, in order to improve our answers to our four basic problems, we create a vast network of sub-problems and preliminary problems – the specialized, technical problems of science and scholarship. A great deal of intellectual activity consists in seeking to solve these limited, technical problems of specialized scientific, academic disciplines. It is, however, of supreme importance – according to universalism – that we do not lose our way within this network, this maze, of sub-problems. If intellectual inquiry is to be rational, it is essential that intellectual priority be given to the four fundamental problems, and to the tasks of proposing and critically assessing possible solutions to them. In order to tackle specialized problems in a rational fashion, in short, it is essential to tackle such problems as sub-problems of the four fundamental problems. Specialized scientists and scholars, in other words, in order to be rational, must also be philosophers or generalists, concerned in their specialized work to help us solve our fundamental problems.

Figure 9.1 gives an indication of the way in which specialized academic disciplines may be conceived, in critical fundamentalist terms, as being designed to help us solve the above four basic problems. As the

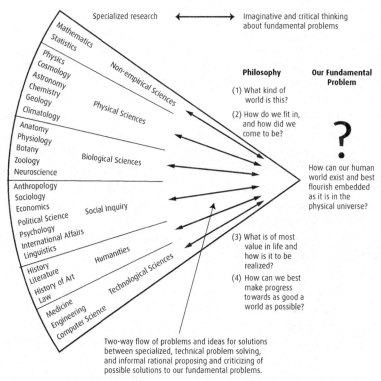

Figure 9.1 Specio-universalism, integrating specialized and fundamental problem solving (Source: author)

figure indicates, it is essential for the intellectual integrity and rationality of intellectual inquiry as a whole that there be a constant two-way flow of information between specialized problem solving and fundamental problem solving.

A major feature of specio-universalism, as depicted in Figure 9.1, is that philosophy is quite different from all the other disciplines of academic inquiry – physics, history, anthropology, sociology and the rest. It is not a specialized discipline at all. It has, as its primary task, to engage in rational, that is, imaginative and critical, thinking about our fundamental problems, intellectual and practical. Philosophy, so conceived, is not the exclusive preserve of trained, expert, academic philosophers; everyone is encouraged to do a bit of philosophy, to engage in a bit of thinking about fundamental problems – academics and non-academics, five-year-olds and ninety-five-year-olds. In so far as there are professional, academic philosophers, their basic job is to encourage non-philosophers, anyone and everyone, to do a bit of philosophical pondering. No professional qualifications are required to make a contribution to philosophy. All that matters is the quality of the contribution, not the academic status or qualifications of the contributor. Philosophy is open to everyone, but seeks to highlight the best of the contributions that have been made over time.

Philosophy differs from other academic disciplines, not only in being open to everyone (the qualified and the unqualified), not only in everyone being encouraged to participate, but also in its relationship to other disciplines, and to other aspects of life. In order to fulfil its vital role of ensuring that all four elementary rules of rational problem solving are implemented by academic inquiry, philosophy must ensure that there is a two-way interaction between specialized problem solving in all the specialized disciplines and the sustained attempts to solve the four fundamental problems of the whole endeavour. Philosophy, according to this view, constantly gives rise to new specialized problems, and is itself profoundly influenced by our success and failure in seeking to solve specialized problems. In so far as professional, academic philosophers do need to have a qualification, it is in order to have an expert knowledge and understanding of one or more specialized fields of study, so that they may actively participate in ensuring that fundamental and specialized problem solving interact with one another.

In the end, of course, it does not matter what we *call* those academics who actively ensure that academia keeps thinking about our fundamental problems, and thus implements all four basic rules of problem solving, as long as the job gets done. There are, however, historical grounds for holding that it is the proper job of philosophy to do it. Once upon a time, in the sixteenth to eighteenth centuries, philosophers instinctively sought to solve fundamental problems – and even tackled specialized problems of science and mathematics as well. This is true, for example, of Descartes and Leibniz; and thinkers we now regard as scientists, such as Galileo or Newton, did not hesitate to tackle fundamental philosophical problems. Since then, however, academic philosophy has dwindled in scope and significance until, in the twentieth century, it reached its nadir of specialized triviality in the form of Oxford, linguistic or analytic philosophy, or of bombastic, anti-scientific obscurity in the form of so-called "Continental" philosophy.[7]

If academic philosophy is to take on the role that speciouniversalism assigns to it, then it must be radically reformed. A revolution in philosophy is required. As it is, academic philosophy has become so much an esoteric, irrelevant speciality that even those all too aware of the failure of academia to keep alive thinking about fundamental problems that cut across all disciplinary boundaries never think to blame *philosophy* for this situation. And yet that is where the blame lies. Philosophy has abandoned its proper responsibility to keep alive, at the heart of academia, sustained, influential thinking about what our fundamental problems are, and how they are to be solved. The situation has become so bad that scarcely anyone thinks that that is what philosophy ought to be doing.[8] And the result is astonishingly damaging. It means that academia fails to put rules (a) and (b) into practice and, as a result, fails to put rule (d) into practice as well. Rule (c) is implemented splendidly. As we shall see, academia today is an intricate maze of ever more specialized research. But, disastrously, rules (a), (b) and (d) are violated. As a result of the retreat of philosophy into specialized triviality, *three* of the four most elementary rules of rational problem solving are violated in a wholesale, structural way, by academia today – and this has very damaging consequences for our world, as we shall see.

In one important respect, specio-universalism, as just characterized, needs to be qualified. It is quite wrong – it may be argued – to suggest that the enterprise of seeking to improve our answers to the above four fundamental questions is somehow exclusively the concern of intellectual inquiry. Literature, theatre, music, art, religion can all be interpreted as being concerned to illuminate our responses to these basic questions – especially the last two questions. Our whole culture can, in other words, be conceived of in universalist terms as being designed, ideally, to help us to discover and create that which is of most value in life. In engaging in our work, in social and political activity, we should ideally – it may be argued – be seeking to develop improved answers in practice to the last two questions, in one way or another. Indeed, in our whole way of life – in our way of being on this earth – we give implicitly our actual answers to such questions, whether we are aware of this or not. And in so far as we seek to improve our lives, we seek to improve the actual answers that we give to these questions, in the fabric of our actions. Universalism, in short, needs to be conceived as a philosophy of life, a social philosophy, a philosophy of culture: universalism interpreted as a philosophy of intellectual inquiry is simply a fragment of all this.

Universalism, as just characterized, may seem at first sight to be a somewhat autocratic, doctrinaire position, in that it seems to determine for us what our problems are and how they should be conceived. For this reason, it may at first sight seem unacceptable. For do not our problems – even our "universal" or fundamental problems – change, quite legitimately, from circumstance to circumstance, from person to person, from culture to culture? Can we really ever know for certain what our fundamental problems are, how they should be conceived?[9]

It will I hope become clear, as the argument unfolds, that my basic purpose in this chapter is to depict – and argue for – a kind of intellectual inquiry specifically designed to offer us maximum help with discovering for ourselves, whoever we may be, what our own unique problems of living are, how we are to conceive of them, and how we are to set about resolving them. My claim is that intellectual inquiry, so designed, is universalism. It is intellectual inquiry so designed that it has the kind of intellectual–institutional structure depicted in Figure 9.1 above, according to which, problems and their discussion are, as it were, hierarchically organized, with four vague, general, fundamental problems at the top, a maze of specific, restricted, precise, specialized problems at the bottom, and in between a continuous range of problems, more and less specific, interconnecting the top and the bottom by means of the relationship "Problem P_1 is more fundamental than problem P_2," or, equivalently, "Problem P_2 is subordinate to problem P_1."[10] A few universal, fundamental problems are needed so that we do not get lost in the maze of restricted, specialized problem solving. These fundamental problems must be formulated informally, imprecisely, without restricting specific presuppositions, just so that all people everywhere, in all societies, cultures and circumstances, can in principle interpret their own more or less specific, basic problems as specific versions or interpretations of the four fundamental problems, as formulated above. Only this can ensure that no one is excluded *a priori* from entering into rational inquiry by their own specific circumstances, view of the world or philosophy of life. In addition, we need discussion of more precise, restricted problems so that we can make progress with solving our problems, as a result of putting into practice the third and fourth of the above four rules of rational problem solving, (c) and (d).

Universalism needs to be implicit in the way in which our own personal thinking and problem solving is organized, so that we may have the best opportunity to understand and learn from others, even from those who think very differently from ourselves – learning from others being essential for the development of our own capacity to recognize and solve our own problems.[11] Universalism needs to be built into education, into the intellectual–institutional structure of scientific, academic inquiry, and generally, into our whole social, political, economic and cultural order, on a worldwide basis, so that learning, understanding and cooperation between people is given every opportunity to flourish.

There is nothing autocratic or doctrinaire in what I am advocating here, just because universalism amounts to a kind of intellectual inquiry, a way of thinking or problem solving which, when put into practice, gives us our maximum chances of discovering for ourselves what our own unique problems are and how they are to be solved, enabling us, ideally, to exploit for this purpose the very best thinking or problem solving that humanity has to offer. The autocratic and doctrinaire, the dogmatic, arise to the extent that *we fail* to put universalism into practice.

A central task of a kind of academic inquiry that puts universalism into practice is, then, actively to promote imaginative and critical – that is, rational – thinking about fundamental problems in personal and social life. What ultimately matters is the quality of the thinking we engage in as we live, guiding our personal and social actions and enabling us, at its

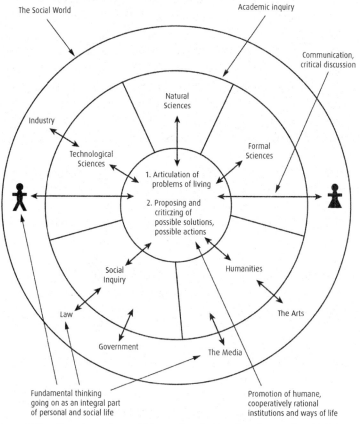

Figure 9.2 Specio-universalist academic inquiry devoted to helping people realize what is of value in life (Source: author)

best, and if all goes well, to resolve conflicts and problems in increasingly cooperative ways so that we may realize what is genuinely of value to us in life. Academic thought has, as its ultimate purpose, to promote personal and social thought guiding personal and social life. In order to highlight this aspect of specio-universalism, here, in Figure 9.2, is another depiction of the view, emphasizing that thinking at its most important and fundamental is the thinking we engage in as we live, academic thought having, as a basic task, to promote the excellence of personal and social thinking guiding personal and social life.

Critical fundamentalist intellectual inquiry can thus incorporate all possible conceptions of the world, all religious views, all philosophies of life, in all possible social and cultural milieux – all possible ways of conceiving of life's problems and how they should be tackled. There is just one proviso: all these diverse views and values, in being plugged into critical fundamentalist inquiry, as it were, must take note of the following basic points: many ways of conceiving of the world, life and its problems, exist and are possible; whoever we may be, our view as to what sort of world this is, and what is of most value in life, is guesswork; we have much to learn from others – especially by taking the achievements and failures, the views, values and arguments of others seriously, by ourselves engaging, with others, in critical fundamentalist inquiry, as we live. Sincere attention to the lives, views and values of others is desirable – and ought to be held to be desirable – within all viewpoints and value systems, since this is absolutely essential for mutual understanding in the world, mutual learning, mutual cooperation, peace, friendship and love. Much of the real richness in life comes from the good things that go on between people; and for these good things to happen, sincere attention to the lives, views and values of others – universalism built into the pattern of our lives, into the structure of society – is essential.

Universalism takes into account the point, stressed especially by Popper in *The Open Society and Its Enemies* (Popper, 1966a), that social, cultural pluralism or diversity is essential for the development of reason and science – the development of what Popper has called *critical rationalism*. I shall even argue, somewhat analogously to Popper, that rational inquiry can be understood as developing as a result of our departure from tribal life – from the human compactness and unity of tribal life. In sharp disagreement with Popper, however, I wish to argue that such things as mutual cooperation, mutual learning, understanding and communication can only flourish within social and cultural diversity if some kind of common unity can be discovered within this diversity. We must be able to agree at some level about what sort of world this is, and what is desirable and of value. Engaging in cooperative intellectual inquiry – the very act of participating in rational discussion – presupposes that it is at least possible to discover or create, at some level, common purposes and assumptions, an agreed framework, an agreed outlook on life and the world. This agreement must, however, accommodate equably the existing differences. It is in order to do justice to this requirement of unity in diversity – essential for cooperative rational discussion and inquiry – that universalism postulates or stipulates the above kind of hierarchical ordering of problems and their attempted resolution. The hierarchical structure of critical fundamentalist inquiry is precisely what we need if we are to discover or create, as readily as possible, just, equable agreement within disagreement, unity within diversity. Agreement can be sought at the fundamental level: disagreement and doubt can then be rationally explored at less fundamental levels, wherever it arises.

In his best epistemological, social and political thought, Popper is centrally concerned to attack authoritarianism, the dogmatic attitude. In *The Open Society and Its Enemies* (Popper, 1966a) this concern takes the form of a mighty onslaught upon those major figures in the history of Western thought who, in Popper's view, have failed to come to terms with the strains of civilization – the strain of living in an open, pluralistic society – and, as a result, have given way to romantic longings for the cohesion of the closed society, the tribal way of life. It is this longing, this potent false nostalgia for a golden past, which Popper argues has led even some of the greatest minds, with the best of intentions, to become the enemies of the open society, the enemies of democracy, reason and pluralism, and as a result, tragically, actually helping totalitarianism and fascism to grow, with all the consequent appalling human suffering of our history.

Popper's diagnosis is of fundamental importance. However, in the midst of his ferocious determination to establish once and for all the intellectual disreputability and appalling potential human destructiveness of views which value the tribal way of life, Popper neglects to consider the possibility that there is indeed much to value, potentially, in the cohesiveness of the tribal way of life which humanity – science, reason and civilization – cannot do without.[12] It is just this possibility that is affirmed here. I shall argue that our departure from the human compactness and unity of tribal life does indeed involve serious loss. Mere pluralism is not enough. It is essential that we develop a common unified view of the world and ourselves through cultural and social diversity if there is to be mutual learning, understanding, and cooperation through diversity – minimal requirements for reason and for civilization. Only universalism can do justice to these apparently conflicting requirements of unity and diversity. In our emergence from tribal life into the modern world a basic task confronting us is to create and develop unity within diversity: only by putting universalism into practice can we achieve this in a just, equable, genuinely rational and humane way. Popper's ideal of the "Open Society" needs to be replaced by the ideal advocated in this essay of the "Fundamental Society". It is precisely our failure to establish universalism on a worldwide basis that is responsible for so much suffering in modern times, and which indeed at present threatens us all. (I refer here to our present worldwide incapacity to cope with fundamental problems posed by such things as the population explosion; the continuing rapid depletion of vital, finite natural resources; widespread poverty and malnutrition in the developing world; global warming; the proliferation, indeed the mere existence, of nuclear weapons, which threatens to engulf us all in the nuclear holocaust. A critical fundamentalist world order is, almost by definition, a world order capable of recognizing its fundamental problems and, where possible, developing and putting into practice, in a cooperative fashion, just, humane, effective solutions.)

Popper's failure to recognize the vital need to create or develop a version of tribal unity within the diversity, complexity and sheer immensity of the modern world, in order to preserve and develop reason, mutual cooperation, humanity and civilization, is intimately connected with his analogous failure to recognize the vital role that fundamental unifying assumptions play in science, and in academic inquiry in general. Scientific, academic inquiry has basic presuppositions about what sort of world this is, and what is important or of value in social life, built into its whole intellectual–institutional structure, built into the priorities for research, built into its implicit methodology. According to universalism, these basic presuppositions need to be explicitly articulated and scrutinized – thus creating a tradition of discussion of presupposed solutions to *fundamental* problems – if scientific, academic inquiry is to be genuinely rational and rigorous, of maximum human value and use. Only by putting universalism into scientific, academic practice can we do justice to – and develop – the inherent rationality, the intellectually progressive character and the human value of the best of scientific, academic work and thought. As we shall see below, Popper fails to characterize adequately the rationality and progressive character of science – in that, for example, he fails to solve the problem of induction – just because he fails to do justice to the need for fundamental metaphysical and evaluative assumptions persisting through scientific revolutions, scientific diversity.

The Open Society and Its Enemies fails to characterize a genuinely rational society; *The Logic of Scientific Discovery* (Popper, 1959) fails to characterize a genuinely rational science: both failures are by-products of Popper's basic failure to articulate and advocate the hierarchical structure of universalism, so essential for genuinely rational, cooperative problem solving and inquiry in life as well as thought.

Having argued that we need to recognize, quite generally, that our thinking goes on in the world, presupposing a view of the world and a view of what is of value in life, I am of course eager to acknowledge that my advocacy of universalism in this chapter is intimately bound up with a view of what sort of world this is and what is of value in life – a broad, general, fundamental answer to the fundamental questions (1) and (3) above. As to the material universe, I hold a view not too dissimilar from the overall conception of the world implicit in much of modern science – a view of the world which does justice to the probable truth of Einstein's remark that "all our science, measured against reality, is primitive and childlike – and yet it is [one of] the most precious thing[s] we have" (quoted in Hoffman, 1973, p. viii). I recognize, of course, the intellectual legitimacy of conceptions of the world – such as animistic and religious views – very different from that of modern science. Critical fundamentalist inquiry acknowledges such rivals, and retreats to a more modest "common sense" view of the world, designed to be indifferent between these rival, explanatorily fundamental views, so that there may be a common, agreed base, in terms of which the merits of the rival explanatorily fundamental views may be discussed. (Universalism recognizes, in other words, that, in certain contexts, and for certain purposes, the epistemologically fundamental may differ from what is presumed to be ontologically and explanatorily fundamental.)

As to that which is of value, I hold that all that is of value in existence has to do with life, and especially, for us, with human life. Enjoyment in living; curiosity and wonder; perceptive awareness, understanding and appreciation of significant and beautiful aspects of the world; kindness, laughter, honesty, friendship, love, intimacy, cooperative creative work, personal responsibility, happiness, fulfilment: these are the kind of things that are of value. For each one of us, this short life is our only opportunity to discover, experience and take part in life of value; all too many people in the modern world – especially the developing world – lack this opportunity. We need to do all we can to change things so that all people everywhere have the opportunity to realize what is of most value in life. Value in the world has much to do with the diversity of life, the unique particularity of each individual life. A uniform world would be a world denuded of value. It is of the essence of value – it is essentially desirable and of value – that there be multiplicity and variety, amongst people, amongst ways of life, amongst societies and cultures. However, if this desirable variety is to flourish in this one crowded world, it is essential that we discover how to cooperate, to learn from and understand each other, in the midst of this variety. And there is a further point. As I have remarked above, much of what is of value in life comes directly from good things that go on between people – mutual understanding and appreciation, sharing, intimacy, cooperative creative work. Such interpersonal or social things, of value in themselves, only become possible in a world full of variety if there is cooperation, communication, learning and understanding amidst variety. Variety is only enriching in so far as there is understanding and learning between people amidst variety. It is to help facilitate all this that I advocate universalism (or critical fundamentalism). Universalism is put forward as a conception of learning and problem solving designed, above all, to help us resolve more adequately the third and fourth of the above fundamental problems.

Amongst other things, universalism amounts to a reply to social and cultural *relativism*. Like relativism, universalism acknowledges the existence and value of social, cultural and intellectual diversity. Unlike relativism, universalism recognizes that we all live in a common world in which we all have a real value, and that we all need to learn from one another so that mutual understanding and cooperation may flourish – so that what is of value in all our lives, potentially and actually, may flourish. The existence of a multiplicity of cultures need not prevent us from recognizing our common humanity, our common value, since it is at least possible for this multiplicity to be interlaced with and unified by a common acceptance of universalism.

Adoption of universalism is especially important for societies and cultures in the developing world. For in learning from the industrially advanced West – in acquiring the science, technology and industry of the West – there is always the grave danger that the indigenous culture and social order will simply be annihilated, as opposed to being helped to develop and flourish.[13] A developing society can only avoid this danger by articulating, at a fundamental level, basic presuppositions, values and problems of the society, so that it becomes possible to discover how to develop these presuppositions and values, to solve these problems, in the new social and cultural circumstances made possible by the importation of Western ideas and techniques. Only in this way can such a society employ these ideas and techniques discriminatingly, for its own best purposes, instead of becoming a hollow imitation of the Western way of life. In addition, of course, the industrially advanced West has a special responsibility, in its interactions with the developing world, to construe its own social and cultural order in universalist (or critical fundamentalist) terms. Only cooperative universalism can enable a mutually desirable kind of learning to go on in both directions.

Analogous considerations arise in connection with education. The most profound, instinctive and passionate fundamentalist thinkers are of course very young children, since all children must, as a practical necessity, arrive at working answers to the four fundamental questions in order to become human. If education is to develop, and not annihilate, instinctively fundamentalist childish thought, then education must itself be organized along universalist (or critical fundamentalist) lines.[14] Only those teachers who learn from their pupils really educate.[15]

To sum up: critical fundamentalist inquiry does justice to the Socratic and Kantian idea that reason forms a basis for the unity of mankind, in such a way as to encourage the flourishing of desirable kinds of diversity within this unity; it might be called "the tribal discussion of humanity". Of course, universalism cannot of itself vanquish tyranny, exploitation, manipulation, war, terrorism, crime. Universalism does, however, hold out the hope that if it is actively promoted in our personal, social, intellectual, economic and political lives wherever possible, then the spirit and practice of mutual cooperation between people may gradually grow, thus enabling us gradually to dismantle those social and cultural arrangements which tend to breed misunderstanding and mistrust, manipulation and exploitation, the use and abuse of power, the dreadful spiral of threat, counter-threat and violence.

So much for my preliminary exposition of universalism. I turn now to a consideration of the rival doctrine of *specialism*.

9.3 Specialism

For most scientists, scholars and educationalists today, specialism is a much more familiar doctrine than universalism: my exposition of specialism can therefore be much briefer. Specialism, unlike universalism, is almost exclusively a view of professional, expert, scientific, academic inquiry – even though this view, being embodied in so much present-day scientific, academic practice, has far-reaching consequences for all our personal, social lives. In complete contrast to universalism, specialism insists that only the specialized, technical problems of the various academic disciplines deserve serious intellectual attention. In order to be capable of serious scientific or scholarly treatment, in other words, a problem must satisfy certain conditions. It must be capable of being given an agreed, precise formulation. The problem must have an objective character, in that experts agree as to how the problem is to be formulated. The nature of the problem must not depend on such subjective, personal or idiosyncratic matters as mood, feelings, personal desires, attitudes or convictions. There must exist agreed procedures for tackling the problem. Above all, there must be general agreement as to what counts as the solution. It must be possible for the problem to receive a definitive solution. Academically respectable problems must, in short, have many of the characteristics of puzzles – chess or crossword puzzles, for example – as emphasized by Kuhn in connection with what he has called "normal" science (Kuhn, 1970a, ch. 4). Such problems arise quite essentially within the context of specialized disciplines, where there are agreed methods, results, assumptions and procedures. It is precisely by excluding all that is vague, ambiguous, controversial, metaphysical or philosophical that such academically respectable problems can be formulated or created. In order to be in a position to understand, solve and assess proposed solutions to such problems, one needs to be an expert, with specialized knowledge of the relevant discipline, its methods and results. It is not in general necessary to have broad intellectual or cultural sympathies and understanding. By and large, ignorance of social, political, religious, moral and philosophical issues lying beyond the scope of his discipline does not in any way hamper or disqualify the expert in his professional work. A "mere" expert or specialist can be as well equipped as anyone to make outstanding contributions to his discipline.

Experts can be in a position to pronounce authoritatively and definitively on matters that fall within the field of their specialized knowledge. In addition, *only* experts can be in a position to make such authoritative pronouncements: the rest of us cannot legitimately challenge or criticize expert judgements unless we too have specialized knowledge. Scientists and scholars are thus fully justified in ignoring criticism of their work and results by "outsiders", by those without expertise. The price that the expert pays, however, in being able to make unassailable, authoritative judgements, is that he must confine himself, *qua* expert, to delivering judgements that lie within the limited sphere of his professional competence – that small part of his discipline about which he does have expert knowledge. He must not in his capacity as expert make pronouncements about broad political, moral, religious and philosophical issues – the immensely complex human, social problems of real life – which, in their very nature, cannot be amenable to specialized, academic treatment.

Specialism may seem to represent an intolerably narrow-minded, dogmatic, scholastic conception of intellectual inquiry. All that is adventurous, imaginative, speculative, free-ranging and creative may seem to be excluded from science and scholarship. Those who defend specialism, however, usually do so in terms of the following kind of argument. It is precisely by eschewing consideration of imaginative, speculative, imponderable issues, and instead concentrating attention on much more limited, specialized "puzzles", capable of definitive solutions, that science and scholarship have made such giant steps forward in recent times. In the end, sustained attention paid to limited, technical problems pays dividends, and may even result in a definite solution to some "profound" philosophical problem. The problem of how the human race has come into existence has been discussed fruitlessly for centuries. Not until the work of Darwin was any real contribution made towards solving this "philosophical" problem. The crucial point about Darwin's contribution, however, so the argument goes, is that it arose out of painstaking attention to highly detailed, limited, specialized problems within zoology and botany.[16]

It is, I hope, obvious from the above that according to specialism, the four basic problems of universalism lie wholly outside the field of reputable science and scholarship. Inevitably these four problems are such that there can be no general agreement as to how they ought to be formulated, or what methods ought to be adopted in seeking to solve them. It is most improbable – perhaps even undesirable – that there should ever be general agreement as to what is to count as a correct, acceptable solution to any of these problems. And it is extremely unlikely that any of them will receive a definitive solution. The four basic problems of universalism satisfy none of the requirements which specialism demands of academically reputable problems. Thus, according to specialism, discussion of these four problems has no place at all within scientific, academic inquiry. Academic inquiry may perhaps produce work that has some bearing on the answers we give to the four basic questions, as in the case of Darwin's work. This comes about, however, as a result of aiming at solutions of exclusively specialist, technical problems. The four basic problems of universalism only have a place within academic inquiry at one remove, as it were, within anthropology, sociology or the history of ideas. A historian of ideas, for example, may quite legitimately discuss the writings of those who have speculated about such problems. Such a historian will, however, be concerned to solve specialized problems within his field, concerning the evolution of ideas. He will not concern himself with the fundamental problems as such – not if he is to continue to function as an intellectually reputable academic.

Extreme versions of specialism – such as logical positivism – condemn the four basic critical fundamentalist problems as metaphysical and evaluative, and therefore strictly meaningless. Less extreme versions of specialism merely place them outside the domain of intellectually respectable scientific, academic inquiry.

9.4 Universalism, specialism and intellectual standards

Universalism and specialism uphold diametrically opposed intellectual standards.

According to universalism, it is absolutely essential for the rationality, intellectual rigour and integrity of intellectual inquiry as a whole that sustained attention be given to the four basic problems. Indeed, this attention needs to be given intellectual priority over all else. All other intellectual activity needs to be subservient to the central and fundamental activity of imaginatively proposing and critically assessing possible answers to the four basic problems. Only in this case can even the most elementary of requirements for rational problem solving be realized.

According to specialism, on the other hand, rationality, intellectual rigour and integrity actually demand that the four "basic" problems of universalism be placed outside the domain of reputable intellectual inquiry. Mature science, authentic scholarship, genuine intellectual progress only really get underway when inconclusive philosophical debate about fundamental issues has been put firmly aside.[17]

One important aspect of this difference in intellectual standards is that universalism and specialism uphold different conceptions of intellectual progress.

According to universalism, intellectual progress is to be conceived in terms of the success that intellectual inquiry has in enabling us to improve our answers to the four fundamental problems, and to improve our capacity to tackle these problems in a rational fashion. One might say that universalism, ultimately, conceives of intellectual progress in personal and social terms – in that what is at issue are in fact the answers that people give to fundamental questions in their lives. Our assessment of intellectual progress will of course depend to some extent on the kind of tentative, broad answers that we give to these questions. Intellectual progress itself is no doubt something absolute and definite; our assessment of intellectual progress, however, is bound to be somewhat tentative, it being possible for there to be a number of different legitimate assessments.

According to specialism, on the other hand, intellectual progress is to be conceived in terms of the success that intellectual inquiry meets with in solving specialized, technical, scientific/academic problems. Progress – or the lack of it – is thus something definite, uncontroversial, something about which there can be general agreement. This is especially true for science. According to specialism, all scientific problems are essentially problems we encounter in seeking to predict more and more phenomena more and more accurately. Thus scientific progress is to be assessed simply in terms of the success we meet with in developing laws and theories which predict more and more phenomena more and more accurately.[18]

9.5 Specialism: its dominance and untenability

Actual scientific, academic inquiry, as it exists at present, and has existed during the last hundred years or so, amounts to an uneasy admixture of universalism and specialism. In many ways, however, specialism predominates.

It must of course be acknowledged that some aspects of scientific, academic inquiry do exemplify critical fundamentalist standards. For example, there can be no doubt that science, technology and scholarship have made great progress when viewed from a universalist (or critical fundamentalist) perspective. The special and general theories of relativity and quantum theory have changed profoundly our conception of the physical universe. The theory of evolution, and subsequent developments since Darwin's day, have done much to improve our understanding of how we fit into the world and have come to be. Our whole conception of the cosmos has been utterly transformed during this period. Technological research has done much, potentially and actually, to provide us with the means to create a better human world. Research in history, archaeology, anthropology – and more questionably, research in other social sciences and humanities – has deepened our understanding of ourselves, our past, our potentialities.

In addition to this there have been many noteworthy "universalist" or "critical fundamentalist" thinkers who have consciously sought to help solve one or other of the four fundamental problems. Almost at random, one might mention Albert Einstein, Sigmund Freud, Erwin Schrödinger, Arthur Eddington, Bertrand Russell, Alfred North Whitehead, Henri Poincaré, Carl Gustav Jung, Erich Fromm, Margaret Mead, Karl Popper, Carl Sagan, Ernst Friedrich Schumacher, Ivan Illich, Thomas Szasz, Friedrich Hayek, Arthur Koestler, George Orwell, Theodore Roszak, Jane Goodall, Thomas Nagel, Lynn Margulis, Roger Penrose, Mark Lynas, Ronald Higgins, and of course many others also of varying repute.[19]

In many ways, however, the influence of universalism on actual scientific, academic practice is submerged beneath the massive influence of specialism on all but a minute proportion of scientific, academic work. Most scientists and scholars are specialists, concerned only to solve specialist problems not consciously conceived of as sub-problems of the four fundamental problems. Almost all scientific, academic publications are concerned with the resolution of specialist problems. Education is shaped primarily by specialist assumptions and standards, especially towards the upper end of the educational ladder, culminating as it does in the extreme specialism of the PhD thesis. Academic appointments, academic honours, academic success, are all judged in terms of specialist standards – apart from quite exceptional cases.

Perhaps most crucially of all, the overall organization, the institutional structure, of scientific, academic inquiry exemplifies specialism rather than universalism. Universities are split up into relatively autonomous faculties: for example, faculties of physical sciences, biological sciences, technology, medicine, humanities or arts. Each faculty is subdivided into a number of relatively autonomous departments corresponding roughly to distinct academic disciplines. On the intellectual level, however, the subdivisions proceed further: each discipline is subdivided into a number of sub-disciplines; a specialist whose field of expertise lies within such a sub-discipline may not even be able to communicate properly – let alone share problems – with colleagues working within the same discipline. Such an expert will communicate almost exclusively with his fellow specialists scattered throughout the world – thus participating in what has been called an "invisible college" (Price, 1961).[20]

The striking point to note about all this is that nowhere is any provision made whatsoever for sustained, explicit, influential discussion of fundamental problems. This does not exist at the level of individual universities; nor does it exist at the level of published intellectual discussion, at the level of "invisible colleges".[21] Scientific, academic inquiry is, in other words, organized overwhelmingly in accordance with the intellectual standards of specialism. And recently, this drive towards more and more specialized research has become much, much worse as a result of the pressures academics are under to publish, or perish.

All this has dire intellectual consequences – especially, of course, if viewed from the perspective of universalism. The remorseless concern to solve exclusively specialist problems for their own sake; the proliferation of specialized disciplines (disciplines within disciplines, the autonomy of each jealously guarded); the accumulation of specialized results and vocabulary; increasingly specialized education (specialist indoctrination); the absence of informed, critical, non-technical discussion of fundamental issues – all these factors combine to make it overwhelmingly difficult for anyone to discover, understand and use the fundamentalist implications of specialized results. Intellectual inquiry becomes increasingly fragmented and incoherent, increasingly unusable from the standpoint of helping us to improve our answers to the four fundamental questions.

That over-specialization can have undesirable consequences has, it is true, been rather widely recognized. This scarcely amounts, however, to a recognition of the inadequacy of specialism. For if we look at what has been done in an attempt to compensate for fragmentation brought about by over-specialization, we find that new interdisciplinary subjects have been created, subjects such as biophysics, biochemistry, mathematical logic, industrial sociology. This typically specialist way of attempting to solve the problem actually, in many ways, serves only to make it worse. In seeking to facilitate communication between disciplines, additional buffer disciplines are created which only have the effect of further obstructing interdisciplinary communication. Thus even those who seek to combat some of the bad consequences of specialism can only adopt specialist methods in seeking to do so – so powerful a hold does specialism exercise over the academic mind – the end result being in consequence the exact opposite of what was originally intended. What cannot be done, of course, is what is needed most: the development of a tradition of influential, informal discussion of fundamental problems, feeding into, and being fed by, diverse specialist discussion. This obvious solution cannot be adopted for the simple reason that it involves violating specialist intellectual standards!

A further powerful indication of the increasing predominance of specialism over universalism is provided by the way in which academic philosophy has developed in recent times. Increasingly, academic philosophers have been concerned to develop philosophy as an academically respectable specialized discipline, with its own particular problems and methods, existing alongside other academic disciplines.

For the vast majority of academic philosophers, progress in philosophy is to be achieved by pinpointing and solving technical problems mainly conceived as problems of "conceptual confusion" requiring "conceptual analysis".[22] Universalism, of course, becomes quite impossible if "philosophy" is pursued in this specialized way. For universalism requires the existence of the Enlightenment conception of philosophy: philosophy conceived as the open, non-professional, unspecialized discussion of fundamental problems, influencing and being influenced by specialized problem solving in all other scientific, academic disciplines. In seeking to develop academically respectable, professional, specialized philosophy, academic philosophers have sabotaged almost all possibility of developing intellectual inquiry in critical fundamentalist directions.

Consider the following specialist account of the way in which intellectual inquiry has developed over the centuries.

Intellectual inquiry begins with myth, religion and philosophy. Originally, philosophy (or perhaps theology or metaphysics) is the queen of the sciences, other intellectual disciplines having only a highly subservient, specialized role to play within philosophy. This state of affairs exists in the thought of ancient Greece, in the thought of mediaeval Europe, and, to some extent, in the thought of seventeenth-century Europe during the so-called scientific revolution. For Kepler, Galileo, Bacon, Descartes, Newton, Spinoza and Leibniz, philosophy and theology represented the primary, central disciplines – so much so that science was known as "natural" or "experimental" philosophy. Gradually, however, successive disciplines emerged out of philosophy, dissociating themselves from the parent discipline of philosophy, intellectual success and progress being essentially bound up with this long process of dissociation. Over the centuries philosophy has given birth to the autonomous disciplines of mathematics, astronomy, physics, logic, biology, history, political science, sociology, psychology, cosmology, linguistics (the last three or four only having become autonomous in the twentieth century). As a result of having bred these autonomous disciplines, philosophy itself has been left in a highly impoverished state. The nature and status of philosophy, in other words, have changed dramatically. Instead of being the queen of the sciences, overarching all other sciences, philosophy has been transformed into a highly specialized, technical, somewhat meagre enterprise, concerned not with improving our knowledge and understanding of the world – for that is the business of the empirical sciences – but rather with clarifying concepts and solving conceptual problems. In line with the general trend, academic philosophy seeks to transform itself into a specialized discipline, dissociated from "philosophy" in the original sense of Plato, Spinoza, Leibniz, Diderot, Voltaire, Hume or Kant.[23]

It must be admitted, I think, that this specialist account of intellectual history does considerable justice to the way intellectual inquiry has *in fact* developed over the centuries. Furthermore, this account is today in practice widely upheld throughout the scientific, academic world as providing us with an adequate description of how intellectual inquiry *ought* to develop. Scientists and scholars have had something like this account in mind in pursuing and developing diverse disciplines. Above all, most contemporary academic philosophers take for granted the conception of modern philosophy that emerges from this account.[24] All of which provides a strong indication of the extent to which specialism has come to be built into the institutional framework of contemporary scientific, academic inquiry.

Universalism, of course, provides us with a quite different picture of how intellectual inquiry ought to develop. If intellectual inquiry begins with myth, religion, philosophy, metaphysics, this is because intellectual inquiry begins quite properly with a concern with the above four fundamental questions. Intellectual progress requires, of course, the development of specialized disciplines concerned to solve diverse subordinate and preliminary problems. It is of crucial importance, however, according to universalism, that this development occurs in such a way that we can, all the more readily, tackle the four fundamental questions in a rational fashion. The development of *autonomous* disciplines – the essential feature of the specialist account – violates the most elementary rules of rational problem solving.

None of the above, however, captures that feature of present-day scientific, academic inquiry which constitutes the most blatant and harmful institutional embodiment of specialism. This feature concerns, not so much the internal intellectual–institutional structure of scientific, academic inquiry, but rather the way in which scientific, academic inquiry is related to society, to life, and to the problem solving that goes on in all our personal and social lives. According to the version of universalism that I wish to defend, the basic task of professional scientific, academic inquiry is to help all of us to recognize and resolve rationally those problems we need to resolve in order to discover and achieve that which is most desirable and of value in life. The basic task of critical fundamentalist academic inquiry is to help us to put universalism into practice in our personal, social lives, and to help us to develop a social order, a world, in which cooperative rational resolving of our most important personal and social life-problems may receive every encouragement. For this goal to be realized, there must be a constant two-way flow of ideas and arguments between discussion of fundamental problems in society, as a part of life, and discussion of fundamental problems within professional scientific, academic inquiry. An intimate, two-way, rational relationship needs to exist between society and science, life and scholarship.

At present this vital rational sociocultural relationship scarcely exists anywhere. This is largely due to the prevalence of specialism, which *prohibits* the above rational social relationship. Specialism demands precisely that scientific, academic inquiry, in order to be intellectually rigorous, must be such that the intellectual domain of scientific, academic inquiry is decisively *dissociated* from the discussion of problems that goes on in society, as a part of life. Scientists and academics, upholding specialist intellectual standards, have done their utmost to develop and preserve this dissociation – in order, from their own standpoint, to preserve rigorous intellectual standards. As a result, the scientific, academic community has betrayed its most profound intellectual purpose (as seen from the perspective of universalism): to help us develop more rational, wiser ways of living, a more rational, wiser world. The result of this betrayal, not surprisingly, is that the production of specialist knowledge flourishes, while wisdom in life, worldwide wisdom, falters.[25]

Of the two views under consideration, it is universalism, and not specialism, which provides us with a rational, intellectually rigorous conception of intellectual inquiry.

In assessing the relative merits of the competing doctrines of universalism and specialism, it is vital to recognize that universalism fully acknowledges the immense value of – indeed the absolute necessity for – specialized scientific, academic work and thought. It is often only by putting into practice the two basic rules of rational problem solving (c) and (d), formulated above in section 9.2, that it is possible to make any headway with improving our solutions to our fundamental problems. Specialized problem solving, specialized scientific, academic work is absolutely essential, according to universalism, for rational problem solving in general.[26] And of course most of the time it is quite specific, highly specialized versions of our four fundamental problems that we need to solve in any case – made quite specific to a time and place, a person or group of persons, a specific need or trouble. The decisive additional point insisted on by universalism is that it is absolutely essential to put into practice rules (a) and (b) too. There must be a sustained rational discussion of our common, fundamental problems both within the scientific, academic community and within society, intimately interconnected with specialized scientific, academic problem solving, if intellectual inquiry is to serve our best interests in a genuinely rigorous, rational fashion. It is legitimate, even desirable, that many individual scientists and scholars be absorbed by the pursuit of highly restricted, specialized topics and problems. What is vital is that the overall intellectual–institutional structure of scientific, academic inquiry, and of society itself, accords with the kind of hierarchical structure required by universalism – sustained, explicit attention being given to fundamental problems. Failure to put into practice – to institutionalize – this vital critical fundamentalist perspective must inevitably lead to the fragmentation and trivialization of intellectual inquiry, and to a general incapacity to tackle cooperatively and effectively mankind's fundamental problems. The institutionalizing of specialism, however, obliges us to neglect the critical fundamentalist perspective. As a result we cease to tackle rationally just those problems it is most important for us to tackle rationally. While diverse sub-problems may be brilliantly tackled, our most general and important problems fall into neglect.

The motivation for insisting that it is of the essence of rationality to articulate our basic problems, and to propose and criticize possible solutions, is really very simple. If we do this, we give ourselves the best chance of seeking to solve those subordinate problems which are relevant to our main objectives. If we do not do this, the chances are that we will become engaged in seeking to solve sub-problems which are entirely misconceived or wholly irrelevant from the standpoint of achieving our basic objectives. Putting specialism into practice, in other words, is almost bound to lead to a mass of problem-solving activity which is misconceived or irrelevant from the standpoint of what matters most in life – a fair comment, I suggest, on a great deal of scientific, academic inquiry as pursued at present.[27]

What if no serious doubts really arose as to how we should answer the fundamental questions: What kind of world is this? How do we fit in? How have we come to be? What is of most value in life and how is it to be realized? How can we help develop a better human world? In that case universalism would be somewhat redundant. Serious doubts presumably would only arise in connection with much more specific, particular issues. But this is not our situation. The above questions are all profoundly problematic, even if many people appear not to recognize the fact. Our greatest uncertainties simply do arise in connection with our most general and important problems. This being the case, it is essential that we give due intellectual emphasis to the critical discussion of these problems, granted that we seek to develop a genuinely rational kind of intellectual inquiry.

Specialism is thus to be rejected, on the grounds that it provides us with a conception of intellectual inquiry that is both irrational and humanly undesirable, these two features indeed being intimately connected. Instead of prompting us to attend to what is most important and problematic, specialism does precisely the opposite!

The harmfulness of specialism does not lie in its tendency to encourage specialized puzzle solving. Universalism, too, insists on the vital importance of such puzzle solving. Nor need the harmfulness of specialism lie primarily in any tendency actively to suppress inquiry into fundamental problems. An upholder of specialism may simply see thought about fundamental problems as yet another specialized intellectual enterprise – grotesquely bankrupt intellectually, it is true, but scarcely deserving to be suppressed for all that. No, the real harmfulness of specialism arises from the fact that it appears to justify the pursuit of specialized problem solving divorced from the consideration of fundamental assumptions and problems. Worse, specialism holds that intellectual integrity and respectability actually demand that fundamental assumptions – vague, conjectural, controversial – be excluded from specialized inquiry. As a result, the adoption of specialism leads to the development of specialized inquiries, within a multitude of diverse disciplines, all of which become immune to elementary, outside, fundamental criticism.

This feature of specialism is responsible for such widespread intellectual corruption in present-day scientific, academic inquiry, that it deserves further comment. The key point that needs to be recognized is that it must always be irrational and undesirable to pursue specialized problems isolated from all consideration of fundamental problems. This is because the whole paraphernalia of specialized problem solving, as described above, actually requires us to give answers to fundamental problems. Choice of problems, formulation of problems, methods of attack, criteria for acceptable solutions, criteria for progress – all these essential features of specialized problem solving implicitly presuppose more or less broad answers to the four basic questions – answers all too likely to be more or less false or unacceptable and standing in need of improvement. If specialized puzzle solving cuts itself off from all critical consideration of fundamental issues (as specialism requires), then such puzzle solving becomes irrational in the straightforward and basic sense that implicit, influential and controversial assumptions are made which are permanently protected from critical assessment. Only by openly acknowledging the basic metaphysical and evaluative presuppositions implicit in specialized puzzle solving can such puzzle solving become genuinely rational.

It is above all the enormous success of *science* – conceived of in traditional empiricist terms – which has seemed to provide the most powerful case for specialism, and for the central assumption that specialized problem solving needs to be dissociated from fundamental assumptions and problems.

According to universalism, a basic task of science is to help us to improve our answers to the question "What kind of world is this?" Thus, according to universalism, a genuinely rational science, putting into practice the two most elementary rules of rational problem solving, gives intellectual priority to the task of proposing and criticizing answers to this question. Proposing and criticizing rival comprehensive metaphysical views about the nature of the universe, the nature of reality, constitutes, in other words, a central intellectual activity of a genuinely rational science.

Metaphysical assumptions at this level will influence drastically more restricted, specialized scientific problem solving – the kind of methods adopted, the kind of theories developed and tested. Thus if we believe ourselves to be in some kind of animistic universe – or in an Aristotelian universe – we will adopt different methods and develop different theories from those which we will adopt and develop if we hold, in Galileo's words, that "the book of Nature is written in the language of mathematics". The success of modern science, according to this standpoint, is due in large measure to the fortunate choice of a comprehensive metaphysical conception of Nature – shared by Kepler, Galileo and their successors – which sets the stage for a characteristic kind of specialized problem solving. According to this critical fundamentalist standpoint, then, science needs to be understood in terms of an interplay between fundamental and specialized problem solving, fundamental ideas and methods evolving with evolving specialized knowledge, this, in part, explaining the explosive growth of scientific knowledge. As our scientific knowledge improves, our knowledge about how to improve our knowledge – our *methods* – improves as well. All this illustrates the four rules of rational problem solving formulated above.[28]

Just this critical fundamentalist conception of science – exemplifying elementary rules of rational problem solving – is, however, rejected absolutely by almost all contemporary scientists and philosophers of science. For, according to traditional empiricist conceptions of science – almost universally taken for granted within the scientific community – it is the essential, defining characteristic of science that scientific theories are selected impartially with respect to empirical success, independently of their compatibility or incompatibility with comprehensive metaphysical assumptions about the nature of the world. Many, of course, acknowledge that simplicity considerations play an important role in the assessment of scientific theories in addition to empirical considerations (for example Mach, Duhem and Poincaré); the decisive point, however, is that biased preference for simple theories in science is not interpreted as committing science to the metaphysical, and possibly false, assumption that the universe itself is simple. According to this traditional empiricist standpoint, science is successful precisely because theories are selected impartially with respect to empirical considerations isolated from all *a priori* metaphysical assumptions about the nature of the world. This was one of Bacon's main points. (Descartes disagreed; but with the downfall of Cartesian science, and the success of Newtonian science, generally and wrongly held to incorporate Baconian inductivism, Cartesian universalism was rejected by the scientific community.) The diverse philosophies of science of inductivism (Bacon and Mill), conventionalism (Duhem and Poincaré) and logical empiricism (Carnap, Hempel and Nagel) all take for granted that in science theories are selected with respect to empirical success alone, unbiased by metaphysical assumptions about the nature of the universe as a whole. Even those thinkers who acknowledge the importance of *a priori* metaphysical ideas (Descartes, Spinoza, Leibniz, Kant) miss the essential point of the critical fundamentalist conception of science outlined above. For instead of emphasizing that our fundamental metaphysical ideas about the nature of the universe are *conjectures*, more or less bound to be false, and therefore needing constant critical scrutiny and development within science, these thinkers, on the contrary, seek to show, in one way or another, that fundamental metaphysical assumptions or principles can be conclusively established by reason, by argument. In effect empiricists and so-called "rationalists" agree on one main point: metaphysical principles, unverifiable by experience, have a legitimate place in science only if they can be conclusively established by reason. Rationalists defend the existence of such principles; empiricists, correctly, reject this possibility. Both parties miss the essential point: metaphysical principles play a decisive role in science; these principles are, however, conjectures, more or less bound to be false. Hence, if science is to be rational, it is essential that these principles be articulated, criticized and developed as an integral part of the scientific enterprise. Even Russell, it should be noted, misses this point. Russell (1948) recognizes that scientific method implicitly makes substantial metaphysical presuppositions about the world; he fails, however, to draw the critical fundamentalist conclusion from this, namely that a genuinely rational science seeks to improve its metaphysical presuppositions, and its methods, as it progresses.[29]

The point is decisively rejected even by Popper. Popper has many critical fundamentalist arguments and remarks to his credit. His book *The Open Society and Its Enemies* (1966a) tackles an issue central to universalism. Popper emphasizes that metaphysical ideas have often played a highly fruitful role in science (1959, pp. 19, 38 and 277–8). He has emphasized the importance of "metaphysical research programmes" for science, some science, in his view, even *amounting* to metaphysical research programmes (for example, in his view, the theory of natural selection) (1976a, pp. 148–51, and sections 33 and 37).[30] He has argued that metaphysical ideas can be assessed rationally, as more or less adequate, tentative solutions to problems (1963, pp. 193–200). He has stressed that intellectual inquiry needs to be organized, not in terms of subject matter and disciplines, but rather in terms of problems and attempts to solve problems (1963, pp. 66–7). He has emphasized that science at its best is cosmology – the attempt, in effect, to answer the question "What kind of world is this?" (1959, p. 15; 1963, p. 136). He has argued for philosophy conceived as a part of our attempt to improve our knowledge and understanding of the world, and against the view that philosophy is merely specialized "puzzle solving", or conceptual analysis (1963, pp. 67–96 and 136). Finally, he has explicitly condemned specialism. Thus, commenting on the attitude of mind of the normal scientist, as described by Kuhn, Popper remarks:

> I admit that this kind of attitude exists: and it exists not only among engineers, but among people trained as scientists. I can only say that I see a great danger in it and in the possibility of its becoming normal (just as I see a great danger in the increase of specialization, which also is an undeniable historical fact): a danger to science and, indeed, to our civilization. (Popper, 1970, p. 53)

Elsewhere, as I have already noted, he remarks:

> If the many, the specialists, gain the day, it will be the end of science as we know it – of great science. It will be a spiritual catastrophe comparable in its consequences to nuclear armament. (Popper, 1976b, p. 296)

Nevertheless, the central tenet of Popper's thought in effect lends strength to a mainstay of specialism, namely traditional empiricism. Much of

Popper's later writings elaborate and apply the main thesis of his first book *The Logic of Scientific Discovery* (1959). There Popper seeks to solve a problem central to traditional empiricism, namely how to demarcate science from metaphysics. Popper's solution, of course, is that a theory, in order to be scientific, must be experimentally falsifiable. A discipline, in order to be scientific, must assess theories solely with respect to empirical considerations, priority being given to those theories which have best survived severe testing and are most amenable to being severely tested. In other words, Popper, along with Bacon, Mill, Duhem, Hempel and others, is centrally concerned to drive a sharp and decisive wedge between the assessment of specialized, partial solutions to scientific problems (laws and theories) and the assessment of solutions to the fundamental problem of science, namely metaphysical answers to the question: What kind of world is this? In *Conjectures and Refutations* Popper makes the matter altogether explicit when he defends *"the principle of empiricism* which asserts that in science, only observation and experiment may decide upon the *acceptance* or *rejection* of scientific statements, including laws and theories" (1963, p. 54). Dramatically and decisively, Popper rejects the basic tenet of the critical fundamentalist conception of science, as outlined above.

However, as we have seen, and as I have argued at greater length elsewhere,[31] this "standard empiricist" viewpoint is unacceptable. The insolubility of the problem of induction as formulated, for example, by Popper, shows clearly that scientific laws and theories – solutions to specialized scientific problems – cannot be assessed solely with respect to empirical success, in an entirely impartial fashion. If we honestly attempted to select theories in this way, we would always be overwhelmed by a vast number of complex, empirically successful theories, and we would fail to select the theories we do actually select in science. In practice, in science selection of theories is permanently biased in the direction of simplicity and unity, even to the extent of overruling mere empirical success. This means that in scientific practice, whether we recognize it or not, we presuppose that the universe has some kind of underlying unified structure (or at least that it behaves as if it had such an underlying structure, to a high degree of approximation). In other words, science is only possible in so far as a more or less specific, tentative answer is given to the question "What kind of world is this?" Much of the success of modern science depends upon the aptness of this answer – so we may well judge. The answer is built into the whole methodology of science. In order to pursue science in a genuinely rational fashion, in a fashion which gives us the best hope of making real progress in improving our knowledge and understanding, we need to propose and criticize modified versions of our answer to the question "What kind of world is this?" as an integral part of science. We need to do this in an attempt to improve further the methods, and the success, of science. We need in short to put universalism into practice. Any attempt, like Popper's, to characterize science in terms of fixed methods which select theories solely with respect to empirical success (and the lack of it) must fail to solve the problem of induction – simply because science, so characterized, violates the two most basic rules of rational problem solving.[32] In addition, the vital capacity of science to develop improved methods with improving knowledge, such an essential feature of scientific progress, must inevitably be missed out.[33] Instead of holding speculation about the ultimate nature of the universe to be metaphysical, philosophical, and thus of questionable scientific status, if not downright unscientific or even meaningless, we need, rather, actively to pursue such speculation, imaginatively and critically, as an integral part of science itself. We need to put into practice the kind of critical fundamentalist way of doing science so brilliantly initiated and exploited by Einstein, in developing the special and general theories of relativity.[34]

That specialized scientific problem solving requires some kind of answer to be given to the question "What kind of world is this?" has been vividly and dramatically demonstrated by Kuhn in his book *The Structure of Scientific Revolutions* (1970a). Kuhn establishes convincingly that the "puzzle solving" of normal science depends upon the acceptance of a paradigm – in effect a *Weltanschauung,* a view of the world, for a given scientific discipline. One might well suppose that Kuhn, having realized this crucial point, this decisive objection to specialism, would go on to defend universalism, and the need for sustained development and criticism of "paradigms" as an integral part of science. Kuhn, of course, does exactly the opposite. Discussion of fundamental issues has, for Kuhn, no place within a "mature" science.[35] Furthermore, for Kuhn, changes of paradigm – scientific revolutions – inevitably involve a breakdown of rationality. Instead of emphasizing that rational assessment of paradigms is essential for the rationality of the whole of science – as universalism does – Kuhn, on the contrary, declares that choice of paradigm in general lies beyond the scope of reason (see Kuhn, 1970a, ch. 12). Kuhn, in short, is quite unable to conceive of non-specialist standards of rationality. *The Structure of Scientific Revolutions* (Kuhn, 1970a) brilliantly reveals the glaring defect of specialism, and yet, perversely, is itself a defence of specialism, of specialist intellectual standards. This provides yet another illustration of the powerful hold that specialism has over the academic mind – especially when one takes into account the great success of Kuhn's book in academic circles.

The profound irrationality of science as depicted by Kuhn in his book can perhaps be brought out by considering the following comparison. Our problem, let us suppose, is to wend our way through an obstacle-strewn path, from *A* to *B* (from ignorance to knowledge). Kuhn's advice is to proceed as follows. Standing at A, arrive at a general idea as to how to get to B (a paradigm); then, with head down – one might almost say with eyes shut – set off, sticking rigidly to this general idea. Even if you bump into a wall, fall into a ditch, or get tangled in brambles (anomalies), nevertheless adhere rigidly to your route (normal science). However, if you seem to have got into permanent difficulties (crisis), you may open your eyes, look around, and hit upon a new route (revolution), which, however, you must stick to as rigidly as before (new phase of normal science).

This blind blundering about may eventually bring you to your goal, *B*. It is hardly, however, the most intelligent, the most rational way to proceed.

A rather more sensible procedure is to keep one's eyes open, and continuously adjust one's route (paradigm) in the light of what one sees and learns on one's way from *A* to *B*. In order to pursue science intelligently and rationally, in other words, we need to reconsider, explicitly and persistently, our most fundamental paradigmatic ideas as an integral part of science. Instead of adhering blindly and dogmatically to some paradigm until our difficulties have become overwhelming and we are forced to reconsider, we need rather to attempt to improve our paradigm even before insoluble empirical problems overwhelm us, taking into account important *a priori* considerations such as simplicity, coherence, unity, intelligibility, comprehensiveness. This was the way Einstein developed the special and general theories of relativity; Einstein was much too intelligent, and much too interested in discovering the "thoughts of God", to follow Kuhn's advice.[36]

To sum up this part of the discussion, specialized scientific problem solving cannot proceed unless some kind of answer is given to the question "What kind of world is this?" This answer is almost bound to be more or less wrong, standing in need of improvement. Hence it needs explicit, sustained, critical discussion.[37] Specialized scientific problem solving dissociated from such critical fundamentalist discussion is irrational, as our glance at Popper's and Kuhn's work has shown.

Analogous considerations arise in connection with all other specialized academic disciplines, and in connection with the other three fundamental problems. Inevitably, in pursuing specialized lines of research, in history, for example, in literary criticism, anthropology, sociology, psychology, medical research or engineering, we presuppose some kind of rough and ready answer to one or other – or to all – of the four fundamental questions, this answer influencing our choice of problems, criteria for successful solutions, and so on. Since such implicit and influential answers are all too likely to be more or less inadequate, it is essential, for rationality, that these answers be explicitly articulated and critically assessed, as an integral part of specialized problem solving.[38]

In recent years a number of writers – so-called "externalist" historians of science and sociologists of knowledge – have argued in effect that specialized scientific, academic problem solving is substantially influenced by the social and cultural circumstances in which it proceeds. Material conditions, religious, political, moral and social ideals, human interests and values of one kind or another, all influence intellectual inquiry.[39] (This may be understood as a generalization of the Kuhnian point that specialized scientific problem solving is influenced by paradigmatic assumptions.) Specialism insists that such non-rational influences must be kept to a minimum, and must be excluded altogether when results are being assessed, if intellectual inquiry is to retain its rationality and objectivity. Universalism, on the other hand, insists that such influences must be openly acknowledged and critically scrutinized if intellectual inquiry is to be rational and objective. If our task is to discover what is of value in life, and to help develop a better human world, then of course our thinking must not be dissociated from our personal and social lives, from our material circumstances, our political, moral and religious ideals, our desires and values. A basic task of intellectual inquiry must be to promote more rational problem solving in life, thus gradually helping us to develop a more rational human world; intellectual inquiry must not merely seek to shield itself from the corrupting influences of an irrational society, as specialism would have it.

Most contemporary externalist historians of science and sociologists of knowledge would probably agree that the specialist programme of excluding social and cultural influences from intellectual inquiry cannot succeed, and is even perhaps incoherent. One might suppose that as a result of recognizing the general untenability of specialism, these writers would advocate and practise universalism. In fact, one finds nothing of the kind. Perversely, like Kuhn, these writers continue to accept and practise specialism – contributing to the highly specialized disciplines of history of science and sociology of knowledge. The main implication of their work is to undermine specialism; but if those who do this work do not themselves see this implication, how can anyone else be expected to see it? Once again we see the extraordinarily powerful hold that specialism has over the contemporary academic mind.[40]

Specialism, then, quite generally, must be rejected. All specialized problem solving dissociated from fundamental problem solving must be held to be seriously irrational.

This simple point has profound and far-reaching implications for the whole of scientific, academic inquiry, and for education. For we have seen that scientific, academic inquiry is on the whole at present organized, institutionalized, along specialist, rather than critical fundamentalist, lines. The urgently needed enterprise of discussing fundamental problems in an informal, informed, critical manner – in a manner capable of influencing, and being influenced by, specialized problem solving – is obstructed by the prevalence of irrational specialist intellectual standards.

9.6 Why does specialism prevail?

If universalism, and not specialism, provides us with a rational conception of intellectual inquiry, why is it that it is specialism which exercises the predominant influence over most actual scientific, academic work?

The question becomes all the more poignant when we realize how little is new or original in the critique of specialism offered here. Writing over *seventy* years ago now, Aldous Huxley said:

> Artistic creation and scientific research may be, and constantly are, used as devices for escaping from the responsibilities of life. They are proclaimed to be ends absolutely good in themselves – ends so admirable that those who pursue them are excused from bothering about anything else. This is particularly true of contemporary science. The mass of accumulated knowledge is so great that it is now impossible for any individual to have a thorough grasp of more than one small field of study. Meanwhile, no attempt is made to produce a comprehensive synthesis of the general results of scientific research. Our universities possess no chair of synthesis. All endowments, moreover, go to special subjects – and almost always to subjects which have no need of further endowment, such as physics, chemistry and mechanics. In our institutions of higher learning about ten times as much is spent on the natural sciences as on the sciences of man. All our efforts are directed, as usual, to producing improved means to unimproved ends. Meanwhile intensive

specialization tends to reduce each branch of science to a condition almost approaching meaninglessness. There are many men of science who are actually proud of this state of things. Specialized meaninglessness has come to be regarded, in certain circles, as a kind of hall-mark of true science. Those who attempt to relate the small particular results of specialization with human life as a whole and its relation to the universe at large are accused of being bad scientists, charlatans, self-advertisers. The people who make such accusations do so, of course, because they do not wish to take any responsibility for anything, but merely to retire to their cloistered laboratories, and there amuse themselves by performing delightfully interesting researches. Science and art are only too often a superior kind of dope, possessing this advantage over booze and morphia: that they can be indulged in with a good conscience and with the conviction that, in the process of indulging, one is leading the "higher life". (Huxley, 1938, pp. 276–7)

In fairness to Huxley – in order to excuse the mildness of his words here – we must remember how long ago this passage was written. Since that time, before the Second World War, everything that Huxley spoke of has of course become much worse.

How and why has this happened? In fact, of course, anyone who has sought to put universalism into practice, and who has explored specialized problems for the light they throw on fundamental problems, will have no difficulty in answering this question. Here, briefly, are seven factors responsible for the ever-increasing tyranny of specialism.

1. We fail to put universalism into practice primarily because, as Huxley points out, we fail to take up a measure of personal responsibility for the world in which we find ourselves. And we fail to take up such personal responsibility because of the enormous difficulties that we must inevitably encounter at present in seeking to do so.

These difficulties have arisen as a kind of unforeseen side-effect of the way in which our human world has evolved throughout recorded history. For consider the way in which the problem arises for those who live in the kind of "human world" experienced by people in prehistorical times – small, closely knit hunting and gathering tribes. In such circumstances, the difficulties that we experience in attempting to assume some personal responsibility for our world do not really arise. Adults, and even children, can without great difficulty assume some measure of personal responsibility for the welfare of the tribe as a whole. All the members of the tribe are known to each individual personally. Relationships of mutual interdependence are experienced daily, on a personal basis, in hunting, gathering food and so on. Obligations, responsibilities towards the tribe, can be experienced in a personal, emotional way, in terms of known individuals, in much the same way as we can experience responsibilities towards our family today. (Perhaps the modern family should be understood as a contraction of the prehistorical tribe.) All members of the tribe have a common outlook on things, a common cosmology and system of values. Thus barriers to intimacy, to mutual understanding, do not arise as a result of differences of outlook and values. Individuals do not face agonizing problems of deciding who they are, how they should live, what there is to give meaning and value to life. On the contrary, the meaning and value of life as lived by the tribe is assured, and is even beyond question, in that no alternative is conceivable. Finally, because of the relative smallness of the tribe, each individual makes a personal impact on the life of the tribe as a whole, and can be well aware of this impact. The tribe, as it were, acknowledges the existence, value and potency of the individual, and is clearly affected by the actions of the individual.[41]

Time passes; agriculture is invented; societies become bigger, more complex, specialized and diversified, requiring much more elaborate, fixed organization. Inter-tribal trade develops; tribes coalesce. Modern methods of travel, transport and communication develop. As a result, our tribe has become the whole human world, humanity, even, perhaps, life on earth in general.

As a result of these historical developments, the task of assuming some personal responsibility for our common human world has been transformed utterly, and has become almost inconceivably more difficult. Our task is not only to take on some responsibility for the welfare of those who are known intimately to us; rather, in addition, our task is to assume some responsibility – at least to some extent – for the welfare of millions upon millions of complete strangers. No doubt our own welfare is closely bound up with the lives, actions and welfare of many of these millions of strangers through international relationships such as trade; such relationships of mutual interdependence are, however, remote, abstract, not experienced daily on a personal basis. We cannot conceivably experience direct, emotional ties with these millions upon millions of strangers as we do with our friends and members of our own family. Millions of our fellow human beings live lives, see the world and have values in many ways very different from our own. Not only does this create barriers to mutual sympathy and understanding; responsible concern to understand others, to enter into their different worlds, must inevitably lead us to question the basic assumptions, practices and values of our own world. The immense diversity of ways of life, cultures, social systems, views of the world and systems of values with which we are confronted in considering our common human life on this planet must inevitably, at some level, plunge us into doubt and indecisiveness about how to live, what to choose, what to believe and value. And finally, when put into the context of the whole human world, our own life and actions must inevitably, and quite properly, seem to shrink almost to a vanishing point. Unless we possess quite exceptional personal power or influence – something that is perhaps inherently undesirable – all that we do with our lives will have almost no kind of impact or effect whatsoever on the human world as a whole. From this standpoint we are, individually, insignificant and impotent – which may not exactly encourage us to conceive of our world and ourselves from such a standpoint.

For all these reasons it is extraordinarily difficult for the individual today to assume some personal responsibility for our common world. In earlier times this failure did not perhaps matter so much since our power to bring about worldwide changes was strictly limited. Quite suddenly, however, we have developed the capacity to make drastic changes to our whole world. As a result, our common evasion of responsibility has become extremely dangerous for us all. Disasters result. War, starvation of millions, immense imbalances of wealth and power on a worldwide basis, the population explosion, reckless squandering of irreplaceable natural resources, international politics conducted like gang warfare, the widespread existence of brutal dictatorships, criminal psychopaths (like Hitler, Papa Doc, Amin and Assad) even seizing and holding power, the worldwide accumulation of armaments, the constant threat of the nuclear holocaust, and above all global warming – all these familiar worldwide dangers and disasters are the direct outcome of our general failure to assume personal, adult responsibility for our world.

The members of a small tribal society can, without great difficulty, confront and tackle common problems of the tribe, in a cooperative, responsible fashion. Tribal meetings can be convened at which everybody can be free to articulate problems, and propose and criticize possible solutions.

In our modern world this cannot be done. The population of the earth cannot hold a meeting to discuss common problems where everyone is free to speak. And yet something like this must exist if general understanding of, and responsibility towards, our common human problems is to develop at the personal level – something that we must develop if we are to be able to cope with the dangers and disasters just indicated. We cannot rely on existing institutions, existing centres of power, existing governments, whether democratic or dictatorial: all this is all too blatantly failing at present to cope adequately, that is, humanely and rationally, with our problems. In the end the point is very simple. In the absence of general understanding of, and responsibility towards, our problems, genuinely democratic governments responsive to public opinion will be unable to act responsibly as far as our most urgent, general, common problems are concerned.[42] Public opinion will not permit it. In a sense, only undemocratic, dictatorial governments, capable of suppressing or ignoring public opinion, will be able to act in such a fashion. Dictatorships, however, put us at the mercy of the decisions and actions of those few individuals who have won the fight for power (thus being, almost inevitably, ruthless and power-mad). Either way it is most unlikely that global problems will be tackled responsibly. For this we need a widespread, even worldwide understanding of, and responsible attitude towards, our basic problems at the personal level. And for this in turn it is essential that we develop a modern, worldwide institutional equivalent of the tribal meeting.

It is in this way, I suggest, that we need to conceive intellectual inquiry: as the open, sustained, responsible discussion of our common problems. Intellectual inquiry needs to be conceived and pursued as the tribal meeting of humanity, permanently in session, open to all, our joint endeavour to develop cooperative, personal responsibility for our common problems. *Something* must be created to replace the tribal meeting. Intellectual inquiry, at its best, constitutes such a replacement: it is from this standpoint that intellectual inquiry needs to be understood, contributed to and judged.

And only universalism can do justice to this conception of intellectual inquiry. This, indeed, *is* universalism: intellectual inquiry conceived as the outcome of our personal, cooperative, responsible attempts to improve our solutions to our fundamental problems.[43]

The difficulty we experience, then, in putting universalism into practice is an important part of the difficulties we experience in seeking to take on a degree of personal responsibility for our shared world. Specialism is, as Huxley correctly points out, an evasion of responsibility, the outcome of a failure to cope with the stress of responsibility. Specialism can even be seen as the outcome of a kind of intellectual or professional tribalism, the specialist's tribe being the "invisible college" of like-minded specialists.

A number of writers have been concerned to emphasize – in terms somewhat analogous to those outlined here – that the blessings resulting from moving from the intimate, coherent tribe to the big, complex, diversified modern world are mixed. These writers all emphasize, in one way or another, that this transition makes possible the development of choice, freedom, reason, science, on the one hand, but can also lead to uncertainty, fear, loneliness, a sense of meaninglessness and impotence, on the other hand.

Thus in *Coming of Age in Samoa* Margaret Mead (1943) tells us that children in Samoa fail to experience anything like the trauma of adolescence so familiar in Western society. She concludes that this is due to the absence in Samoa of the problem of choosing between rival ways of life and values. Adolescent trauma, then, is due to the great difficulties that we experience in coming to terms with cultural diversity in our society – in turn due, without doubt, to a general failure of our culture to cope adequately with this central problem of diversity. As I have already indicated, in *The Open Society and Its Enemies* Popper (1966a) argues that the open society – the society in which diverse ways of life are tolerated – is essential to our humanity, our reason, our civilization. It is only with the existence of social diversity that we can begin to doubt, to criticize, to learn, and perhaps to make progress. In Popper's view, rationality is to be understood primarily in terms of the capacity to doubt, to criticize and thus to learn; criticism, however, is only really possible if a plurality of views and ways of life coexist in society. Thus, for Popper, rationality is to be understood primarily in social terms, arising as a result of social developments – the development of social and cultural diversity, and a tradition of criticism.[44] The development of the open society makes possible the development of both freedom and reason. Popper is at pains to emphasize, however, the price we pay for these developments, the strain that civilization puts upon us. It is indeed a major thesis of *The Open Society and Its Enemies* that the uncertainties, the emotional stress, created by our movement towards the open society, can be so great that we long passionately for a return to the simplicities and certainties of the monolithic closed society. This anti-rational, anti-humanitarian longing is responsible for the totalitarianism of both left and right. The difficulties that confront us in coming to terms with the open society are indeed, according to Popper, so extreme that even many of our greatest thinkers in the past have failed to surmount them: Heraclitus, Plato, Aristotle, Hegel and Marx all in one way or another, in Popper's view, sought to return us to the closed society. Many of our greatest philosophers and rationalists have been *enemies* of the open society.[45]

It is scarcely surprising, then, that adolescents, emerging from the "closed" society of the family into our quasi-"open" society, should experience difficulties. The problems of adolescence need to be understood in philosophical or rationalistic terms, in terms of emotional reactions to an intense awareness of possibilities and uncertainties, and not merely in terms of some psychological theory of emotional development.

Isaac Bashevis Singer, in his novels and short stories, has given us a wonderfully vivid and perceptive account of the enormous difficulties we encounter in emerging from a closed society.[46] In *The Manor* (1965a) and *The Estate* (1975b) Singer provides us with a wholly convincing picture of the confusion, the sense of loss, that overwhelmed those enlightened Jews who, towards the end of the nineteenth century, emerged from the highly traditional, almost mediaeval, Jewish communities still existing then in Poland. Singer's writings are especially noteworthy for the fact that many of his protagonists are themselves deeply conscious of the problem, and not merely affected emotionally by it without any understanding of its nature. Singer is concerned to show us, in a fictional form, individuals grappling passionately with the task of pursuing critical fundamentalist intellectual inquiry. Singer's vivid and honest imagination takes us to the heart of the problems of our civilization.

Essentially the same problems – explored by both Popper and Singer – have also been discussed by Erich Fromm, for example, in his *The Fear of Freedom* (1960; see also Fromm, 1963). Finally, Peter Gay, in his marvellous book *The Enlightenment: An Interpretation*, provides us with a haunting account of the anguish experienced by the thinkers of the Enlightenment in attempting to come to terms with their doubts, their scepticism, as they emerged from the religious tribalism of contemporary Christianity (Gay, 1973, vol. 1, pp. 59–71).

It is, I hope, clear that all these writers are concerned essentially with the same problem: the difficulties we encounter in coming to terms with something that is essentially desirable, namely social and cultural diversity. One disastrous consequence of specialism is that it disrupts understanding of problems as fundamental as this; the problem is scattered amongst a number of disparate disciplines, and lost sight of. Instead of discussion being organized around the problem, so that contributions such as those of the above writers can fruitfully interact with, and supplement, each other, discussion is organized instead within the disciplines: anthropology, epistemology, political philosophy, history, psychology, history of ideas, fiction. As a result, we fail to discover the interconnections between the contributions; we fail to improve our understanding of the underlying problem. We fail to understand the problems of adolescence as those of moving from a closed to an open society – in part philosophical problems. We fail to appreciate the social, cultural and personal implications of Popper's philosophical and epistemological discussions. We fail to grasp the universal significance of Singer's fiction. We do not see that Popper and Singer are concerned with essentially the same problem. Fromm may be dismissed as pursuing the pseudoscientific discipline of sociocultural-psychoanalytic psychology, instead of being understood as contributing to our understanding of the problems discussed in *The Open Society and Its Enemies*, and in Feyerabend's "Problems of Empiricism" (1965).

Specialism thus prevents us from *seeing* our fundamental problems. As a result, we fail to see the urgent need to improve our thinking at this level, and the considerable difficulties that arise in connection with this task.

2. It is in the nature of universalism to raise questions and doubts that can be highly awkward for those who wield power in society. In particular, of course, universalism challenges all those who claim to have authoritative answers to fundamental problems – religious and secular centres of power and influence in society. Universalism calls into question cherished beliefs and values, and thus also is liable to collide with public opinion. Powerful social forces, then, will inevitably discourage the development of critical fundamentalist intellectual inquiry – as Socrates, Galileo and Spinoza, for example, found out. Only a society which had, quite generally, taken universalism to heart would encourage the development of critical fundamentalist intellectual inquiry; but of course no such society has as yet come into existence.

The case of specialism is, however, quite different. Specialist scientists and scholars may well be quite content to let non-academic authorities decide fundamental issues, scientific, academic inquiry confining itself to solving those specialized, technical problems whose solutions are required by those who wield power in society. Critical fundamentalist issues in any case lie beyond the reach of specialist intellectual standards and concerns. Specialism thus robs the scientist and scholar of the capacity, from a professional standpoint, to criticize fundamentalist decisions made on the basis of power in society – except where those in authority are foolish enough to transgress specialist standards and results.

3. Specialism is especially appealing to those who uphold what may be called "oracular" conceptions of reason – according to which, reason, ideally, is something that reaches decisions authoritatively for us,

rather than being something which helps us to decide.[47] Any academic who upholds such an oracular conception of reason (or method) quite clearly cannot pursue inquiry in accordance with universalism. For that would involve the appalling prospect of academic professors deciding authoritatively for the rest of us how fundamental questions are to be answered, how we should live, what is of value in life, what our life problems are and how they are to be resolved. This is a modern version of Plato's vision of the dictatorship of the philosopher-king. If one is to hold on to an oracular conception of reason or method (which includes of course appeals to evidence), without incurring the charge of becoming a Platonic philosopher-king, a prophetic dictator, it becomes necessary to restrict the field of one's research to some specialized, limited, factual, value-neutral domain over which, a bit more plausibly, one may seek to claim authoritative expertise. One must become, in other words, a specialist, and implement specialism.

The proper response to this is, of course, to reject all such oracular conceptions of method (whether empiricist or rationalist) in all fields, in the contexts of both specialized and universalist inquiry. Even the most secure results of specialized research ought not to be held to be so secure that they are inherently immune to all possibility of being criticized usefully by non-specialists. And when it comes to exploration of fundamental questions, there may well be some individuals who deserve greater attention than others, and who are, to that extent, experts, but such individuals cannot conceivably be held to speak authoritatively for the rest of us about fundamental issues, as if Platonic prophets or popes whose pronouncements must be taken on trust.

Specialists can claim to reach their results by employing such an oracular conception of reason (which includes, of course, appeals to evidence), and on that basis, can claim full authoritative status for their pronouncements, immune to the sceptical doubts of the ignorant. Of course, even specialists should not think that the results of their specialist research is immune, in an *a priori* way as it were, from non-specialist criticism. Oracular conceptions of reason (of method), which claim to bequeath indubitability to results obtained by their means, need to be rejected in all fields, including specialized fields of research (even mathematics, where the boldest claims for certainty are made). It is, of course, crucial that such oracular conceptions of reason are rejected when it comes to universalism – otherwise we would have the appalling prospect of professors deciding authoritatively for the rest of us how we should live, what is of value in life, what our life problems are and how they are to be resolved. This is Plato's vision of the dictatorship of the philosopher-king.

Even if we do identify reason with some set of rules, laws, methods or criteria which, in some restricted field perhaps, dictate decisions to us, we ought always to be aware that it is our own decision to adopt these laws, methods, etc. Genuine rationality involves being able to choose and develop such laws to suit our purposes. Universalism is correct in insisting that genuine rationality involves recognizing that ultimately *we* choose and decide. And choose and decide fallibly.

4. According to specialism, the expert is entirely entitled to pronounce authoritatively on matters relating to his discipline, in a manner which ignores the contributions, the criticisms, of non-experts. This is because, according to specialism, only specialized considerations can be relevant for an assessment of specialized results. Only the expert can be competent to contribute to a specialized discipline. There can be no doubt that being able to pronounce authoritatively in this kind of way is something that is deeply appealing to many. Universalism, however, deprives the expert of this deeply appealing authoritative immunity from outside criticism. Basic assumptions about the nature of the world, and about the meaning and value of life, must inevitably, according to universalism, pervade specialist work. It cannot be correct for experts to decide for the rest of us what these assumptions should be. It is thus entirely proper that non-experts should be able to challenge and contribute to critical fundamentalist assumptions implicit in specialized work. It is indeed important that experts do listen to non-expert comments and criticisms concerning fundamental assumptions, since it is all too easy for the expert to forget the prevalence and influence of such assumptions amidst his technical work – losing sight of the wood for the trees.

There is, of course, a very serious problem here, which confronts academic inquiry pursued along the lines of universalism. It is the problem of distinguishing between honest, potentially significant and fruitful criticism of academic work, and criticism that is dishonest, dogmatic, pathologically ignorant, merely destructive, and even possibly in the pay of corporations and other interests out to discredit authentic results, as when special interests seek to cast doubt on the reality of human-induced climate change to protect profits from oil and coal. Legitimate, authentic non-specialist comment and criticism must be distinguished from illegitimate, inauthentic comment and criticism. And academia must discover how to maintain this all-important and highly problematic distinction, in order to preserve and enhance the intellectual integrity of academic thought.

5. Increasingly, during the last fifty to one hundred years, scientific, academic work has become something that is engaged in as a profession, a career, rather than out of amateur love. The scientific, academic enterprise has become increasingly institutional and bureaucratic in character. All this favours, and almost requires, specialism. For these factors require that scientific, academic work can be assessed in a definite, agreed way, sound work being distinguishable from unsound work in an uncontroversial manner. Promotions, funding of research work, professional status, management of research – these career and institutional matters all favour the adoption of definite, agreed specialist intellectual standards. Sustained inquiry into fundamental problems is much more difficult to professionalize and institutionalize. Crucial institutional questions such as whose work is to receive funds, to be taught, to be rewarded with promotion and academic honours, become almost impossible to decide in a standard, bureaucratic manner.[48]

6. Once a conception of intellectual inquiry has become established – built into the institutional and bureaucratic structure of intellectual inquiry – all sorts of mechanisms tend to preserve this institutionalized conception. Education will tend to indoctrinate pupils and students in this conception. Only those who conform to the standards of the conception will be able to do research work, publish, obtain academic jobs. Only that work which conforms to the accepted standards will be published, and will be accepted on publication. Even those who disagree with the institutionalized viewpoint will be obliged to pay lip service to it, simply in order to teach, publish and do research. As a result, the public face of scientific, academic inquiry will come overwhelmingly to conform to the general viewpoint, and it will seem increasingly absurd to call this viewpoint into question. Once specialism is established institutionally, in short, no problem arises as to why this viewpoint should persist.

7. Specialism receives support from all those scientists and scholars keen to promote their own careers. An excellent way to ensure that one's career flourishes is to found, develop and contribute to a new discipline, a new sub-discipline or speciality. This can be done by founding, with others, a new journal devoted to the new specialized field of study; it can be done by establishing research groups devoted to the new speciality, acquiring students who seek to gain PhDs in the discipline. With others,

experts in the new discipline, one can hope to win grants for research; one may even be able to create new departments or research institutions devoted to the discipline. For academic success, creating the *impression* of authentic scientific or scholarly activity may be more important than producing work that is of genuine intellectual value. In short, alongside authentic intellectual reasons for creating specialized fields of research, there are also merely personal, social or institutional reasons having to do with the desire, or the need, to make a success of one's career.

These, then, are some of the factors responsible for the failure to put universalism into practice – responsible for a pervasive corruption of intellectual standards.

9.7 Universalism, knowledge and wisdom

If universalism were to be put into practice we would expect intellectual inquiry as a whole to give priority to our most general and important problems – specialized problems being chosen and tackled in order to help us solve the former.

The result to be expected from putting specialism into practice is, however, the exact opposite. Although specialized, technical problems may well be tackled with brilliance and great success, from the standpoint of what matters most in life the vast industry of specialized problem solving may well seem largely irrelevant. Most specialized problem solving will be unrelated to our fundamental problems. Specialized problems will not be understood or tackled as subordinate problems to fundamental problems. Instead of illuminating our understanding of how fundamental problems may be solved, intellectual inquiry will tend to do the exact opposite. We will tend to be overwhelmed by a vast maze of specialized disciplines, jargon and results. It is not just intellectual inquiry as a whole that will suffer as a result. *We* will suffer.

Our capacity to think and act intelligently, in response to our basic problems, will be sabotaged. Experts will become, not our servants, but our masters.

Some years ago, in his Nobel Peace Prize lecture, Martin Luther King declared:

> Modern man has brought this whole world to an awe-inspiring threshold of the future. He has reached new and astonishing peaks

of scientific success. He has produced machines that think and instruments that peer into the unfathomable ranges of interstellar space. He has built gigantic bridges to span the seas and gargantuan buildings to kiss the skies. His airplanes and spaceships have dwarfed distance, placed time in chains, and carved highways through the stratosphere. This is a dazzling picture of modern man's scientific and technological progress.

Yet, in spite of these spectacular strides in science and technology, and still unlimited ones to come, something basic is missing. There is a sort of poverty of the spirit which stands in glaring contrast to our scientific and technological abundance. The richer we have become materially, the poorer we become morally and spiritually. We have learned to fly the air like birds and swim the sea like fish, but we have not learned the simple art of living together as brothers. (In Haberman, 1972, pp. 333–4)

The predominance of specialist intellectual inquiry plays its part, I suggest, in the development of the "glaring contrast" to which Martin Luther King here refers: the achievement of specialist knowledge at the expense of the achievement of wisdom.

Consider the following analogy. Our problem, let us suppose, is to build a house. On the one hand, we may tackle this problem in a critical fundamentalist manner. We propose and criticize possible solutions to our basic problem – thus developing an overall plan. In order to solve our basic problem, however, a host of specialized, technical, subordinate problems need to be solved. Bricks need to be made; so, too, slates, doors, window frames, windows, beams, plaster, floorboards, and so on. Foundations need to be dug and cemented. All the various parts need to be assembled properly, in conformity with the plan, to build the house. Plumbers need to put in pipes, tanks, sinks, a bath; electricians need to wire the house; and so on. An intricate maze of highly specialized, technical problems need to be solved by an army of experts if the house is to be built. Equally, however, if the house is to be built, it is absolutely essential that the specialized problem solving be properly coordinated so that it all gives rise to a solution to the fundamental problem – to build a house. There needs to be a constant two-way flow of information between problem solving at the fundamental level, and at the specialized level. Failure to solve certain specialized problems may necessitate a revision of the basic plan.

This common-sense, critical fundamentalist approach is in complete contrast to a specialist approach. According to specialism, building a house only involves solving specialized, technical problems. The fundamental problem – what kind of house do we want? – is not a problem that the building trade can take seriously. (It is meaningless, subjective, incapable of being decisively solved, philosophical, evaluative or religious; in any case, not a matter for the trade to concern itself with professionally.) The building trade needs to concern itself with specialized, technical "puzzles" – manufacturing bricks, mortar, cement, slates, floorboards, windows, wiring, pipes, and so on. Progress in the building trade is to be judged in terms of how well these specialized puzzles are being solved.

The outcome of all this will of course be ever increasing piles of completely unusable bricks, slates, wire, pipes, etc. – and no house will be built at all. And if we complain, we will no doubt be met with indignation in that each specialist has indeed performed his task with skill and expertise.

The "house" that intellectual inquiry as a whole should help us build is, I suggest, a life of value – a rich and fulfilling life, a life in which we can share friendship, love, happiness, beauty, creative work, joy in being alive. Our "fundamental" problems are the problems we encounter in our lives in seeking to discover, experience, participate in and help create that which is of value. The basic rationale for the whole of intellectual inquiry is to help us to articulate and solve these fundamental problems of living. All intellectual problems are subordinate to these fundamental personal and interpersonal problems of living. The problems of mathematics, logic, philosophy, theoretical physics, cosmology, molecular biology, neurology – all these need to be understood as sub-problems of our fundamental personal and interpersonal problems of living.

It is, of course, not the case that intellectual inquiry is pursued only for pragmatic reasons, as a means to the realization of non-intellectual, practical ends. Intellectual inquiry is also pursued for its own sake. Intellectual inquiry is, in other words, itself a part of life, enriching life directly when pursued for its own sake, like music or poetry. It is, for example, of the essence of life of value that we are perceptive and curious about our surroundings – in touch with our environment. "Pure" research in physics, say, or cosmology, geology, history or anthropology, amounts simply to a cooperative following-up of such personal perceptiveness and curiosity. From the standpoint of pure intellectual inquiry, it is the curiosity, the imaginative explorations, the thoughts and feelings, the knowledge and understanding, the intellectual honesty and passion, the problem solving, of people in society as a part of life, that really matters. It is our shared exploration of our world, as an aspect of life of value, that is important. It is the personal knowledge and understanding of our world that we have ourselves developed, integrated into our lives, that really matters. Pure intellectual inquiry is, in other words, at the most fundamental level, personal and interpersonal in character, a part of life. The impersonal or institutional aspects of pure intellectual inquiry exist simply as a means to an end: to aid personal and interpersonal curiosity, wonder, knowledge and understanding, as a vital aspect of our personal and social lives. Thus both "pure" and "pragmatic" intellectual inquiry seek to contribute to the richness, the value, of our shared lives here on earth. In both cases, what ultimately matters is the value of our personal and interpersonal lives.

The fundamental aim of intellectual inquiry, we may say, is to enhance our personal and interpersonal wisdom – our capacity to discover and achieve what is of value in life both for ourselves and for others. All intellectual problems are problems subordinate to our basic life-problems of wisdom. Of the four fundamental problems formulated above, it is the third and fourth that are the most fundamental, the first and second being pursued as a part of our concern to discover and achieve what is of value in life.

If intellectual inquiry is to meet with success in helping us to discover and achieve what is of value in life, then it must of course be generally understood to have this basic purpose. Education must enable us to come to understand and use intellectual inquiry in this kind of way,[49] so that we discover fruitful interconnections between our personal problems and "impersonal", "objective" intellectual problems, our own personal, childish wonderings about the nature of the universe, for example, illuminating and being illuminated by the "official" wonderings of Kepler, Newton, Faraday or Einstein, or our personal problems of adolescence illuminating and being illuminated by the philosophical, social problems of the open society discussed, for example, by Popper, Fromm, Mead and Singer. Intellectual inquiry must itself be organized in such a way as to be amenable to this kind of understanding and use. Above all, scientists and scholars must be fundamentally concerned to develop intellectual inquiry in such a way that it is designed to help us build our "houses" of wisdom with our lives. All this is essential if intellectual inquiry is to be developed as the tribal discussion of humanity, designed to help us create more valuable lives, a better human world.

When viewed from this perspective of the philosophy of wisdom,[50] what present-day scientific, academic inquiry produces is, in terms of our analogy, more like an unusable, chaotic heap of bricks, slates, window frames and pipes, than something out of which we can build a habitable house. What confronts us is an immense pile of specialized jargon-ridden disciplines pursuing specialized intellectual problems dissociated from our problems of living, there being little indication as to how the non-specialist is to find his way through all this to discover and achieve what is of most value in life. Scientific, academic inquiry is not pursued, understood, taught or organized in accordance with the overall assumption that what ultimately matters is personal and social wisdom.

In so far as a basic organizing assumption is built into present-day scientific, academic inquiry, it is that the aim of such inquiry is to improve objective, impersonal, institutional *knowledge*, not personal and social *wisdom*. Intellectual priority is not given to our problems of living, to the difficulties, frustrations and sufferings that we encounter in our lives in attempting to discover and achieve what is of value in life: on the contrary, intellectual priority is given to impersonal problems of knowledge encountered by the various academic disciplines in seeking to describe, predict and explain phenomena. Even the social sciences give intellectual priority to problems of knowledge as they arise within sociology, psychology, and so on, rather than to the problems encountered by people in their lives. Intellectual progress is assessed, not in terms of the success we meet with in achieving what is of value in life, but rather in terms of the success achieved in acquiring academic knowledge. Intellectual progress is conceived as being decisively dissociated from human, social progress.

Impersonal, academic problems of knowledge may of course be tackled out of a concern to develop knowledge which can subsequently be used or applied to help solve human, social problems. The all-important point, however, is that these problems of knowledge are neither understood nor tackled as intellectually subordinate to our more fundamental problems of living, but are, on the contrary, decisively dissociated from these. If science is to be of human value, it tends to be argued, it is essential that science acquires reliable, objective, impersonal factual knowledge, this in turn requiring – so the argument goes – that the problems of knowledge be tackled in a way which is decisively dissociated from the problems of life.

From the standpoint of developing a kind of intellectual inquiry designed to help us achieve what is of value in life, however, all this is irrational in a quite elementary fashion, and for precisely the reasons emphasized throughout this essay. Granted that the fundamental task of intellectual inquiry is to help us solve those personal, social problems of living we encounter in seeking to achieve what is of value in life, elementary rules of rational problem solving require us to give intellectual priority to the task of articulating these personal, social problems of living, and proposing and criticizing possible (and actual) solutions to them. Rationality also requires, of course, that we develop a multitude of subordinate, specialized problems – for example, technological problems, scientific problems, problems of knowledge and understanding. It is absolutely essential for rationality, however, that these specialized problems are understood as subordinate, the enterprise of tackling them being set within the framework of the more fundamental intellectual activity of proposing and criticizing possible solutions to our problems of living.

The philosophy of knowledge is, as I have said, at present almost universally taken for granted by the academic community, and is built into the whole institutional structure of the scientific, academic enterprise. As a result, the elementary irrationality of this philosophy has damaging repercussions for the whole of intellectual inquiry, and indeed for the whole modern world, all our lives here on earth. Both "applied" and "pure" intellectual inquiry, it should be noted, are damaged by the general acceptance of the philosophy of knowledge.[51]

On the one hand we may – with Bacon, Comte, Bernal and Ravetz, for example – be concerned primarily with the capacity of intellectual inquiry to help us solve our practical social problems. If so, then according to the philosophy of wisdom, intellectual priority needs to be given to articulating these problems, and proposing and criticizing possible solutions. Solutions to practical social problems are appropriate personal, social *actions*. Hence, according to the philosophy of wisdom, the fundamental intellectual task of intellectual inquiry is to develop imaginatively and assess critically possible and actual personal, social actions. The development of knowledge and technology needs to be rationally subordinated to the more fundamental intellectual activity of proposing and criticizing social actions.

The philosophy of knowledge, however, gives intellectual priority to the development of knowledge divorced from a concern with our social problems. New knowledge leads to the development of new technology which is then applied in ways which help, we may hope, to solve these problems. The crucial point, however, is that intellectual priority is given to the task of proposing and criticizing *claims to knowledge* – laws, theories, experimental results – instead of possible social actions.

Inevitably, as a direct result of giving intellectual priority to the development of knowledge rather than to proposing and criticizing possible solutions to social problems, intellectual inquiry must (1) fail to help us solve all those major social problems which require new social actions, policies and institutions for their resolution rather than new knowledge and technology; (2) fail to help us give priority to the development of new knowledge and technology most needed for the resolution of urgent social problems; (3) fail to help us use such knowledge and technology, where developed, to maximum advantage in a rational fashion to help solve social problems; (4) fail to help us anticipate and prevent new knowledge and technology being used in socially harmful ways; (5) fail to help us anticipate and refrain from engaging in intrinsically harmful scientific research; and (6) fail to concentrate intellectual attention on our most urgent social problems.[52] These six kinds of failure are all immediate consequences of the fundamental failure to give intellectual priority to rational human, social problem solving. As long as our thinking about the world and ourselves is dominated by the philosophy of knowledge, it is almost inevitable that the social ills of the modern world will arise, even if almost everyone acts with good will.

On the other hand we may – with Kepler, Spinoza, Einstein and Popper, for example – be concerned primarily with the "intrinsic" or cultural value of intellectual inquiry, intellectual inquiry pursued for its own sake. If so, then we need to recognize – as emphasized by the philosophy of wisdom – that it is knowledge and understanding achieved by people that ultimately matters. "Pure" intellectual inquiry, conceived of in impersonal or institutional terms, is of value in so far as it helps us to achieve that which really has value – our personal knowledge and understanding of our world, our personal curiosity, perceptiveness, capacity to discover that which is of significance in our surroundings, and the extent to which all this enriches our life. The problems of "pure" intellectual inquiry are, in other words, at the most fundamental level, personal and interpersonal problems, problems that we encounter in seeking to enhance our personal knowledge and understanding of the world, our personal perception and appreciation of what is significant and of value in existence. As Einstein once remarked: "Knowledge exists in two forms – lifeless, stored in books, and alive in the consciousness of men. The second form of existence is after all the essential one; the first, indispensable as it may be, occupies only an inferior position" (Einstein, 1973, p. 80). To this I would only add that from the standpoint of "pure" intellectual inquiry it is perhaps the activity, as a part of life, of imaginatively exploring the world, following up our passionate curiosity, the lively encountering of aspects of reality, that is essentially of value. And just as the professional, specialized, institutionalized activities associated with music are designed, ideally, to further our making and enjoying of music, so too the professional, specialized, institutionalized activities associated with science are, ideally, designed to further our exploration and enjoyment of our world.

All this is in marked contrast with the views of those who, like Popper (1972) and Ziman (1968), emphasize the fundamental importance of "objective knowledge", of "knowledge without a knowing subject", of "public knowledge", or of "institutional knowledge", conceived as ends in themselves, rather than as means to the achievement of the end of life of value, via enhancement of personal awareness of the world. In insisting that "pure" science be dissociated from life, intellectual progress being understood in wholly objective, impersonal or institutional terms, the philosophy of knowledge misses out precisely that which matters most, our personal apprehension of the world. As a result of putting this philosophy of impersonal knowledge into practice, a disastrous split develops between the way we personally apprehend or conceive of the world, and the way "science" apprehends or conceives of the world. We fail to exploit science in order to enrich and extend our personal vision of things; and we fail to develop science in such a way that it is amenable to such exploitation. We fail to discover how to use scientific theories as spectacles through which we may, conjecturally, view the world. Instead of emphasizing the priority of the personal problems of understanding we need to solve in order to make such a use of scientific theories, the problems are dismissed as "subjective", the development of impersonal knowledge embodied in scientific theories becoming an end in itself. As a result we become blind to – or ignore – the profound discrepancies that exist between the world as conceived by us in life, and the world as conceived, impersonally, by science. A kind of advanced intellectual schizophrenia in our thinking develops. Theoretical physics, for example, ceases to be, with Einstein, a personal "attempt conceptually to grasp reality as it is thought independently of its being observed" (Schilpp, 1969, p. 80), and becomes merely the impersonal, institutional, ritualistic prediction of phenomena, 'the whole thing ... a wretched bungle' which can 'only claim the interest of shopkeepers and engineers'.(Einstein, in Przibram, 1967, p. 39). Personal awareness of what is significant and of value in existence, intellectual passion, curiosity, wonder, all degenerate into nothing more than the possession of information and expert skills, the accumulation of dry knowledge of fact. As a result of dissociating "pure" intellectual inquiry from life, we lose sight of the value which intellectual inquiry has when pursued for its own sake.

Above all, and quite generally, as a result of engaging in, and thinking in terms of, intellectual inquiry as in the first instance the pursuit of impersonal knowledge, we lose sight of those problems which, quite fundamentally, create the need for intellectual inquiry, and which intellectual inquiry ought fundamentally to be helping us to solve. By giving priority to the pursuit of impersonal knowledge, we fail to emphasize the fundamental character of the personal and social problems of our pluralistic world. Intellectual inquiry must then fail to enhance our common understanding of these problems and our common capacity to develop more adequate resolutions to them. Conceiving of things in terms of the pursuit of impersonal knowledge, we fail entirely to see the urgent need to develop intellectual inquiry as the critical fundamentalist tribal discussion of humanity, as a vital part of all our lives, as *a personal and social reality, as a part of the world*, designed to help us create wiser ways of living, wiser institutions, a wiser world.

Whereas the philosophy of wisdom, in short, in subordinating intellectual inquiry to the needs of life of value, does justice to both the pragmatic and the cultural aspects of intellectual inquiry, in a unified way,[53] the philosophy of knowledge fails to do justice to both aspects.

Specialism is a relatively recent phenomenon, a general intellectual malaise that has progressively overtaken scientific, academic inquiry during the last hundred years or so, and especially during the last fifty years. The natural philosophers of the seventeenth century, the *philosophes* of the eighteenth century, and many scientists, philosophers and social thinkers of the nineteenth century had no difficulty in conceiving and pursuing intellectual inquiry in broadly critical fundamentalist terms (even if epistemological and methodological misconceptions prevented them from having a full understanding of the rationale for universalism indicated here).

I have argued in this last section that there is nevertheless an even deeper intellectual and humanitarian malaise inherent in scientific, academic inquiry, which cannot by any means be construed as a relatively recent phenomenon. On the contrary, it goes back to the origins of modern science some four hundred years ago and can even be traced back to the ancient Greeks of over two thousand years ago. It is built into the very foundations of the Western tradition. It can be put like this. Intellectual inquiry has been pursued in accordance, not with the philosophy of wisdom, but rather with the philosophy of knowledge. Instead of problems (3) and (4) of section 9.2 being taken as fundamental, problems (1) and (2) being tackled as an aspect of, and subordinate to, problems (3) and (4), on the contrary, scientific, academic inquiry has been devoted primarily to solving problems (1) and (2), solutions to aspects of these problems incidentally helping people in social life to develop improved answers to problems (3) and (4) (or so it is hoped). Instead of problems (3) and (4) being held to constitute the central problems of intellectual inquiry, on the contrary, these problems have been ostracized from rational inquiry, relegated to the domain of the personal and the political, solutions to them being determined by such "irrational" factors as subjective emotion and motivation, political power, market forces.

But if present-day scientific, academic inquiry really is damagingly irrational in the quite elementary and fundamental way indicated, how, it may be asked, is it possible? How can such a wholesale, fundamental irrationality have been tolerated for so long? It is not difficult to understand why in the seventeenth century questions concerning the value of life should not have been open to rational discussion: the combined power of church and state made it impossible. (One only has to remember the difficulties encountered by Galileo, Descartes and others in seeking to establish the principle that relatively neutral problems concerning the nature of the material universe should be open to non-authoritarian, rational discussion, to realize that any attempt to establish an analogous principle in connection with problems concerning the meaning and value of life was, at the time, out of the question.) The *philosophes* of the eighteenth century sought to devote reason to the enhancement of human enlightenment, human progress; unfortunately, and understandably, being over-impressed by Bacon and Newton, they failed to emphasize, clearly and unambiguously, that intellectual priority needs to be given to wisdom rather than to knowledge. Romantic writers of the late eighteenth and nineteenth centuries can be interpreted as emphasizing the priority of questions concerning life of value. Unfortunately, in doing so, they abandoned "reason" under the mistaken impression that reason is relevant only for the acquisition of impersonal knowledge of truth, and that it involves the repression of personal feelings, desires and imagination. The question we need to ask is this: Why were these past failures not put right in the twentieth century? Why have we still not put them right, in the twenty-first century? A major part of the answer is, I suggest, the increasing prevalence of specialism, which has cancelled the very possibility of critical, influential discussion of fundamentals. Indeed, the existing fundamental disorganization of contemporary scientific, academic inquiry, with its elevation of knowledge above wisdom, is just what one would expect from putting specialism into practice – as the house analogy indicates. Indeed the pursuit of knowledge dissociated from the pursuit of wisdom is itself the outcome of a kind of specialism – the tackling of impersonal, objective or institutional problems of knowledge dissociated from those more fundamental personal and interpersonal problems that face us in our search for what is, or can be, of value in existence. This elementary irrationality inherent in our official, public thinking about the world and ourselves is at the root of our present failure, as indicated by Higgins and others, to tackle our fundamental problems effectively and humanely. It is this that is responsible for the "glaring contrast" noticed by Martin Luther King.

It must be admitted that in recent years many more voices have been raised against rampant specialization. *Interdisciplinary, cross-disciplinary, transdisciplinary, multidisciplinary* and *post-disciplinary* are now buzzwords to an extent that was hardly the case a decade or so ago. These buzz terms now almost denote specialized fields of study in their own right. But this seems at most to involve encouraging interdisciplinary research, setting up interdisciplinary courses of various kinds, and perhaps creating interdisciplinary research groups, centres and journals. What it does not involve is transforming the overall structure and character of universities so that an arena is created for the sustained, informal, imaginative and critical discussion of fundamental problems – discussion that influences and is influenced by more specialized research. Nor does it involve the kind of radical transformation of the whole relationship between the university and society that is required by universalism.

Another indication of attempts to develop a more socially responsive science is the emphasis that research funding bodies give to the importance of *impact* – that is, *social* impact – in deciding what research projects to support. But here again this development is little more than a botched version of what is really needed. Demanding that research projects, in order to gain financial support, must have *impact*, may well lead to research being funded that has considerable social impact that is of little real value, or even impact that is harmful. On the other hand, research projects that have no immediate impact whatsoever, but nevertheless are of great potential intellectual value, or potential social value that may only come to fruition decades into the future, are likely to receive no funding at all. Impact may actually degrade the intellectual and social value of science. Holding impact to be important is no substitute for recognizing the inherently and profoundly problematic character of the aims of science, imbued as they are with problematic assumptions concerning metaphysics, values and social use, there thus being the need to subject aims to sustained imaginative and critical exploration as an integral part of scientific inquiry itself. Research scientists and groups, funding bodies, governments and industry may make decisions as to which research projects are, and are not, supported financially. Such decision-making needs to be bathed, however, in sustained imaginative and critical exploration of problematic possibilities. This does not take place at present because the current orthodox conception of science does not permit it, or fails to insist that scientific rigour requires it to take place. What we need, in short, is the implementation of a new conception of science, a new conception of rational inquiry – universalism, or critical speciofundamentalism. *Impact* is no substitute.

In some respects, things have got worse in the last few decades. There has been the growth of various anti-rationalist creeds within sections of academia: post-modernism, the "strong programme" within the sociology and history of science, social constructivist views about scientific knowledge. This is in turn has led some scientists and philosophers of science to defend orthodox conceptions of science and reason. It all came to a head with the publication of Alan Sokal's spoof article, "Transgressing the Boundaries" (reprinted in Sokal and Bricmont, 1998). The so-called "Science Wars" that resulted have amounted to little more than a distraction from what really does need to be debated:[54] What kind of inquiry can best help us learn how to realize what is of value in life, for ourselves and others? What kind of inquiry can best help humanity learn how to create a better world – or at least learn how to avoid some of the worst possible future worlds?[55]

My own specialized field of research (in so far as I have one), namely philosophy of science, has suffered what I can only see as a degrading splintering into diverse specialized sub-disciplines. When I began my academic career, in the mid 1960s, the debate between Popper and Kuhn was all the rage. There was the idea that both history and philosophy of science together have the profound task of understanding how scientific progress has come about – this astonishing record of progress in knowledge and technology across generations and centuries that has transformed the world. There was even the idea that much may be learnt from scientific progress about how progress may be achieved in other areas of human life, where it falters and where it is urgently needed – above all, social progress towards a better, wiser world. This optimistic and inspiring nascent research programme was then dealt a series of hammer blows. First, adherents of the Edinburgh "strong programme", and of social constructivism more generally, denied that there is any such thing as scientific progress – or at least abolished the idea from history of science as illegitimate "Whiggishness". Then philosophers of science fell to the lure of specialism. Philosophy of science, instead of becoming what was needed, the more general philosophy of inquiry, degenerated into the philosophy of physics, biology, chemistry, neuroscience, computing, geology, microbiology, etc., etc. The fundamental problems of the discipline disappeared from view, and an ever growing army of specialists fell upon an ever expanding domain of specialized puzzles.[56]

Academic philosophy, more generally, has perhaps improved somewhat in recent decades. The worst excesses of ordinary language philosophy have died away, and many academic philosophers strive to engage with serious problems that arise in connection with serious issues: injustice, environmental degradation, war, science, politics, how to live. Old habits of thought nevertheless linger on. Open any issue of *Mind* (a leading journal of philosophy) and one can still find papers published in the tradition of analytic philosophy. I cannot help but note that philosophers of science, and philosophers more generally, have by and large remained uninterested in my attempts to draw their attention to the argument that the aims and methods of academia are profoundly and damagingly irrational when judged from the standpoint of helping to promote human welfare.[57]

Another regrettable development is the loss of what one might call the idea of the liberal university. When I started out as a young academic, in the mid 1960s, the idea was still around that a university should concentrate on, first, hiring good people, and then, second, giving them the freedom to teach and do research as they themselves saw fit, the job of the administration being to provide support for these two essential university activities. In the UK at least, research assessments, committee work, loss of tenure, short-term contracts, restricted funding and increasing power of the administration seem together to have all but destroyed this idea of the liberal university.

On the other hand, there have been a number of recent developments (at the time of writing) in the UK – and no doubt elsewhere – which can perhaps be interpreted as constituting first steps towards putting universalism, or critical specio-fundamentalism, into academic practice. A number of new departments, institutes and centres have been created devoted to policy and peace studies. Growing concern about environmental problems, especially those associated with climate change, have led to the founding of new institutions which seek to bring specialists together to engage in relevant interdisciplinary research, and to communicate with government, the media and the public. Thus at the University of Cambridge there is the Cambridge Environmental Initiative (CEI), launched in December 2004, which brings together diverse specialized fields of research to work on environmental problems, and holds seminars and public lectures to put research scientists in touch with one another, and with the public. A similar coordinating, interdisciplinary initiative exists at the University of Oxford. It is called the School of Geography and the Environment, and it was founded in 2005 (under another name). At University College London (UCL), my own university, there is a recent and very active initiative called the Grand Challenges programme, which seeks to bring together a wide range of specialists to work on seven broad themes all having to do with human wellbeing. A policy document produced in 2011 is called "The Wisdom Agenda".[58] It seeks to "deliver a culture of wisdom" at UCL. There is also the Tyndall Centre for Climate Change Research, founded by twenty-eight scientists from ten different universities or institutions in 2000. It is based in six British universities, has links with six others, and is funded by three research councils, NERC, EPSRC and ESRC (environment, engineering, and social and economic research). In recent years many scientists have become concerned to involve the public in debate about questions of science policy.[59] There is now an active movement which seeks to promote public engagement with science.[60]

In 2015, *Nature* produced a special issue devoted to interdisciplinarity. One article argues that it is vital to "bridge the divide between the biophysical and the social sciences" in order to tackle global problems, but goes on to stress just how difficult this is to do (Brown, Deletic and Wong, 2015). The authors suggest five principles which help such interdisciplinary work succeed. Another contribution discusses the problems that confront efforts to set up interdisciplinary research (Rylance, 2015). Proponents of interdisciplinarity complain that funding, career prospects and status all favour disciplinary rather than interdisciplinary work. Much depends, however, on how close or distant from one another the component disciplines of interdisciplinary work are – the more distant, the greater the obstacles to success. Especially difficult is the task of combining science and humanities. Another contribution attempts to assess the extent to which interdisciplinarity is increasing. It turns out that there is, at least, an increase in papers that mention "interdisciplinary" in the title (Van Noorden, 2015). Throughout this special issue of *Nature*, it is assumed that we need a kind of science – a kind of academic inquiry – well designed from the standpoint of helping us solve problems in the real world. It is argued that interdisciplinarity will suffice to provide what is required. But it will not. We need to instigate sustained imaginative and critical thinking about our fundamental problems – intellectual and humanitarian – at the heart of the academic enterprise, this to be conducted in such a way that it interacts both with more specialized academic problem solving, and with thinking in the social world, guiding personal, social, institutional, national and global life.

There are, in short, a few scattered signs that the revolution, from specialism to universalism, or from knowledge to wisdom, is already under way. It will need, however, much wider cooperative support – from scientists, scholars, students, research councils, university administrators, vice chancellors, teachers, the media and the general public – if it is to become anything more than what it is at present, a few fragmentary, scattered changes intended to put right quite specific perceived defects in the status quo. What we need is a high-profile campaign, in the public eye, concerned to make out the case for a comprehensive revolution in our universities so they come to put universalism, or the philosophy of wisdom, into academic practice. If this revolution ever comes about it will be comparable in its long-term impact to that of the Renaissance, the scientific revolution or the Enlightenment. The outcome will be that we will at last have what we so urgently need, institutions of learning and research rationally organized and devoted to helping us realize what is of value in life – helping us make progress towards as good a world as possible.

10
Karl Popper and the Enlightenment Programme

10.1 Karl Popper's most significant contribution

Karl Popper's most significant contributions are contained in his first four books: *The Logic of Scientific Discovery*, *The Open Society and Its Enemies*, *The Poverty of Historicism*, and *Conjectures and Refutations*.

It is important to appreciate the existence of a central backbone of argument running through these four books. In *The Logic of Scientific Discovery* (1959), Popper argues (as we all know) that all scientific knowledge is irredeemably conjectural in character, it being impossible to verify theories empirically. Science makes progress by proposing bold conjectures in response to problems, which are then subjected to sustained attempted empirical refutation. This falsificationist conception of scientific method is then generalized to form Popper's conception of (critical) rationality, a general methodology for solving problems or making progress. As Popper puts it in *The Logic of Scientific Discovery*, "inter-subjective *testing* is merely a very important aspect of the more general idea of inter-subjective *criticism*, or in other words, of the idea of mutual rational control by critical discussion" (Popper, 1959, p. 44, n*1).[1]

In order to make sense of the idea of *severe* testing in science, we need to see the experimentalist as having at least the germ of an idea for a rival theory up his sleeve (otherwise testing might degenerate into performing essentially the same experiment again and again). This means experiments are always *crucial experiments*, attempts at trying to decide between two competing theories. Theoretical pluralism is necessary for science to be genuinely empirical. And, more generally, in order to *criticize* an idea, one needs to have a rival idea in mind. Rationality, as construed by Popper, requires plurality of ideas, values, ways of life. Thus, for Popper, the rational society *is* the open society. Given pre-Popperian conceptions of reason, with their emphasis on proof rather than criticism (and associated plurality of ideas), the idea that the rational society is the open society is almost a contradiction in terms. There is thus a very close link between *The Logic of Scientific Discovery*, on the one hand, and *The Open Society and Its Enemies*, *The Poverty of Historicism*, and *Conjectures and Refutations* on the other.

And the direction of argument does not go in just one direction, from *The Logic of Scientific Discovery* to *The Open Society and Its Enemies*: it goes in the other direction as well. For in *The Open Society and Its Enemies* Popper argues that rationality, and scientific rationality as well, need to be conceived of in social and institutional terms (1966a, vol. 2, pp. 217–20) (and the argument is echoed in *The Poverty of Historicism*, in connection with a discussion about the conditions required for scientific progress to be possible [1961, pp. 154–7]). *The Open Society and Its Enemies*, *The Poverty of Historicism*, and *Conjectures and Refutations* illuminate and enrich the doctrines of *The Logic of Scientific Discovery*.

10.2 The Enlightenment Programme

Much of the importance of Popper's first four books stems from the fact that they constitute a major contribution to what may be called "The Enlightenment Programme" – the basic idea of the eighteenth-century Enlightenment, especially the French Enlightenment, of *learning from scientific progress how to achieve social progress towards an enlightened world*.[2] Popper's work does much to revitalize and improve on the version of the Enlightenment Programme that we have inherited from the eighteenth-century Enlightenment, from Voltaire, Diderot, Condorcet and the other *philosophes*.[3] But, as we shall see – and this is the main point of this chapter – Popper's version of the Enlightenment Programme, despite its great virtues, is still defective, and needs further improvement.

But before I discuss Popper's contribution, I want first to say a few words about how profoundly important the basic Enlightenment idea is of learning from scientific progress how to achieve social progress towards an enlightened world.

Science has made astonishing progress in improving knowledge. But social progress towards an enlightened world seems much more problematic. This discrepancy would be reason enough to take very seriously indeed the Enlightenment idea of seeing whether we can learn from scientific progress how to achieve greater social progress towards an enlightened world. But what makes this Enlightenment idea so much more important and urgent, for our times, is that all too often, and tragically, modern science and technology have actually been implicated in some of our worst human disasters. Modern science and technology have undoubtedly done much to help relieve human suffering and enhance the quality of human life; but there is still our terrible record of unnecessary human suffering and death, our record of man-made disasters during the past hundred years or so: horrifyingly destructive wars; the terrifying threat posed by modern armaments, conventional, chemical, biological and nuclear; vast inequalities in wealth around the globe; explosive population growth; the destruction of natural habitats and the rapid extinction of species; pollution of earth, sea and air, the latter leading to the impending devastation of climate change. And a crucial point to note about these global problems is that they have been made possible, have even, in a perfectly legitimate sense, been *caused*,[4] by the advent of modern science and technology. Without the amazing success of modern science and technology, they would not have happened. There is nothing surprising about this. New scientific knowledge and technological know-how enormously increase our power to act: in the absence of enlightenment, of wisdom, our new power to act will sometimes have good consequences, but will also, as often as not, have bad consequences, whether intended, as in war, or unintended, as in global warming. Before the advent of modern science, lack of wisdom, of enlightenment, did not matter too much; we lacked the power to do too much damage to ourselves and the planet. But now that we do have modern science, and the unprecedented powers that it has given us, lack of enlightenment puts us into a position of unprecedented peril. It may even be that our very survival depends on humanity learning a bit more wisdom. Instead of blaming science for our troubles, as many do, we need, rather, to see whether we can learn from the astonishing and dangerous success of science about how to acquire a bit more global wisdom.

The eighteenth-century *philosophes* interpreted the basic Enlightenment idea as requiring that the social sciences be developed alongside the natural sciences. Francis Bacon had already argued that, in order to better the lot of humanity, it is essential to improve our knowledge of the natural world. The *philosophes*, understandably enough, came to the conclusion that, in order to make social progress it is, if anything, even more important to improve knowledge of the social world. So they set about creating and developing the social sciences: economics, psychology, history, anthropology, sociology, political science. This was continued throughout the nineteenth century by such men as Saint-Simon,

Comte, Mill and Marx until, by the mid-twentieth century, departments of these various social sciences, as conceived of by the eighteenth century, had been created in universities all over the world.[5] It is hardly too much to say that academic inquiry, as it exists today, is the outcome of developing and institutionalizing the scientific revolution of the seventeenth century, and the Enlightenment Programme of the eighteenth century.

The Enlightenment was of course opposed by the Romantic movement. This put its faith in emotion, imagination, spontaneity, inspiration, art, genius, and opposed the Enlightenment faith in science and reason. This Romantic opposition to the Enlightenment, to science and reason, is still influential today in such fields as politics, education, anti-science movements, the arts. And it is still influential in some parts of academic inquiry, in such areas as philosophy, cultural studies, anthropology. Postmodernism comes out of the Romantic movement.[6]

But in objecting to the rationalism of the Enlightenment, Romanticism entirely missed the point. For the Traditional Enlightenment, inherited from the eighteenth century, suffers, not from too much reason, but from not enough. It amounts to a characteristic kind of irrationality masquerading as rationality. In developing the basic Enlightenment idea intellectually, the *philosophes* botched the job; and unfortunately it is this botched, irrational version of the Enlightenment Programme that we now have built into the intellectual and institutional structure of the academic enterprise. Academic inquiry today, when judged from the standpoint of helping us create a better world, is damagingly irrational in a wholesale, structural way.[7]

There are three steps that need to be got right to put the basic Enlightenment idea into practice correctly:

(i) The progress-achieving methods of science need to be correctly identified.
(ii) These methods need to be correctly generalized so that they become fruitfully applicable to any worthwhile, problematic human endeavour, whatever the aims may be, and not just applicable to the endeavour of improving knowledge.
(iii) The correctly generalized progress-achieving methods then need to be exploited correctly in the great human endeavour of trying to make social progress towards an enlightened, civilized world.

Unfortunately, the *philosophes* of the Enlightenment got all three points disastrously wrong. They failed to capture correctly the progress-achieving methods of natural science (in that they defended inductivist, or at least verificationist, conceptions of science); they failed to generalize these methods properly; and, most disastrously of all, they failed to apply them properly so that humanity might learn how to become more civilized or enlightened by rational means. Instead of applying the generalized progress-achieving methods of science to *social life itself*, so that social progress might be achieved, the *philosophes* sought to apply scientific method merely to *social science*. Reason (as construed by the *philosophes*) got applied, not to the task of making social progress towards an enlightened world, but to the task of making intellectual progress towards greater knowledge about the social world. Social inquiry was developed, not as social methodology or social philosophy, but as social science.

That the *philosophes* made these blunders in the eighteenth century is forgivable; what is unforgivable is that these blunders still remain unrecognized and uncorrected today, over two centuries later. Instead of correcting the blunders, we have allowed our institutions of learning to be shaped by them as they have developed throughout the nineteenth and twentieth centuries, so that now the blunders are an all-pervasive feature of academia, as we shall see in more detail in a moment.

10.3 Popper's contribution to the Enlightenment Programme

As I have already indicated, Popper made enormously important improvements to the Enlightenment Programme inherited from the eighteenth century. Inductivism and verificationism become falsificationism. Traditional conceptions of reason, with all the emphasis on proof and justification, become critical rationalism. These improvements, at steps (i) and (ii) of the Enlightenment Programme, become all-important when it comes to step (iii): the application of reason to social life, to politics, to problems of living, to political philosophy. As a result of bringing about a revolution in our conception of scientific method, and of rationality more generally, Popper in effect transforms the very idea of "the rational society", so that this ceases to be something that is morally and politically abhorrent, and becomes both highly desirable, and achievable, instead.

Given traditional, pre-Popperian conceptions of science and reason, which tend to see science as establishing secure knowledge of truth by means of evidence, and tend to see reason as establishing truth by means of deductive argument, the "rational society" can only be a society determined, or at least severely constrained, by "the rules of reason". Reason becomes a kind of tyrant. Individual liberty, diversity of views and ways of life, wayward imagination, disagreement and protest would all be suppressed by the iron rule of reason and logic. Granted such verificationist, authoritarian conceptions of reason,[8] the "rational society" can only be regarded as a kind of nightmarish totalitarian state, the very opposite of democracy and liberalism.[9] No wonder the Romantics protested.

But Popper's revolutionary ideas about science and reason change all this dramatically. First, granted Popper's falsificationist conception of scientific method, imagination plays a crucial role in science. Imagination is needed to dream up new wild speculations, subsequently to be submitted to ferocious attempts at empirical refutation. Second, plurality of conflicting theories is absolutely essential for scientific progress, not only to increase the store of theories to be submitted to attempted refutation, but in order to ensure that theories are severely tested in the first place. As I have already mentioned, in order to make sense of the idea of severe testing, we need to see the experimentalist as having at least the germ of an idea for a rival theory up his sleeve (otherwise testing might degenerate into performing essentially the same experiment again and again). This means experiments are always *crucial* experiments, attempts at trying to decide between two competing theories. Theoretical pluralism is necessary for science to be genuinely empirical.[10]

Both these points carry over when Popper's falsificationist conception of scientific method is generalized to form critical rationalism. Reason, quite generally, is at a loss without imagination. Imagination is required to dream up possible solutions to problems, which can then be submitted to severe criticism. Again, plurality of views is an essential ingredient of Popper's conception of reason. Criticism can only deliver a good idea as to how to solve a problem if there is a plurality of ideas to criticize in the first place. And merely in order to criticize an idea, one needs to have some kind of rival idea in mind, at least as a possibility.

Rationality, as construed by Popper, requires plurality of ideas, values, ways of life; the freedom to imagine, to criticize authority, dogma and received opinion. It demands sustained tolerance of diversity of views and ways of life, together with the existence of traditions of criticism, so that good ideas may be selected from a pool of not-so-good ideas. Reason, as Popper emphasizes, needs to be seen in social, political and institutional terms (Popper, 1966a, ch. 24; 1961, section 32). Thus, granted Popper's revolutionary conceptions of scientific method and reason, the "rational society" is not some kind of totalitarian society, but just the opposite, the "open society" – a society that tolerates doubt, diversity of views and ways of life, and criticism, and sustains individual liberty, reasonableness, humanity, justice and democracy. Reason, instead of being the enemy of freedom, individuality, imagination, democracy and justice, becomes the friend of these things, indeed essential for their preservation and development. As Popper puts it in a stray remark tossed out during the course of developing the argument: "We have to learn the lesson that intellectual honesty is fundamental for everything we cherish" (Popper, 1966a, vol. 2, p. 59).

In *The Open Society and Its Enemies*, Popper depicts an epic struggle between those who have sought to help sustain and promote the open society (i.e. the rational society), and those who have opposed it. And he shows how even some of the greatest thinkers of the past have been beguiled by false ideas of science and reason into arguing for the closed society, above all, Plato and Marx.

In these ways, the path Popper pursues, from his conjectural, falsificationist conception of science to its generalization to form critical rationalism, and its application to some of the most urgent and profound political and social problems of our times, represents an immensely valuable rediscovery and transformation of the eighteenth-century Enlightenment Programme. Popper's contribution is important and profound; but it is nevertheless defective. It needs further improvement. (Popper's followers, in so far as they refuse to consider the need for further improvement, do Popper a great disservice – and humanity, of course, an even greater disservice. At the heart of Popper's thought there is the insight that scientific method, and reason, rightly understood, deliver, not certainty, but rather uncertain progress, improvement, development, growth.[11] We betray Popper's philosophy quite fundamentally if we do not take it as a set of proposals, suggestions and arguments urgently in need of further development.)

In what follows I indicate how Popper's version of the Enlightenment Programme needs further improvement in two stages, which I shall discuss under the headings "The Improved Popperian Enlightenment" and "The New Enlightenment".

10.4 The Improved Popperian Enlightenment

A basic inadequacy of Popper's version of the Enlightenment Programme is that it depicts social inquiry, in highly traditional terms, to be social science, indeed pro-naturalist social science, with methods akin to those of the natural sciences. Popper's criticisms of some traditional views associated with social inquiry – his criticisms of historicism, historicist social science and Utopian social engineering – are excellent and decisive. But he does not carry through this criticism of traditional views far enough; he fails to correct the greatest blunder of the eighteenth-century Enlightenment, namely the mistake of applying reason to social *science* rather than to social *life*!

The basic Enlightenment idea, after all, is to learn from scientific progress how to make social progress towards an enlightened world. Putting this idea into practice involves getting appropriately generalized progress-achieving methods of science *into social life itself*! It involves getting progress-achieving methods into our institutions and ways of life in addition to science, into government, industry, agriculture, commerce, international relations, the media, the arts, education. But in sharp contrast to all this, the Traditional Enlightenment has sought to apply generalized scientific method, not to social *life*, but merely to social *science*! Instead of helping humanity learn how to become more civilized by rational means, the Traditional Enlightenment has sought merely to help social scientists improve knowledge of social phenomena (this knowledge then being applied to help solve social problems). The outcome is that today academic inquiry devotes itself to acquiring knowledge of natural and social phenomena, but does not attempt to help humanity learn how to become more civilized. Instead of social inquiry having, as its basic task, to promote cooperatively rational tackling of problems of living in the social world, its primary task, rather, is to acquire knowledge of social phenomena. Instead of being social methodology or social philosophy, social inquiry is pursued as social *science*.

This is the blunder that Popper simply reproduces, and fails to correct. Popper, one might almost say, argues for the open society, but fails to argue for open social inquiry, for a kind of open inquiry devoted to promoting the open society by rational means.

In order to correct this third, monumental and disastrous blunder, we need, as a first step, to bring about a revolution in the nature of academic inquiry, beginning with social inquiry and the humanities. Social inquiry needs to be, not social *science*, but rather social *methodology* or social *philosophy*, concerned to promote rational tackling of problems of living in the social world.[12]

Let us now see, in a little more detail, what would result from correcting this third, monumental blunder of the Traditional Enlightenment. What we need to do is to see what results from applying the progress-achieving rules of reason (arrived at by generalizing the progress-achieving methods of science) to social *life* rather than to social *science*, to the task of making *social progress* towards a civilized world rather than to the task of making *intellectual progress* towards better knowledge of social phenomena.

In order to make clear what is at stake here, I need to appeal to the four rules of rational problem solving of the last chapter. I shall call these rules, which constitute an improved version of Popper's critical rationalism, "problem-solving rationality":

1. Articulate, and try to improve the articulation of, the problem to be solved.
2. Propose and critically assess possible solutions.
3. When necessary, break up the basic problem to be solved into a number of preliminary, simpler, analogous, subordinate or specialized problems (to be tackled in accordance with rules 1 and 2), in an attempt to work gradually towards a solution to the basic problem to be solved.
4. Interconnect attempts to solve basic and specialized problems, so that basic problem solving may guide, and be guided by, specialized problem solving.[13]

Popper's critical rationalism consists of rules 1 and 2; problem-solving rationality improves on this by adding on rules 3 and 4, which become relevant when we are confronted by some especially recalcitrant problem – such as the problem of understanding the nature of the universe, or the problem of creating a civilized world – which can only be solved gradually and progressively, bit by bit, and not all at once.[14] Popper was too hostile to specialization to emphasize the need for rule 3; he did not appreciate that the evils of specialization can be counteracted by implementing rule 4.

It might seem that in moving from scientific method to critical and problem-solving rationality we lose the idea of learning from experience; but this is not so. Problem-solving rationality, as enshrined in the above four rules, is a method of learning from experience. Experience is what we acquire through trying out various possible solutions to the problem we wish to solve, and discovering that these possibilities more or less fail. Consider, for example, a problem of action, a technological or political problem, perhaps: in criticizing a proposed solution we may well appeal to the (adverse) outcome of attempting to put the solution into practice; that is, we appeal to experience. Experience, in this broad sense, is what we acquire through trying to do things, trying to solve problems: it is a generalization of the notion of experience as this arises in connection with science – observation and experimentation. Problem-solving rationality might also be called "problem-solving empiricism"; it is as much a generalization of scientific empiricism as it is of scientific rationality.

These four rules, though by no means sufficient for rationality,[15] are certainly necessary for it. No mode of inquiry can hope to be rational which systematically violates any of these rules. In a moment we shall see that academic inquiry as it exists in the main at present, devoted to the pursuit of knowledge, systematically violates *three* of these four elementary, almost banal, entirely uncontroversial, rules of reason.

Two preliminary points now need to be made.

First, in order to create a more civilized, enlightened world, the problems that we need to solve are, fundamentally, problems of *living* rather than problems of *knowledge*. It is what we *do* (or refrain from doing) that matters, and not just what we *know*. Even where new knowledge or technology is needed, in connection with agriculture or medicine for example, it is always what this enables us to *do* that solves the problem of living.

Second, in order to make progress towards a sustainable, civilized world we need to learn how to resolve our conflicts in more cooperative ways than at present. A group acts cooperatively in so far as all members of the group share responsibility for what is done, and for deciding what is done, proposals for action, for resolution of problems and conflicts, being judged on their merits from the standpoint of the interests of the members of the group (or the group as a whole), there being no permanent leadership or delegation of power.[16] Competition is not opposed to cooperation if it proceeds within a framework of cooperation, as it does ideally within science. There are of course degrees of cooperativeness, from its absence, all-out violence, at one extreme, through settling of conflicts by means of threat, agreed procedures such as voting, via bargaining, to all-out cooperativeness at the other extreme. If we are to develop a sustainable, civilized world we need to move progressively away from the violent end of this spectrum towards the cooperative end.

Granted, then, that the task of academic inquiry is to put the four rules of problem-solving rationality into practice in such a way as to help humanity learn how to make progress towards a civilized, enlightened world, the primary intellectual tasks must be:

1. To articulate, and try to improve the articulation of, those social problems of living we need to solve in order to make progress towards a better world.

2. To propose and critically assess possible, and actual, increasingly cooperative social actions – these actions to be assessed for their capacity to resolve human problems and conflicts, thus enhancing the quality of human life.

These intellectually fundamental tasks are undertaken by social inquiry and the humanities, at the heart of the academic enterprise. Social inquiry also has the task of promoting increasingly cooperatively rational tackling of problems of living in the social world – in such contexts as politics, commerce, international affairs, industry, agriculture, the media, the law, education.

Academic inquiry also needs, of course, to implement the third rule of rational problem solving; that is, it needs:

3. To break up the basic problems of living into preliminary, simpler, analogous, subordinate, specialized problems of knowledge and technology, in an attempt to work gradually towards solutions to the basic problems of living.

But, in order to ensure that specialized and basic problem solving keep in contact with one another, the fourth rule of rational problem solving also needs to be implemented; that is, academic inquiry needs:

4. To interconnect attempts to solve basic and specialized problems, so that basic problem solving may guide, and be guided by, specialized problem solving.

In Figure 9.2 in Chapter 9 I have tried to depict the kind of inquiry that would emerge as a result of putting the above four rules of rational problem solving into academic practice, as just indicated. I will give some further details below.

There are a number of points to note about this "rational problem solving" conception of academic inquiry. Social inquiry is not, primarily, social science; it has, rather, the intellectually basic task of engaging in, and promoting in the social world, increasingly cooperatively rational tackling of conflicts and problems of living.[17] Social inquiry, so conceived, is actually intellectually more fundamental than natural science (which seeks to solve subordinate problems of knowledge and understanding). Academic inquiry, in seeking to promote cooperatively rational problem solving in the social world, must engage in a two-way exchange of ideas, arguments, experiences and information with the social world.

The thinking, the problem solving, that really matters, that is really fundamental, is the thinking that we engage in, individually, socially and institutionally, as we live; the whole of academic inquiry is, in a sense, a specialized part of this, created in accordance with rule 3, but also being required to implement rule 4 (so that social and academic problem solving may influence each other). Academic inquiry, on this model, is a kind of peoples' civil service, doing openly for the public what actual civil services are supposed to do, in secret, for governments. Academic inquiry needs just sufficient power to retain its independence, to resist pressures from government, industry, the media, religious authorities and public opinion, but no more. Academia proposes to, argues with, learns from, attempts to teach and criticizes all sectors of the social world, but does not instruct or dictate. It is an intellectual resource for the public, not an intellectual bully.

The basic intellectual aim of inquiry may be said to be, not knowledge, but wisdom – wisdom being understood to be the desire, the active endeavour and the capacity to realize what is desirable and of value in life, for oneself and others.[18] Wisdom includes knowledge, know-how and understanding, but goes beyond them in also including the desire and active striving for what is of value; the ability to experience value, actually and potentially, in the circumstances of life; the capacity to help realize what is of value for oneself and others; the capacity to help solve those problems of living that need to be solved if what is of value is to be realized; the capacity to use and develop knowledge, technology and understanding as needed for the realization of value. Wisdom, like knowledge, can be conceived of not only in personal terms but also in institutional or social terms. Thus, the basic aim of academic inquiry, according to the view being indicated here, is to help us develop wiser ways of living; wiser institutions, customs and social relations; a wiser world.

So far academic inquiry has been characterized as having the task of helping humanity learn how to tackle its problems of living more rationally; nothing has been said about learning from experience. But, as I indicated above, the four rules of reason that we are considering are also rules for learning from experience; this has a vital role to play in the conception of inquiry we are considering. What we learn as a result of attempting to put into practice some proposed solution to a problem of living is of course all important for learning how to build a better world. A vital task for academic inquiry (especially for history) is to monitor the successes and failures of our past attempts at solving problems of living. As far as possible we should try to ensure that our failed social experiments, our failed attempts at solving social problems, are performed only *in imagination*, and not *in practice* in the real world, so that we only suffer the consequences of failure in imagination, and not in reality. But however vivid, far-seeing and accurate our imagination may be, failure in practice will always happen, and we should seek to learn all we can from it for future actions. To this extent, the conception of inquiry we are considering can be regarded as a kind of empiricism. In two crucial respects, however, it differs from what is usually meant by empiricism. First, what is learned is how to do things, how to realize what is of value, how to live, and not, primarily, what we learn in the context of science: knowledge of fact. And second, as I have already remarked, "experience" means something like "what we acquire as a result of attempting to do things, attempting to realize what is of value", and not, primarily, what it means in the context of science: observation and experiment. (This latter meaning is a specialized version of the former meaning.)

It is important to appreciate that the conception of academic inquiry that we are considering is designed to help us to see, to know and to understand, for their own sake, just as much as it is designed to help us solve practical problems of living. It might seem that social inquiry, in articulating problems of living and proposing possible solutions, has only a severely practical purpose. But engaging in this intellectual activity of articulating personal and social problems of living is just what we need to do if we are to develop a good empathic or "personalistic" understanding of our fellow human beings (and of ourselves) – a kind of understanding that can do justice to our humanity, to what is of value, potentially and actually, in our lives. In order to understand another person *as a person* (as opposed to a biological or physical system) I need to be able, in imagination, to see, desire, fear, believe, experience and suffer what the other person sees, desires, etc. I need to be able, in imagination, to enter into the other person's world; that is, I need to be able to understand his problems of living as he understands them, and I need also, perhaps, to understand a more objective version of these problems. In giving intellectual priority to the tasks of articulating problems of living and exploring possible solutions, social inquiry thereby gives intellectual priority to the development of a kind of understanding that people can acquire of one another that is of great intrinsic value. In my view, indeed, personalistic understanding is essential to the development of our humanity, even to the development of consciousness. Our being able to understand each other in this way is also essential for cooperatively rational action.

And it is essential for science. It is only because scientists can enter imaginatively into each other's problems and research projects that objective scientific knowledge can develop. At least two rather different motives exist for trying to see the world as another sees it: one may seek to improve one's knowledge of the other person; or one may seek to improve one's knowledge of the world, it being possible that the other person has something to contribute to one's own knowledge. Scientific knowledge arises as a result of the latter use of personalistic understanding – scientific knowledge being, in part, the product of endless acts of personalistic understanding between scientists (with the personalistic element largely suppressed so that it becomes invisible). It is hardly too much to say that almost all that is of value in human life is based on personalistic understanding.[19]

The basic intellectual aim of the kind of inquiry we are considering is to devote reason to the discovery of what is of value in life. This immediately carries with it the consequence that the arts have a vital *rational* contribution to make to inquiry, as revelations of value, as imaginative explorations of possibilities, desirable or disastrous, or as vehicles for the criticism of fraudulent values through comedy, satire or tragedy. Literature and drama also have a rational role to play in enhancing our ability to understand others personalistically, as a result of identifying imaginatively with fictional characters – literature in this respect merging into biography, documentary and history. Literary criticism bridges the gap between literature and social inquiry, and is more concerned with the *content* of literature than the means by which it achieves its effects.

Another important consequence flows from the point that the basic aim of inquiry is to help us discover what is of value, namely that our feelings and desires have a vital rational role to play within the intellectual domain of inquiry. If we are to discover for ourselves what is of value, then we must attend to our feelings and desires. But not everything that feels good is good, and not everything that we desire is desirable. Rationality requires that feelings and desires take fact, knowledge and logic into account, just as it requires that priorities for scientific research take feelings and desires into account. In insisting on this kind of interplay between feelings and desires on the one hand, knowledge and understanding on the other, the conception of inquiry that we are considering resolves the conflict between rationalism and romanticism, and helps us to acquire what we need if we are to contribute to building civilization: mindful hearts and heartfelt minds.

This, then, in bare outline, is the kind of academic inquiry that would have emerged from the eighteenth-century Enlightenment if the third great blunder of the Enlightenment had not been made.

But the blunder was made. Instead of the progress-achieving methods of science (generalized to become problem-solving rationality) being applied to *social life*, scientific method was applied to the task of developing *social science* alongside natural science. The outcome is what we have (by and large) today: a kind of inquiry that gives intellectual priority to the task of acquiring *knowledge*, this knowledge, once acquired, being subsequently and secondarily applied to help solve social problems. Rule 3 of problem-solving rationality is put into practice to splendid effect: the outcome is the maze of specialized disciplines of the formal, natural, social and technological sciences, and of scholarship, that go to make up much of academic inquiry today. But rules 1, 2 and 4 are violated. Academic inquiry today, restricted primarily to solving problems of knowledge, is so irrational, in a wholesale and structural way, that *three* of the four most elementary rules of reason conceivable are violated. Rule 1 is violated because academia can articulate problems of *knowledge* but cannot, at a fundamental level, articulate problems of *living*. Rule 2 is violated because academia can propose and critically assess possible solutions to problems of *knowledge* – theories, observational and experimental results, factual claims of all kinds – but cannot propose and critically assess possible solutions to problems of living – proposals for action, policies, political programmes, political philosophies, philosophies of life. All these latter do not state matters of fact; they embody proposals as to what we should do, how we should live, what we should seek to change and create; they incorporate such things as values, human hopes and fears, policies, strategies for living: they do not constitute potential contributions to knowledge, and are thus excluded from a kind of inquiry devoted to the pursuit of knowledge. Once rules 1 and 2 are violated, rule 4 is necessarily violated as well.

This wholesale, structural irrationality of academic inquiry as it mostly exists today is no mere formal matter. It has far-flung, long-term damaging consequences. It means that knowledge and technological know-how are pursued dissociated intellectually from a more fundamental concern to promote increasingly cooperatively rational tackling of conflicts and problems of living. As I have pointed out (in section 10.2), it is this that is at the root of most of our current global problems.

A kind of inquiry that pursues knowledge and technological know-how, and fails to give intellectual priority to the tasks of articulating our problems of living, and proposing and criticizing possible solutions (thus violating three of the four most elementary rules of reason conceivable), must inevitably tend to *create* the kind of global problems we face today, the outcome of possessing much recently acquired power to act, without the power to act wisely. And the more successful such "knowledge-inquiry" is, so the greater the human suffering it is likely to lead to.

Reason is far too important "for everything we cherish" for it to be tolerable that it should be systematically violated in this way.

Here, then, is a major failing of Popper's version of the Enlightenment Programme. Instead of arguing for the need to reject knowledge-inquiry and replace it with the kind of academic inquiry indicated above, "wisdom-inquiry" as it may be called, Popper defends knowledge-inquiry, and even defends pro-naturalist social science.[20]

10.5 The New Enlightenment, step (i): from falsificationism to aim-oriented empiricism

I come now to a rather more radical revision of Popper's version of the Enlightenment Programme. This begins with a revision of step (i) of the programme. Popper's falsificationism is untenable, and needs to be replaced by a conception of scientific method that I have called *aim-oriented empiricism* (AOE). The reason for this revision, already discussed in Chapters 2, 5 and 7, can be summarized as follows. Science only considers (and only accepts) theories that are sufficiently simple, unified or explanatory, and this means that the methods of science make a persistent metaphysical assumption about the universe, to the effect that it has a simple, unified, explanatory dynamic structure. That such a persistent metaphysical assumption is made by science, as a part of (conjectural) scientific knowledge, contradicts, and refutes, falsificationism. An improved conception of scientific method is required.

In *The Logic of Scientific Discovery* (1959), Popper claims that the more falsifiable a theory is, so the greater its degree of simplicity. (There is a second method for assessing degrees of simplicity, in terms of number of observation statements required to falsify the theories in question, but Popper stresses that if the two methods clash, it is the first that takes precedence.) It is easy to see that Popper's proposal fails. Given a reasonably simple scientific theory, T, one can readily increase the falsifiability of T by adding on an independently testable hypotheses, h_1, to form the new theory, $T + h_1$. This new theory will be more falsifiable than T but, in general, will be drastically less simple. And one can make the situation even worse, by adding on as many independently testable hypotheses as one pleases, h_2, h_3 and so on, to form new theories $T + h_1 + h_2 + h_3 + ...$, as highly empirically falsifiable and as drastically lacking in simplicity as one pleases.[21] Thus simplicity cannot be equated with falsifiability.

And there is a further, even more devastating point. Popper's methodological rules favour $T + h_1 + h_2 + h_3$ over T, especially if h_1,

h_2 and h_3 have been severely tested, and corroborated. But in scientific practice, $T + h_1 + h_2 + h_3$ would never even be considered, however highly corroborated it might be if considered, because of its extreme lack of simplicity or unity, its grossly *ad hoc* character. There is here a fundamental flaw in the central doctrine of *The Logic of Scientific Discovery*.

Later, in *Conjectures and Refutations* (1963), Popper put forward a new methodological principle which, when added to those of the earlier book, succeeds in excluding theories such as $T + h_1 + h_2 + h_3$ from scientific consideration. This principle states that a new theory, in order to be acceptable, "should proceed from some *simple, new, and powerful, unifying idea* about some connection or relation (such as gravitational attraction) between hitherto unconnected things (such as planets and apples) or facts (such as inertial and gravitational mass) or new "theoretical entities" (such as field and particles)" (Popper, 1963, p. 241). $T + h_1 + h_2 + h_3$ does not "proceed from some *simple, new and powerful, unifying idea*", and is to be rejected on that account, even if more highly corroborated than T.

But the adoption of this "requirement of simplicity" (as Popper calls it) as a basic methodological principle of science, has the effect of permanently excluding from science all *ad hoc* theories (such as $T + h_1 + h_2 + h_3$) that fail to satisfy the principle, however empirically successful such theories might be if considered. This amounts to assuming permanently that the universe is such that no *ad hoc* theory, that fails to satisfy Popper's principle of simplicity, is true. It amounts to accepting, as a permanent item of scientific knowledge, the substantial metaphysical thesis that the universe is non-*ad hoc*, in the sense that no theory that fails to satisfy Popper's principle of simplicity is true. But this clashes with Popper's criterion of demarcation: that no unfalsifiable, metaphysical thesis is to be accepted as a part of scientific knowledge.[22]

It is, in fact, important that Popper's criterion of demarcation is rejected, and the metaphysical thesis of non-*ad hocness* is explicitly acknowledged to be a part of scientific knowledge. The thesis, in the form in which it is implicitly adopted at any given stage in the development of science, may well be false. Scientific progress may require that it be modified. The thesis needs to be made explicit, in other words, for good Popperian reasons, namely, so that it can be critically assessed, and perhaps improved. As long as Popper's demarcation criterion is upheld, the metaphysical thesis must remain implicit, and hence immune to criticism.[23]

Popper's falsificationism can be modified, however, so that substantial metaphysical theses, implicit in methods that exclude *ad hoc* theories, are made explicit within science, and are acknowledged to be basic items of (conjectural) scientific knowledge, thus becoming open to critical scrutiny and revision. The outcome is a more rational, a more intellectually rigorous kind of science, just because substantial, influential and problematic metaphysical theses, implicit in the methods of science, become explicitly criticizable and improvable.

The moment we acknowledged that there is a persistent metaphysical thesis implicit in the methods of science, two new problems leap to our attention. What, precisely, does this metaphysical thesis assert? And on what grounds is it to be (conjecturally) accepted as a part of scientific knowledge? AOE is put forward as the solution to these two problems.

As far as the first of the above two problems is concerned, a wide range of metaphysical theses are available. At one extreme, we might adopt a metaphysical thesis that excludes only utterly silly theories; at the other extreme, we might adopt the thesis that the universe is physically comprehensible in the sense that it has a unified dynamic structure, some yet-to-be-discovered unified physical "theory of everything" being true – a thesis that I shall call "physicalism". We might even adopt some specific version of physicalism, which asserts that the underlying physical unity is of a specific type: it is made up of a unified field perhaps, or a quantum field, or empty, curved, topologically complex space-time, or a quantum string field. Other things being equal, the more specific the thesis (and thus the more it excludes) so the more likely it is to be false, whereas the more unspecific it is so the more likely it is to be true.

As far as the second of the above two problems is concerned, there are three considerations that we can appeal to, wholly Popperian in spirit if not to the letter of Popperian doctrine:

(1) If some metaphysical thesis, M, is implicit in some scientific methodological practice, then science is more rigorous if M is made explicit, since this facilitates criticism of it, the consideration of alternatives.
(2) A metaphysical thesis may be such that its truth is a necessary condition for it to be possible for us to acquire knowledge: if so, accepting the thesis explicitly can only help, and cannot undermine, the pursuit of knowledge of truth.
(3) Given two rival metaphysical theses, M_1 and M_2, it may be the case that M_1 supports an empirical scientific research programme that has apparently met with far greater empirical success than

any rival empirical research programme based on M_2: in this case we may favour M_1 over M_2, at least until M_2, or some third thesis, M_3, shows signs of supporting an even more empirically progressive research programme.

Two difficulties arise, however, when one attempts to use (2) and (3) to select the best available metaphysical thesis from the infinitely many options available. As far as (2) is concerned, any thesis sufficiently substantial to exclude empirically successful crackpot theories from science is such that acquisition of knowledge might still be possible even if the thesis is false. On the other hand, any thesis such that its truth is necessary for knowledge to be acquired is much too insubstantial to exclude crackpot theories. As far as (3) is concerned, given any metaphysical thesis, M, that supports a non-crackpot empirically progressive scientific research programme, we can mimic this with a crackpot M* that supports a crackpot empirically progressive research programme, with a series of crackpot theories, T_1^*, T_2^*, ..., these theories becoming progressively more and more empirically successful, and closer and closer to exemplifying M*.

These two difficulties can be overcome, however, if physics is construed as adopting a hierarchy of metaphysical conjectures concerning the comprehensibility and knowability of the universe, these conjectures becoming more and more insubstantial as one ascends the hierarchy, more and more likely to be true (see Figure 2.1 in Chapter 2). At the top of the hierarchy, there is the conjecture that the universe is such that some (conjectural) knowledge of our local circumstances can be acquired, sufficient to make life possible. This, and the next conjecture down are, I argue, to be accepted as permanent items of scientific knowledge, in accordance with (2), on the grounds, that is, that such acceptance can only help, and cannot hinder, the search for factual knowledge whatever the universe may be like. At level 4 the conjecture to be adopted is, I argue, physicalism. At level 5 there is the less precise conjecture that the universe is comprehensible in some way or other. At level 3 there is the best currently available more or less specific version of physicalism, which I call the current "metaphysical blueprint". Examples from the history of physics include the following: the universe consists of (a) corpuscles, which interact by contact; (b) point atoms, which interact by means of forces; (c) a unified classical field; (d) a unified quantum field; (e) empty, curved, topologically complex space-time; and (f) a unified quantum string field. At level 2 are currently accepted fundamental physical theories, and at level 1 there are empirical data. Two considerations govern acceptance of metaphysical conjectures from level 3 to level 6. Any such conjecture must, as far as possible, (A) exemplify, be a precise version of, and imply, the next conjecture up in the hierarchy, (B) be more empirically fruitful than any rival conjecture, in that it is a part of an empirical research programme that seems to be more empirically progressive than any rival research programme, in accordance with (3) above. Two considerations also govern acceptance of testable fundamental dynamical physical theories. Such a theory must be such that (i) it, together with all other accepted fundamental physical theories, exemplifies, or is a special case of, the best available metaphysical blueprint (at level 3) to a sufficiently good extent, and (ii) it is sufficiently successful empirically (where empirical success is to be understood, roughly, in a Popperian sense).

This hierarchical view of AOE overcomes the two difficulties, indicated above, roughly as follows. Only the top two theses are accepted as a result of an appeal to (2); theses at levels 3 to 5 are accepted as a result of (a) an appeal to (3), and (b) compatibility with the top two theses at levels 7 and 6; this suffices to exclude aberrant rivals at levels 3 to 5 (which might be construed to support aberrant, empirically progressive research programmes). For further details of how AOE overcomes the two difficulties indicated above, and for further details of the view itself, see previous chapters.[24]

A basic idea of AOE is to channel or direct criticism so that it is as fruitful as possible, from the standpoint of aiding progress in knowledge. The function of criticism within science is to promote scientific progress. When criticism demonstrably cannot help promote scientific progress, it becomes irrational (the idea behind (2) above). In an attempt to make criticism as fruitful as possible, we need to try to direct it at targets that are the most fruitful, the most productive, to criticize (from the standpoint of the growth, the improvement of knowledge). This is the basic idea behind the hierarchy of AOE. Conjectures at all levels remain open to criticism. But, as we ascend the hierarchy, conjectures are less and less likely to be false; it is less and less likely that criticism, here, will help promote scientific knowledge.[25] The best currently available level 3 conjecture is almost bound to be false: the history of physics reveals, at this level, as I have indicated above, that a number of different conjectures have been adopted and rejected in turn. Here, criticism, the activity of developing alternatives (compatible with physicalism) is likely to be immensely fruitful for progress in theoretical physics. Indeed, in Chapter 3, and elsewhere,[26] I argue that this provides physics with a rational, though fallible and non-mechanical method for the discovery of new fundamental physical theories, a method invented and exploited by Einstein in discovering special and general relativity, something which Popper has argued is not possible (see Popper, 1959, pp. 31–2). Criticizing physicalism, at level 4, may also be fruitful for physics, but (the conjecture of AOE is that) this is not as likely to be as fruitful as criticism at level 3. (Elsewhere I have suggested alternatives to physicalism: see Maxwell, 2004, pp. 198–205.) And, as we ascend the hierarchy (so AOE conjectures), criticism becomes progressively less and less likely to be fruitful. Against that, it must be admitted that the higher in the hierarchy we need to modify our ideas, so the more dramatic the intellectual revolution that this would bring about. If physicalism is rejected altogether, and some quite different version of the level 5 conjecture of comprehensibility is adopted instead, the whole character of natural science would change dramatically; physics, as we know it, might even cease to exist.

The biggest change, in moving from falsificationism to AOE, has to do with the role of metaphysics in science, and the scope of scientific knowledge. According to falsificationism, untestable metaphysical theses may influence scientific research in the context of discovery, and may even lead to metaphysical research programmes; they cannot, however, be a part of scientific knowledge itself. But according to AOE, the metaphysical theses at levels 3 to 7 are all a part of current (conjectural) scientific knowledge. In particular, physicalism is. According to AOE, it is a part of current (conjectural) scientific knowledge that the universe is physically comprehensible – certainly not the case granted falsificationism.

Another important change has to do with the relationship between science and the philosophy of science. Falsificationism places the study of scientific method, the philosophy of science, outside science itself, in accordance with Popper's demarcation principle. AOE, by contrast, makes scientific method and the philosophy of science an integral part of science itself. The activity of tackling problems inherent in the aims of science, at a variety of levels, and of developing new possible aims and methods, new possible more specific or less specific philosophies of science (views about what the aims and methods of science ought to be) is, according to AOE, a vital research activity of science itself. But this is also philosophy of science, being carried on within the framework of AOE.[27]

AOE differs in many other important ways from Popper's falsificationism (see chapter 2 and Maxwell, 1998). Nevertheless the impulse, the intellectual aspirations and values, behind the hierarchical view of AOE are, as I have tried to indicate, thoroughly Popperian in character and spirit. The whole idea is to turn implicit assumptions into explicit conjectures in such a way that criticism may be directed at what most needs to be criticized from the standpoint of aiding progress in knowledge, so that conjectures may be developed and adopted that are the most fruitful in promoting scientific progress, at the same time no substantial conjecture, implicit or explicit, being held immune from critical scrutiny.[28]

10.6 The New Enlightenment, step (ii): from critical to aim-oriented rationalism

Falsificationism is defective because it fails to identify the problematic aim of science properly, and thus fails to specify the need for science to improve its aims and methods as it proceeds. Critical rationalism is defective in an analogous way. It does not make improving aims and methods, when aims are problematic, an essential aspect of rationality.

If, however, we take AOE as our starting point, and generalize that, the outcome is different. It is not just in science that aims are problematic;[29] this is the case in life too, either because different aims conflict, or because what we believe to be desirable and realizable lacks one or other of these features, or both. Above all, the aim of creating global civilization is inherently and profoundly problematic. Furthermore, it is not just science that "represses" problematic aims (see Maxwell, 2002a); many other institutional and traditional endeavours repress problematic aims and acknowledge ostensibly unproblematic, token aims instead. Quite generally, then, and not just in science, whenever we pursue a problematic aim we need first to acknowledge the aim; then we need to represent it as a hierarchy of aims, from the specific and problematic at the bottom of the hierarchy, to the general and unproblematic at the top. In this way we provide ourselves with a framework within which we may improve more or less specific and problematic aims and methods as we proceed, learning from success and failure in practice what it is that is both of most value and realizable. Such an "aim-oriented" conception of rationality is the proper generalization of the aim-oriented, progress-achieving methods of science.[30]

Any conception of rationality which systematically leads us astray must be defective. But any conception of rationality, such as Popper's critical rationalism, which does not include explicit instructions for the *improvement* of aims, must systematically lead us astray. It will do so whenever we fail to choose that aim that is in our best interests or, more seriously, whenever we misrepresent our aim – as we are likely to do whenever aims are problematic. In these circumstances, the more "rationally" we pursue the aim we acknowledge, the worse off we will be. Systematically, such conceptions of rationality, which do not include provisions for improving problematic aims, are a hinderance rather than a help; they are, in short, defective.[31]

AOE and its generalization, aim-oriented rationality (AOR), incorporate all the good points of Popper's falsificationist conception of science and its generalization, critical rationalism, indicated above, but also improve on Popper's notions, in being designed to help science and other worthwhile endeavours progressively improve problematic aims and methods.

10.7 The New Enlightenment, step (iii): from knowledge to wisdom

I come now to step (iii) of the New Enlightenment Programme. The task, here, is to help humanity gradually get more AOR into diverse aspects of social and institutional life – personal, political, economic, educational, international – so that humanity may gradually learn how to make progress towards an enlightened world. Social inquiry, in taking up this task, needs to be pursued as social *methodology* or social *philosophy*. What the philosophy of science is to science, as conceived by AOE, so sociology is to the social world: it has the task of helping diverse valuable human endeavours and institutions gradually improve aims and methods so that the world may make social progress towards global enlightenment. (The sociology of science, as a special case, is one and the same thing as the philosophy of science.) And a basic task of academic inquiry, more generally, becomes to help humanity solve its problems of living in increasingly rational, cooperative, enlightened ways, thus helping humanity become more civilized. The basic aim of academic inquiry becomes, as I have already said, to promote the growth of *wisdom*. Those parts of academic inquiry devoted to improving knowledge, understanding and technological know-how contribute to the growth of wisdom. The New Enlightenment Programme thus has dramatic and far-reaching implications for academic inquiry, for almost every branch and aspect of science and the humanities, for its overall character and structure, its overall aims and methods, and its relationship to the rest of the social world. I have spelled out in some detail what these implications are in a number of publications.[32]

As I have already remarked, the aim of achieving global civilization is inherently problematic.[33] This means, according to AOR, that we need to represent the aim at a number of levels, from the specific and highly

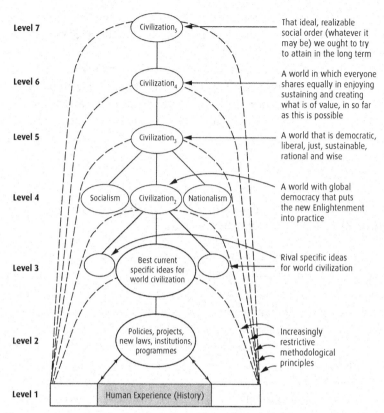

Figure 10.1 Aim-oriented rationality applied to the task of making progress towards a civilized world (Source: author)

problematic to the unspecific and unproblematic. Thus, at a fairly specific level, we might, for example, specify civilization to be a state of affairs in which there is an end to war, dictatorships, population growth, extreme inequalities of wealth, and the establishment of democratic, liberal world government and a sustainable world industry and agriculture. At a rather more general level, we might specify civilization to be a state of affairs in which everyone shares equally in enjoying, sustaining and creating what is of value in life *in so far as this is possible*. Figure 10.1 depicts a cartoon version of what is required, arrived at by generalizing and then reinterpreting Figure 2.1 (see Chapter 2).

As a result of building into our institutions and social life such a hierarchical structure of aims and associated methods, we create a framework within which it becomes possible for us progressively to improve our real-life aims and methods in increasingly cooperative ways as we live. Diverse philosophies of life – diverse religious, political, economic and moral views – may be cooperatively developed, assessed and tested against the experience of personal and social life. It becomes possible progressively to improve diverse *philosophies of life* (diverse views about what is of value in life and how it is to be realized), much as *theories* are progressively and cooperatively improved in science.

AOR is especially relevant when it comes to resolving conflicts cooperatively. If two groups have partly conflicting aims but wish to discover the best resolution of the conflict, AOR helps in requiring of those involved that they represent aims at a level of sufficient imprecision for agreement to be possible, thus creating an agreed framework within which disagreements may be explored and resolved. AOR cannot, of itself, combat non-cooperativeness, or induce a desire for cooperativeness; it can, however, facilitate the cooperative resolution of conflicts if the desire for this exists. In facilitating the cooperative resolution of conflicts in this way, AOR can, in the long term, encourage the desire for cooperation to grow (if only because it encourages belief in the possibility of cooperation).

10.8 Objections

I now consider, briefly, some objections that may be raised against my claim that the "New Enlightenment" improves on the Popperian version of the Enlightenment Programme.

It may be objected that the Traditional Enlightenment does not dominate current academic inquiry to the extent that I have assumed. But grounds for holding that it does are given in chapter six of my *From Knowledge to Wisdom*. There I looked at the following: (1) books about the modern university; (2) the philosophy and sociology of science; (3) statements of leading scientists; (4) physics abstracts; (5) chemistry, biology, geo and psychology abstracts; (6) journal titles and contents; (7) books on economics, sociology and psychology; (8) philosophy. In 1984, the year *From Knowledge to Wisdom* was published, there can be no doubt whatsoever that the Traditional Enlightenment (or "the philosophy of knowledge", as I called it in the book) dominated academic inquiry.

Have things changed since then? The revolution advocated by *From Knowledge to Wisdom*, and argued for here, has not occurred. There is still, amongst the vast majority of academics today, no awareness at all that a more intellectually rigorous and humanly valuable kind of inquiry than that which we have at present, exists as an option. In particular, social inquiry continues to be taught and pursued as social *science*, and not as social *methodology*. Recently I undertook an examination, at random, of thirty-four introductory books on sociology, published between 1985 and 1997. Sociology, typically, is defined as "the scientific study of human society and social interactions" (Tischler, 1996, p. 4), as "the *systematic, sceptical study of human society*" (Macionis and Plummer, 1997, p. 4), or as having as its basic aim "to understand human societies and the forces that have made them what they are" (Lenski et al., 1995, p. 5). Some books take issue with the idea that sociology is the *scientific* study of society, or protest at the male-dominated nature of sociology (for example, Abott and Wallace, 1990, pp. 3 and 1). Nowhere did I find a hint of the idea that a primary task of sociology, or of social inquiry more generally, might be to help build into the fabric of social life progress-achieving methods, generalized from those of science, designed to help humanity resolve its conflicts and problems of living in more cooperatively rational ways than at present.

The tackling of problems of living rather than problems of knowledge does of course go on within the academic enterprise as it is at present constituted, within such disciplines as economics, development studies, policy studies, peace studies, medicine, agriculture, engineering, and elsewhere. But this does not tell against the point that the primary task of academic inquiry at present is, first, to acquire knowledge and technological know-how, and then, second, to apply it to help solve problems of living. It does not, in other words, tell against the point that it is the Traditional Enlightenment that is the dominant influence on the nature, the aims and methods, the whole character and structure of academic inquiry.

It may be objected that it is all to the good that the academic enterprise today does give priority to the pursuit of knowledge over the task of promoting wisdom and civilization. Before problems of living can be tackled rationally, knowledge must first be acquired.[34]

I have six replies to this objection.

First, even if the objection were valid, it would still be vital for a kind of inquiry designed to help us build a better world to include rational exploration of problems of living, and to ensure that this guides priorities of scientific research (and is guided by the results of such research).

Second, the validity of the objection becomes dubious when we take into account the considerable success people met with in solving problems of living in a state of extreme ignorance, before the advent of science. We still today often arrive at solutions to problems of living in ignorance of relevant facts.

Third, the objection is not valid. In order to articulate problems of living and explore imaginatively and critically possible solutions (in accordance with Popper's conception of rationality) we need to be able to act in the world, imagine possible actions and share our imaginings with others: in so far as some common sense knowledge is implicit in all this, such knowledge is required to tackle rationally and successfully problems of living. But this does not mean that we must give intellectual priority to acquiring new relevant knowledge before we can be in a position to tackle rationally our problems of living.

Fourth, simply in order to have some idea of what kind of knowledge or know-how it is *relevant* for us to try to acquire, we must first have some provisional ideas as to what our problem of living is and what we might do to solve it. Articulating our problem of living and proposing and critically assessing possible solutions needs to be intellectually prior to acquiring relevant knowledge simply for this reason: we cannot know what new knowledge it is *relevant* for us to acquire until we have at least a preliminary idea as to what our problem of living is, and what we propose to do about it. A slight change in the way we construe our problem may lead to a drastic change in the kind of knowledge it is relevant to acquire: changing the way we construe problems of health, for example, to include *prevention* of disease (and not just curing of disease), leads to a dramatic change in the kind of knowledge we need to acquire (the importance of exercise, diet, etc.). Including the importance of avoiding *pollution* in the problem of creating wealth by means of industrial development leads to the need to develop entirely new kinds of knowledge.

Fifth, relevant knowledge is often hard to acquire; it would be a disaster if we suspended life until it had been acquired. Knowledge of how our brains work is presumably highly relevant to all that we do, but clearly, suspending rational tackling of problems of living until this relevant knowledge has been acquired would not be a sensible step to take. It would, in any case, make it impossible for us to acquire the relevant knowledge (since this requires scientists to act in doing research). Scientific research is itself a kind of action carried on in a state of relative ignorance.

Sixth, the capacity to act, to live, more or less successfully in the world, is more fundamental than (propositional) knowledge. Put in Rylean terms, "knowing how" is more fundamental than "knowing that" (Ryle, 1949, ch. 2). All our knowledge is but a development of our capacity to act. Dissociated from life, from action, knowledge stored in libraries is just paper and ink, devoid of meaning. In this sense, problems of living are more fundamental than problems of knowledge (which are but an aspect of problems of living); giving intellectual priority to problems of living quite properly reflects this point.[35] The point made above in section 10.4 deserves to be re-emphasized: a kind of inquiry that gives priority to tackling problems of knowledge over problems of living violates the most elementary requirements of rationality conceivable. If the basic task is to help humanity create a better world, then the problems that need to be solved are, primarily, problems of living, problems of action, not problems of knowledge. This means that to comply, merely, with Popper's conception of critical rationalism (or problem-solving rationality) discussed above, the basic intellectual tasks need to be (1) to articulate problems of living, and (2) to propose and critically assess possible solutions, possible more or less cooperative human *actions*. (1) and (2) are excluded, or marginalized, by a kind of inquiry that gives priority to the task of solving problems of knowledge. The result will be a kind of inquiry that fails to create a reservoir of imaginative and critically examined ideas for the resolution of problems of living, and instead develops knowledge often unrelated to, or even harmful to, our most basic human needs.

It may be objected that in employing AOR in an attempt to help create a more civilized world, in the way indicated above, the New Enlightenment falls foul of Popper's strictures against Utopian social engineering (Popper, 1966a, vol. 1, ch. 9; 1961, pp. 64–92). I have three replies to this objection. First, to the extent that piecemeal social engineering, of the kind advocated by Popper, is indeed the rational way to make progress towards a more civilized world, this will be advocated by the New Enlightenment. Second, when we take into account the unprecedented *global* nature of many of our most serious problems, indicated at the beginning of this chapter (the outcome of solving the problem of acquiring scientific knowledge but failing to solve the problem of becoming wiser), we may well doubt that piecemeal social engineering is sufficient. Third, Popper's distinction between piecemeal and Utopian social engineering is altogether too crude: it overlooks entirely what has been advocated here, aim-oriented rationalistic social engineering, with its emphasis on developing increasingly cooperatively rational resolutions of human conflicts and problems in full recognition of the inherently problematic nature of the aim of achieving greater civilization.[36]

All those to any degree influenced by Romanticism and what Isaiah Berlin has called the counter-Enlightenment will object strongly to the idea that we should learn from scientific progress how to achieve social progress towards civilization; they will object strongly to the idea of allowing conceptions of rationality, stemming from science, to dominate in this way, and will object even more strongly to the idea, inherent in the New Enlightenment, that we need to create a more aim-oriented rationalistic social world.

Directed at the Traditional Enlightenment, objections of this kind have some validity; but directed at the New Enlightenment, they have none. As I have emphasized elsewhere, AOR amounts to a synthesis of traditional rationalist and romantic ideals, and not to the triumph of the first over the second. In giving priority to the realization of what is of value in life, and in emphasizing that rationality demands that we seek to improve aims as we proceed, the New Enlightenment requires that rationality integrates traditional Rationalist and Romantic values and ideals of integrity. Imagination, emotion, desire, art, empathic understanding of people and culture, the imaginative exploration of aims and ideals, which tend to be repudiated as irrational by traditional Rationalism, but which are prized by Romanticism, are all essential ingredients of aim-oriented rationality. Far from crushing freedom, spontaneity, creativity and diversity, AOR is essential for the desirable flourishing of these things in life.[37]

Finally, it may be objected that science is too different from political life for there to be anything worthwhile to be learnt from scientific success about how to achieve social progress towards civilization.[38] (a) In science there is a decisive procedure for eliminating ideas, namely empirical refutation: nothing comparable obtains, or can obtain, in the political domain. (b) In science, experiments or trials may be carried out relatively painlessly (except, perhaps, when new drugs are being given in live trials); in life, social experiments, in that they involve people, may cause much pain if they go wrong, and may be difficult to stop once started. (c) Scientific progress requires a number of highly intelligent and motivated people to pursue science on the behalf of the rest of us, funded by government and industry; social progress requires almost *everyone* to take part, including the stupid, the criminal, the mad or otherwise handicapped, the ill and the highly unmotivated; and in general there is no payment. (d) Scientists, at a certain level, have an agreed, common objective: to improve knowledge. In life, people often have quite different or conflicting goals, and there is no general agreement as to what civilization ought to mean, or even whether it is desirable to pursue civilization in *any* sense. (e) Science is about fact, politics about value, the quality of life. This difference ensures that science has nothing to teach political action (for civilization). (f) Science is male-dominated, fiercely competitive and at times terrifyingly impersonal (Harding, 1986); this means it is quite unfit to provide any kind of guide for life.

Here, briefly, are my replies. (a) Some proposals for action can be shown to be unacceptable quite decisively as a result of experience acquired through attempting to put the proposal into action. Where this is not possible, it may still be possible to assess the merits of the proposal to some extent by means of experience. If assessing proposals for action by means of experience is much more indecisive than assessing scientific theories by means of experiment, then we need, all the more, to devote our care and attention to the former case. (b) Precisely because experimentation in life is so much more difficult than in science, it is vital that in life we endeavour to learn as much as possible from (i) experiments that we perform in our imagination, and (ii) experiments that occur as a result of what actually happens. (c) Because humanity does not have the aptitude or desire for wisdom that scientists have for knowledge, it is unreasonable to suppose that progress towards global wisdom could be as explosively rapid as progress in science. Nevertheless progress in wisdom might go better than it does at present.

(d) Cooperative rationality is only feasible when there is the common desire of those involved to resolve conflicts in a cooperatively rational way. (e) Aim-oriented rationality can help us improve our decisions about what is desirable or of value, even if it cannot reach decisions for us. (f) In taking science as a guide for life, it is the progress-achieving methodology of science to which we need to attend. It is this that we need to generalize in such a way that it becomes fruitfully applicable, potentially, to all that we do. That modern science is male-dominated, fiercely competitive, and at times terrifyingly impersonal should not deter us from seeing what can be learned from the progress-achieving methods of science – unless, perhaps, it should turn out that being male-dominated, fiercely competitive and impersonal is essential to scientific method and progress. (But this, I submit, is not the case.)

10.9 Implications for academic inquiry

Popper's version of the Enlightenment Programme, as enshrined in his first four books, has major implications, still unacknowledged by many, for a wide range of human endeavours, such as politics, education, the arts, philosophy and the humanities. But when developed further, in ways indicated above, the Popperian Enlightenment has even more fruitful, dramatic widespread implications. It is hardly too much to say, in my view, that the upshot of the argument is that we require a social and cultural revolution as substantial and dramatic, perhaps, as that of the Reformation, the Scientific Revolution, or the eighteenth-century Enlightenment itself. This revolution involves changing the Traditional Enlightenment and the Romantic opposition so that these become unified in the New Enlightenment; it involves appropriately modifying all those activities and institutions affected by the Traditional Enlightenment and the Romantic opposition so that they come to embody the New Enlightenment: science, art, politics, education, medicine, philosophy, law, industry, agriculture, education.

In particular, it involves changing academic inquiry so that, instead of being shaped by the Traditional Enlightenment (modified somewhat by Popper) and the Romantic opposition, as at present, it comes to be shaped by the New Enlightenment. I conclude this chapter with a list of twenty-three structural changes that need to be made to academic inquiry if it is to come to embody the New Enlightenment. The upshot would be universities rationally devoted to helping us realize what is of value in life, rationally devoted to helping us make progress towards as enlightened a world as possible.

How academic inquiry must change to put the New Enlightenment into practice

1. There needs to be a change in the basic intellectual *aim* of inquiry, from the growth of knowledge to the growth of wisdom – wisdom being taken to be the capacity and active endeavour to realize what is of value in life, for oneself and others, and thus including knowledge, understanding and technological know-how (but much else besides).
2. There needs to be a change in the nature of academic *problems*, so that problems of living are included, as well as problems of knowledge – the former being treated as intellectually more fundamental than the latter.
3. There needs to be a change in the nature of academic *ideas*, so that proposals for action are included as well as claims to knowledge – the former, again, being treated as intellectually more fundamental than the latter.
4. There needs to be a change in what constitutes intellectual *progress*, so that progress-in-ideas-relevant-to-achieving-a-more-civilized-world is included as well as progress in knowledge, the former being indeed intellectually fundamental.

5. There needs to be a change in the idea as to where inquiry, at its most fundamental, is located. It is not esoteric theoretical physics, but rather the thinking we engage in as we seek to achieve what is of value in life. Academic thought is a (vital) adjunct to what really matters, personal and social thought active in life.
6. There needs to be a dramatic change in the nature of social inquiry (reflecting points 1 to 5). Economics, politics, sociology, and so on, are not, fundamentally, *sciences*, and do not, fundamentally, have the task of improving knowledge about social phenomena. Instead, their task is threefold. First, it is to articulate problems of living, and propose and critically assess possible solutions, possible actions or policies, from the standpoint of their capacity, if implemented, to promote wiser ways of living. Second, it is to promote such cooperatively rational tackling of problems of living throughout the social world. And third, at a more basic and long-term level, it is to help build the hierarchical structure of aims and methods of aim-oriented rationality into personal, institutional and global life, thus creating frameworks within which progressive improvement of personal and social life aims-and-methods becomes possible. These three tasks are undertaken in order to promote cooperative tackling of problems of living – but also in order to enhance empathic or "personalistic" understanding between people as something of value in its own right. Acquiring knowledge of social phenomena is a vital but subordinate activity, engaged in to facilitate the above three fundamental pursuits.
7. Natural science needs to change, so that it includes at least three levels of discussion: evidence, theory and research aims. Discussion of aims needs to bring together scientific, metaphysical and evaluative consideration in an attempt to discover the most desirable and realizable research aims. It needs to influence, and be influenced by, exploration of problems of living undertaken by social inquiry and the humanities, and the public.
8. There needs to be a dramatic change in the relationship between social inquiry and natural science, so that social inquiry becomes intellectually more fundamental from the standpoint of tackling problems of living, promoting wisdom. Social inquiry influences choice of research aims for the natural and technological sciences, and is, of course, in turn influenced by the results of such research. (Social inquiry also, of course, conducts empirical research, in order to improve our understanding of what our problems of living are, and in order to assess policy ideas whenever possible.)

9. The current emphasis on specialized research needs to change so that sustained discussion and tackling of broad, global problems that cut across academic specialities is included, both influencing and being influenced by, specialized research.
10. Academia needs to include sustained imaginative and critical exploration of possible futures, for each country, and for humanity as a whole, with policy and research implications being discussed as well.
11. The way in which academic inquiry as a whole is related to the rest of the human world needs to change dramatically. Instead of being intellectually dissociated from the rest of society, academic inquiry needs to be communicating with, learning from, teaching and arguing with the rest of society – in such a way as to promote cooperative rationality and social wisdom. Academia needs to have just sufficient power to retain its independence from the pressures of government, industry, the military and public opinion, but no more. Academia becomes a kind of civil service for the public, doing openly and independently what actual civil services are supposed to do in secret for governments.
12. There needs to be a change in the role that political and religious ideas, works of art, expressions of feelings, desires and values have within rational inquiry. Instead of being excluded, they need to be explicitly included and critically assessed, as possible indications and revelations of what is of value, and as unmasking of fraudulent values in satire and parody – vital ingredients of wisdom.
13. There need to be changes in education so that, for example, seminars devoted to the cooperative, imaginative and critical discussion of problems of living are at the heart of all education from the age of five onwards. Politics, which cannot be taught by knowledge-inquiry, becomes central to wisdom-inquiry, political creeds and actions being subjected to imaginative and critical scrutiny.
14. There need to be changes in the aims, priorities and character of pure science and scholarship, so that it is the curiosity, the seeing and searching, the knowing and understanding of individual persons, that ultimately matters, the more impersonal, esoteric, purely intellectual aspects of science and scholarship being means to this end. Social inquiry needs to give intellectual priority to helping empathic understanding between people to flourish (as indicated in point 6 above).

15. There need to be changes in the way mathematics is understood, pursued and taught. Mathematics is not a branch of knowledge of anything actual at all. Rather, it is concerned to explore problematic *possibilities*, and to develop, systematize and unify problem-solving methods (Maxwell, 2010c).
16. Literature needs to be put close to the heart of rational inquiry, in that it explores imaginatively our most profound problems of living and aids personalistic understanding in life by enhancing our ability to enter imaginatively into the problems and lives of others.
17. Philosophy needs to change so that it ceases to be just another specialized discipline and becomes instead that aspect of inquiry as a whole that is concerned with our most general and fundamental problems – those problems that cut across all disciplinary boundaries. Philosophy needs to become again what it was for Socrates: the attempt to devote reason to the growth of wisdom in life.
18. Academic contributions need to be written in as simple, lucid, jargon-free a way as possible, so that academic work is as accessible as possible across specialities and to non-academics.
19. There needs to be a change in views about what constitute academic contributions, so that publications which promote (or have the potential to promote) public understanding as to what our problems of livings are and what we need to do about them are included, in addition to contributions addressed primarily to the academic community.
20. Every university needs to create a seminar or symposium devoted to the sustained discussion of fundamental problems that cut across all conventional academic boundaries, global problems of living being included as well as global problems of knowledge and understanding.

The above changes all come from my "from knowledge to wisdom" argument spelled out in detail elsewhere. The following three institutional innovations do not follow from that argument but, if implemented, would help wisdom-inquiry to flourish.[39]

21. Natural science needs to create committees, in the public eye, and manned by scientists and non-scientists alike, concerned to highlight and discuss failures of the priorities of research to respond to the interests of those whose needs are the greatest – the poor

of the earth – as a result of the inevitable tendency of research priorities to reflect the interests of those who pay for science, and the interests of scientists themselves.
22. Every national university system needs to include a national shadow government, seeking to do, virtually, free of the constraints of power, what the actual national government ought to be doing. The hope would be that virtual and actual governments would learn from each other.
23. The world's universities need to include a virtual world government which seeks to do what an actual elected world government ought to do, if it existed. The virtual world government would also have the task of working out how an actual democratically elected world government might be created.

Notes

Introduction

1. Aim-oriented empiricism was first put forward in *The Rationality of Scientific Discovery* (Maxwell, 1974). Anyone interested in the way the view has evolved over the years should have a look at this two-part paper.
2. In my *In Praise of Natural Philosophy* (2017b, ch. 7), I make the point that, once it is apparent that the intellectual aims of science are problematic, and hence need to be improved as science proceeds, it becomes obvious that the philosophy of science – the study of what the aims and methods of science ought to be – needs to be an integral, influential part of science itself, if science is to be rigorous. I there go on to generalize this to all worthwhile human endeavours with problematic aims: rationality requires that the philosophy of a life endeavour with a problematic aim must be an influential part of the life endeavour itself (politics, law, international relations, finance, education).
3. For a discussion of these contradictory impulses to be found in Popper's work, together with relevant references to Popper's writings, see Maxwell (2016a).
4. This may be no more than truth at the empirical level, at the level of empirical predictions.

Chapter 1

1. Critical remarks about Wittgenstein are scattered throughout Popper's works. For an extended criticism of Wittgenstein's *Tractatus* see Popper (1966a, vol. 2, pp. 296–9).

Chapter 2

1. The version of AOE defended here is a simplification and improvement of the version expounded in Maxwell (1998), in turn an improvement of versions of the view expounded in Maxwell (1972a, 1974, 1979, 1984a, 1993a, 1997a). For summaries of Maxwell (1998), see Maxwell (1999a, 2000b, 2002a, 2002b). For more recent expositions, see Maxwell (2007, chs. 5, 9, 14; 2011b; 2013), and especially Maxwell (2017a, 2017b).
2. The standard model is the current quantum field theory of fundamental particles and the forces between them.
3. See Popper (1959, 1963, 1983).
4. See Lakatos (1970, 1978).
5. For discussion of Popper, see Schilpp (1974); Levinson (1982); Hacohen (2000); Catton and Macdonald (2004); O'Hear (2004); Keuth (2005); Jarvie, Milford and Miller (2006); Gattei (2009); Rowbottom (2011); Shearmur and Stokes (2016). For discussion of Kuhn, see: Horwich (1993), Hoyningen-Huene (1993), Bird (2000), Nickles (2003), Gattei (2008), D'Agostino (2010), Wray (2011), Marcum (2015). For discussion of Lakatos, see: Cohen, Feyerabend and Wartofsky (1976); Larvor (1998); Kadvany (2001); Kampis, Kvasz and Stöltzner (2002). For discussion that compares and contrasts the work of Popper, Kuhn and Lakatos, see Lakatos and Musgrave (1970), Stove (1982), Nola and Sankey (2000), Fuller (2003), Agassi (2014). See also Lakatos and Feyerabend (1999), and for a review that includes in a nutshell the main point of the present chapter, Maxwell (2000c).

6. I exempt my own work from consideration – work that so far has been almost entirely ignored by mainstream philosophy of science.
7. I have in mind such publications as Holton (1973), Feyerabend (1978), Glymour (1980), Van Fraassen (1980), Laudan (1984), Watkins (1984), Hooker (1987), Hull (1988), Howson and Urbach (1993), Kitcher (1993), Musgrave (1993), Dupré (1995), McAllister (1996), Cartwright (1999), Lipton (2004) and, more recently, Achinstein (2010), Craver and Darden (2013), Dawid (2014), Sober (2015), Schrenk (2016), Scerri (2016).
8. At the time of writing (2016), the metaphysics of science seems to have become a topic of growing interest: see Rescher (2001); Lange (2009); Kincaid, Ladyman and Ross (2013); Morganti (2013); Mumford and Tugby (2013); Schrenk (2016). None of these works defends, or even expounds, a role for metaphysics in science of the kind required by aim-oriented empiricism. The closest I have come to discovering an exception to this is a book by Craig Dilworth (2007) entitled *The Metaphysics of Science*. Dilworth expounds only a very inadequate version of the view, and fails to provide the key argument for the view, as I have shown elsewhere (see Maxwell, 2009d). Dilworth seems to have derived the idea from my work: several times in his book he refers to Maxwell (1984a), where aim-oriented empiricism is expounded and argued for.
9. For Popper's replies to such criticisms, see Popper (1972, ch. 1; 1974, sections II and III; 1983, introduction and ch. 1).
10. I stress this point because the one Popperian who has taken note of my criticisms of Popper, David Miller, has accused me, incorrectly, of defending "fallibilist justificationism" (see Miller, 1994, p. 37). Miller repeats the criticism in his book *Out of Error* (2006, p. 94). For my refutation of the charge, see Maxwell (2006).
11. Popper discusses such "silly" rival theories in volume 1 of *Postscript* (Popper, 1983, pp. 67–71). He argues that they deserve to be rejected on the grounds that they create more problems than they solve, in particular problems of explanation. This is a relevant consideration granted dressed falsificationism, but not granted bare falsificationism. He also argues that it does not matter if such "silly" theories become potential rivals, since it can be left to scientists themselves to criticize them. But this ignores the fact that it is precisely Popper's methodology which should be providing guidelines for such criticism. Far from condemning such a "silly" theory as worthy of rejection, bare falsificationism holds such a theory to be better than the accepted theory (if it has greater empirical content, is not falsified where the accepted theory appears to be, and some of the excess content of the "silly" theory is corroborated). Popper fails to appreciate that it is his methodology, not he himself, which needs to declare that silly theories are indeed "silly". The fact that his methodology declares these silly theories to be highly acceptable is a devastating indictment of his methodology. To argue that these silly theories, refuting instances of his methodology, do not matter and can be discounted, is all too close to a scientist arguing that evidence, that refutes his theory, should be discounted, something which Popper resoundingly condemns. The falsificationist stricture that scientists should not discount falsifying instances, ought to apply to methodologists as well!
12. In fact even the methodological rules of bare falsificationism are such that persistent application of these rules commits one to making implicit metaphysical assumptions (which may be false). Bare falsificationism, as formulated by Popper, requires of an acceptable theory that it is strictly universal in that it makes no reference to any specific time, place or object. This makes it impossible for science to discover that the laws of nature just are different within specific space-time regions, or that there is a specific object with unique dynamical properties. There is no scope, within bare falsificationism, for the rejection of these metaphysical theses, even though circumstances could conceivably arise such that progress in knowledge would require this. (AOE, by contrast, allows for this remote possibility: that which is dogmatically upheld by bare falsificationism becomes criticizable granted AOE.) Popper recognizes that the methodological rule requiring any theory to be strictly universal does have a metaphysical counterpart (1959, sections 11 and 79), but fails to appreciate how damaging this is for falsificationism.
13. It may be objected that persistent acceptance of unified theories in physics when endlessly many empirically more successful disunified rivals are available does not mean that physics makes a metaphysical assumption about the underlying unity of the universe. Unified theories are accepted over disunified rivals because they are better verified – or better corroborated, as Popper would say. But whether unified theories are more likely to be true – other things being (more or less) equal – depends on what kind of universe we are in. In a unified universe, unified theories will, no doubt, be more likely to be true, but in a disunified universe, the opposite

holds – and we have no valid reasons to hold that the former is more likely to be true than the latter. For a decisive refutation of this and other objections to AOE, and to the key argument for AOE, see Maxwell (2015a).
14. To say that M_1 "supports" an empirically successful research programme is to say that the programme develops a succession of theories, each empirically more successful than its predecessors, in a Popperian sense, and each being closer to exemplifying, to being a precise, testable instantiation of, M_1 than its predecessors.
15. Smart (1963) has used the term "physicalism" to stand for the view that the world is made up entirely of physical entities of the kind postulated by fundamental physical theories – electrons, quarks and so on. As I am using the term, "physicalism" stands for the very much stronger doctrine that the universe is physically comprehensible, that it is such that some yet-to-be-discovered, unified "theory of everything" is true.
16. For much more detailed arguments for accepting metaphysical theses at levels 7 to 3, see Maxwell (2017a, ch. 9).
17. This talk of "justifying" may seem thoroughly un-Popperian in character, but it is not. What is at issue is not the justification of the truth, or probable truth, of some thesis, but only the justification of *accepting* the thesis (granted our aim is truth). Within Popper's falsificationism, there is just such a "justification" for accepting highly falsifiable (and unfalsified) theories: such theories, being most vulnerable to falsification, facilitate the discovery of error, and thus give the most hope of progress (towards truth). Acceptance of such theories is justified (according to falsificationism) because it promotes error detection and progress. This Popperian justification justifies acceptance of that theory most likely to *false*! Nothing could demonstrate more starkly that justifying *acceptance* (in the interest of discovering truth) is not at all the same thing as justifying the *truth* of a theory or proposition.
18. For further details of how AOE overcomes the two difficulties indicated earlier in this chapter, and for further details of the view itself, see Maxwell (1998, ch. 5; 2004a, chs. 1 and 2, and appendix; 2007, ch. 14; 2011b; 2013; 2017b; and especially 2017a).
19. For a more detailed rebuttal of this objection, see Maxwell (2004a, pp. 207–10; 2017a, ch. 9).
20. See Maxwell (1998, pp. 78–89, 159–63, and especially 217–23; 2017a, ch. 11; 2017b, ch. 5).
21. Elsewhere I have suggested an alternative to physicalism (see Maxwell, 2004, pp. 198–205; 2017b, ch. 5, section 10).
22. In holding that metaphysical theses and philosophies of science are an integral part of science itself, AOE implies that Popper's principle of demarcation (Popper, 1963, ch. 11) is to be rejected. Popper's demarcation proposal, apart from being untenable, is in any case too simplistic, in that it reduces to one a number of distinct demarcation issues. Popper rolls into one the distinct tasks of demarcating (a) good from bad science, (b) science from non-science, (c) science from pseudoscience, (d) rational from irrational inquiry, (e) knowledge from mere speculation, (f) knowledge from dogma (or superstition, or prejudice, or popular belief), (g) the empirical from the metaphysical, and (h) factual truth from non-factual (analytic) truth. (a) to (d) involve demarcating between disciplines, whereas (e) to (h) involve demarcating between propositions.
23. For a very much more detailed exposition of this solution to the problem of simplicity, together with an account of the way in which great unifying theories of physics illustrate the solution, see Maxwell (1998, chs. 3 and 4). See also Maxwell (2004a, pp. 160–74; 2007, pp. 373–86; 2017a, ch. 5).
24. For a more detailed discussion of the solution to the problem of verisimilitude, see Maxwell (2017a, ch. 8).
25. It may be objected that if T is assumed to be the true *unified* theory of everything, no meaning can be given to the idea that theoretical physics is making progress, by means of a succession of false theories, to a more or less *disunified* theory of everything. But T does not need to be assumed to be unified; all that is required is that T is such that the notion of "partial derivation" from T makes sense. For further discussion of the inability of any standard empiricist view such as falsificationism to solve the problem of verisimilitude, and the ability of AOE to solve the problem, see Maxwell (1998, pp. 70–2, 211–17 and 226–7).
26. For further discussion of the method of discovery provided by AOE, see Maxwell (1974, part II; 1993a, part III; 1998, pp. 159–63 and 219–23; 2017a, ch. 11; and especially 2017b, ch. 5).
27. Thus one of the aims of geology is to improve knowledge about how and when specific rock strata have been formed – knowledge about the history of a particular object, the Earth. This is very different from the aims of theoretical physics, in seeking to discover the unified laws

28. that govern all phenomena. The aims of evolutionary biology, again, have a historical aspect to them, and are thus quite different from the aims of theoretical physics.
28. Chemistry presupposes quantum theory. Thus an item (quantum theory) that is a component of level 2 as far as physics is concerned, is a high-level presupposition as far as chemistry is concerned, implicit in the basic aims of theoretical chemistry. In short, what is low down in the hierarchy for physics, is high up in the hierarchy for chemistry.
29. For more on the implications of AOE for scientific method, see Maxwell (2004a, ch. 2; 2017a, chs. 10–12).
30. This theme is developed in my "Unification and Revolution: A Paradigm for Paradigms" (Maxwell, 2014a).
31. See Maxwell (2014a) for further criticisms of Kuhn, especially in connection with "incommensurability", and for suggestions as to how Kuhn's view can be modified to move it in the direction of AOE.
32. The phrase "underlying dynamic unity in nature", I hope it is clear, is to be interpreted as appealing to the thesis that the universe is such that there is a yet-to-be-discovered, physical "theory of everything" that is both unified and true.
33. See Lakatos (1970, 1978). For Feyerabend's argument that severe testing requires the development of rival theories, see Feyerabend (1965).
34. Granted Lakatos's overall view, the research programme of science cannot have a hard core, for then, in order to ensure Popperian severe testing, there would need to be a rival research programme with a rival hard core – and that would mean the original research programme was not the whole of science. Actually, Lakatos is not quite consistent here; after the sentence quoted in the text, Lakatos goes on: "Such methodological rules may be formulated, as Popper has pointed out, as metaphysical principles. For instance, the *universal* anti-conventionalist rule against exception-barring may be stated as the metaphysical principle: 'Nature does not allow exceptions'" (1970, p. 132). That this admission is damaging for Popper's bare falsificationism was pointed out in note 12; it is equally damaging for Lakatos's version of Popperianism.
35. I say "not straightforwardly empirical" because both physicalism and the best available blueprint are themselves accepted on the grounds that they support a more empirically progressive research programme than any rival theses. Long-term empirical considerations influence choice of theses at levels 3 and 4, while at the same time these theses can lead to the rejection of potentially empirically successful theories that clash too severely with them (i.e. that are too severely *ad hoc*).
36. The Popperian and Lakatosian demand that theories be strictly universal, places weak but rigid constraints on what theories are acceptable; the demand of AOE that theories accord, as far as possible, with physicalism and the best available blueprint, places strong, but flexible and revisable constraints on what theories are acceptable.
37. For further details and discussion, see Maxwell (1998, pp. 172–80).
38. For an account of the discovery of parity non-conservation, and of the decisive character of the experiments refuting parity conservation, see Franklin (1990, pp. 6–36 and 151–2). See also Franklin (1986).
39. In order to preserve parity in the teeth of the experimental results, one would have had to argue that one or other auxiliary theory – quantum theory, the theory of weak interactions, the theory of nuclear structure or the theoretical description of the experiment – simply did not apply to this specific experiment. But to do that would have amounted to turning one or other of these auxiliary theories into a highly *ad hoc, disunified* theory – and that would clash with AOE. The requirement of overall theoretical unity (plus the experimental result) demanded that parity be rejected.

Chapter 3

1. See Maxwell (1972a) for my criticisms of Popper which led me to develop aim-oriented empiricism. See Maxwell (1979; 1993a, pp. 61–79) for early attempts at employing aim-oriented empiricism to solve the problem of induction. This is a theme which will be taken up in Chapter 7. Finally, see Maxwell (2012) for an account of how I came to develop aim-oriented empiricism as a result of pondering problems faced by Popper's philosophy of science.

2. It may be asked how it is possible for Einstein to be the first to exploit aim-oriented empiricism explicitly in scientific practice if what I have argued in Chapter 2 is correct, and aim-oriented empiricism is inherent in all of science. The answer is straightforward. Actual scientific practice is massively influenced by the long-standing conviction of the scientific community that science ought to proceed in accordance with standard empiricism. The result is that scientific practice is a mixture of aim-oriented empiricism and standard empiricism. Aim-oriented empiricism is implemented in a surreptitious, hypocritical fashion, overlaid by the conviction that science ought to proceed in accordance with standard empiricism. As a result, physicalism and more specific metaphysical blueprints are not acknowledged within the intellectual domain of scientific knowledge, and this sabotages the possibility of putting the rational method of discovery of aim-oriented empiricism into sustained scientific practice. Explicit scientific exploitation of aim-oriented empiricism is frustrated if not prohibited. (See Maxwell [1976b and 1984a] for further discussion of this point.) Einstein's great lucidity about fundamental matters led him to put aim-oriented empiricism into scientific practice unconstrained by hypocritical allegiance to standard empiricism.
3. Gerald Holton comes the closest to interpreting Einstein in the way that I do. One difference, of course, is that Holton espouses his "themata" conception of science and not aim-oriented empiricism (see Holton, 1973).
4. See Maxwell (1979, pp. 647–8; 1988, p. 42; 1998, pp. 80–2; 2017b, ch. 5, sections 5 and 6).
5. Einstein put it very clearly in a paper published in 1911. He states that as a result of assuming the equivalence of acceleration and gravitation for all phenomena "we obtain a principle that, if it really is correct, possesses great heuristic significance. For by means of theoretical considerations of processes that take place relative to a uniformly accelerated reference system, we obtain conclusions about the course of processes in a homogeneous gravitational field" (quoted in Stachel, 2007, p. 85).
6. For a more detailed discussion of the role played by the rotating disc in the genesis of general relativity, see Stachel (1980, 2007).
7. Einstein had to labour long and hard to transform the initial insight of 1911 or 1912, that gravitation is due to the curvature of space-time, into the field equations of general relativity, first formulated in their final form in 1915. For a magnificent detailed account by a number of authors of Einstein's creation of general relativity, see Renn (2007). See also Pitts (2016). For much earlier accounts, see Pais (1982, chs. 11 and 12) and Norton (1984). Authoritative and detailed as all these accounts are, none tells the story of Einstein's discovery of general relativity as the implementation of aim-oriented empiricism, as I have tried to do, all too briefly, here.
8. Good expositions of general relativity are to be found in Friedman (1983), Schutz (1988) and Misner et al. (1973). For a lively, non-technical account of the genesis of the theory and its subsequent applications to astrophysics and cosmology, with the emphasis on accounts of the physicists involved, see Ferreira (2014).
9. The best "blueprint" for physics is the best available idea as to how the universe is physically comprehensible, a vital element of theoretical scientific knowledge according to aim-oriented empiricism (see Chapters 2 and 5).
10. Lucid summaries of these papers are to be found in Lanczos (1974). They are reproduced, translated into English, in Stachel (1998).
11. He did this, too, in the introduction to his 1905 paper introducing the revolutionary idea of light quanta. He there makes clear that there is a fundamental clash between the idea of the particle, associated with theories of matter, and the field idea of Maxwell's theory of electrodynamics (see Stachel, 1998, pp. 177–8).
12. Elsewhere, I have suggested that the way to implement this method of discovery is to take, as P_1, the general idea of deterministic dynamic space-time geometry (from general relativity), and to take, as P_2, the general idea of ontological probabilism (from quantum theory), the task then being to create unified probabilistic dynamic geometry, P_3 (see Maxwell, 1985, pp. 40–1).
13. All Einstein's great contributions to physics arose out of tackling aspects of the clash between Newton and Maxwell – between the classical theories of gravitation and electromagnetism. In seeking to unify general relativity and classical electrodynamics, Einstein was, in a sense, still tackling the problem of his youth, updated by the replacement of Newton's theory of gravitation with his own theory, general relativity.
14. For an account of the great battle between Einstein and Bohr (and others) see Kumar (2008).
15. Interpreting the ψ-function as describing quantum reality directly, measurement having no fundamental role in the theory, has the consequence that quantum superpositions never

disappear. Not only does Schrödinger's cat persist as a superposition of being dead and alive, we persist as a superposition of observing the cat dead and alive when we open the box and look. These consequences might seem sufficient to rule out this interpretation of the theory. They did not prevent Hugh Everett from putting forward this interpretation of quantum theory long ago, in 1957. Recently, the Everett version of quantum theory has become almost fashionable: see Wallace (2012) for a recent exposition and defence. For critical assessments, see Bacciagaluppi and Ismae (2015) and Maxwell (2017b, ch. 5, section 6).

16. See also Maxwell (1993c; 1994a; 1995; 1998, ch. 7; 2004b; 2011a). See also Chapter 8 of the present volume.
17. The six fatal defects of OQT that I have indicated all stem from the failure of OQT to solve the fundamental quantum wave/particle problem. We can draw an important conclusion from this point: no version of quantum theory (QT) is acceptable which fails to solve the wave/particle problem. This constitutes a decisive objection to the currently fashionable Everett or "many-worlds" interpretation of QT: it provides no solution to the wave/particle problem. It does not tell us what sort of physical entity in space and time the electron, photon or atom is. The probabilistic version of QT, to be outlined in Chapter 8, here scores a striking victory over Everett QT. Probabilistic QT provides us with a very natural solution to the wave/particle problem, as we shall see in Chapter 8.
18. "Perfection of means and confusion of goals seem – in my opinion – to characterize our age" Einstein (1973, p. 337). One can regard this state of affairs as the result of the failure of our age to develop and implement a kind of rational inquiry designed to help us improve our goals, informed by aim-oriented rationalism and the philosophy of wisdom, themselves the outcome of generalizing Einstein's way of doing physics (see Maxwell, 1976b, 1984a, 2014b).
19. Perhaps, more modestly, we should say that only the top two theses of aim-oriented empiricism are synthetic *a priori* conjectures – theses we will never abandon because accepting them can only help, and cannot hinder, the search for knowledge whatever the universe may be like, and even though we have no grounds whatsoever to hold that they are true. (That these propositions are conjectures means that they are not synthetic *a priori* propositions in Kant's sense.) It is just about conceivable that we might discover that the universe is not comprehensible, and still live; it is inconceivable that we will discover that it is not partially knowable, and not meta-knowable (see Figure 2.1 in Chapter 2).
20. Reasons for accepting partial knowability and meta-knowability will be spelled out in more detail in Chapter 7. See also Maxwell (2017a, especially ch. 9; and 2017b, appendix 2).
21. In my view, the most important implications of the new way of doing physics created by Einstein in developing special and general relativity lie in fields far beyond that of theoretical physics: see the Prologue and Chapters 9 and 10 of the present volume, and Maxwell (1976a, 1984a or 2007, 2004a, 2014b, 2016b, 2017a, 2017b). See also my articles summarizing aspects of the argument in various ways (Maxwell, 1977a, 1984b, 1986, 1991, 1992a, 1992b, 1999b, 2000a, 2003a, 2003b, 2003c, 2009a, 2012, 2013), where I attempt to spell out the implications of aim-oriented empiricism for science as a whole, for technological research, social inquiry, scholarship, education and global problems confronting humanity.

Chapter 4

1. This argument was expounded in Chapter 2, section 2.3. It is taken up again in Chapter 5. For an earlier account, see Maxwell (1998, 47–56).
2. For more recent discussions of diverse aspects of the problem, see Weber and Lefevere (2014), Schurz (2014, 2015), Votsis (2015), Cohen (2015), Sterkenburg (2016), Niiniluoto (2016), and Dasgupta (2016).
3. I exempt my own earlier work on the problem (Maxwell, 1998, chs. 3 and 4; 2004a, appendix, section 2; 2007, pp. 373–86; 2011b; 2013; 2017a; 2017b).
4. If the theory is formulated as a set of differential equations, then what is invariant throughout the possible phenomena to which the theory applies is what is asserted by the physically interpreted set of differential equations. Laws specifying precisely how diverse physical states evolve in space and time may be quite diverse in character: what matters is that they are all solutions of the same set of differential equations.

5. Counting entities is rendered a little less ambiguous if a system of M particles is counted as (a somewhat peculiar) field. This means that M particles all of the same kind (i.e. with the same dynamic properties) are counted as one entity. In the text I continue to adopt the convention that M particles all the same dynamically represent one kind of entity, rather than one entity.
6. For accounts of the locally gauge invariant structure of quantum field theories, see Aitchison and Hey (1982, part III), Moriyasu (1983), and Griffiths (1987, ch. 11). For introductory accounts of group theory as it arises in physics, see Isham (1989) or Jones (1990).
7. For accounts of spontaneous symmetry breaking, see Moriyasu (1983), Mandl and Shaw (1984), and Griffiths (1987, ch. 11).
8. This account of unity radically simplifies and improves on the account given in Maxwell (1998, chs. 3 and 4).
9. I am grateful to Jos Uffink for drawing my attention to the two objections just discussed.
10. For further discussion of simplicity, and how terminological simplicity can be related to unity, see Maxwell (1998, pp. 110–3 and 157–9).
11. For a discussion of such "approximate derivations", the conclusion being strictly incompatible with the premises, see Maxwell (1988, 211–7).
12. See also Chapters 5 and 7, and Maxwell (1998, 2004a, 2017a, 2017b).

Chapter 5

1. The date of my first publication arguing for aim-oriented empiricism (see Maxwell, 1974).
2. See especially Maxwell (1998); see also Maxwell (1976a, ch. 6; 1984a, chs. 5 and 9, or 2007, chs. 5, 9 and 14; 1993a; 1997a; 1999a; 2000b; 2001a, ch. 3 and appendix 3; 2002d; 2004a; 2008; 2009d; 2010a, ch. 5; 2011b; 2012; 2013; 2014a; 2016a; 2017a; 2017b).
3. Philosophers of science who have praised or criticized aim-oriented empiricism include Kneller (1978, pp. 80–7, 90–1), Harré (1986, pp. 26–32), Midgley (1986), Chakravartty (1999), Chart (2000), Juhl (2000), Shanks (2000), Smart (2000), Weinert (2000), McHenry (2000, 2009a), Roush (2001), Muller (2004, 2008), Schiff (2005), Miller (2006, pp. 92–4), MacIntyre (2009), Rogers (2009), Vicente (2010), Pandit (2010). For my replies to criticisms, see Maxwell (2001c, 2006, 2009b, 2010b).
4. An earlier version of this chapter was delivered as a talk at a conference on "Induction, Confirmation and Science" at the London School of Economics on 10 March 2007. It was then perhaps rather more paradoxical to talk of "scientific metaphysics" than it is now, at the time of writing (2016). This is because, in recent years, as I mentioned in Chapter 2, note 8, something approaching a research industry in the philosophy of science has emerged devoted to the topic "the metaphysics of science" (see note 7 below and associated text). "Scientific metaphysics" has rather lost its paradoxical air.
5. It is widely appreciated that some metaphysical theses have influenced science in the context of discovery, in influencing scientists to try to develop certain sorts of theories, and to ignore others (see, for example, Watkins, 1958; Popper, 1959, p. 278). What is being argued here is very different. I argue that there are metaphysical theses, neither falsifiable nor verifiable, which are an integral part of theoretical scientific knowledge, more firmly established, indeed, than such highly corroborated theories as quantum theory and general relativity. Science is more rigorous if this is acknowledged rather than denied. All this differs dramatically even from Popper's later views concerning the important role that metaphysical research programmes play in science (see Popper, 1976a, sections 33 and 37; 1982b, sections 20–8; 1983, section 23). Popper held on to his demarcation criterion to the end, and never wavered from holding metaphysical theses, however scientifically fruitful, to be "unscientific" (for discussion of this point, see Chapter 4 and Maxwell, 2012).
6. For discussion of the claim that Kuhn and Lakatos defend versions of SE, see Chapter 2 and Maxwell (1998, p. 40).
7. For recent contributions to the "metaphysics of science", see note 8 of Chapter 2, and Maudlin (2007); Ladyman and Ross (2007); Chakravartty (2007); Lange (2009); Kuhlmann (2010); Ladyman (2012); Kincaid, Ladyman and Ross (2013); Morganti (2013); Mumford and Tugby (2013); Robus (2015); Brown (2016); Andersen and Arenhart (2016). Some of this work argues for conjectural essentialism, the anti-Humean view that theoretical physics should be interpreted as seeking to discover, not just physical laws, or regularities in phenomena, but

rather "necessitating properties" possessed by fundamental physical entities which determine, necessarily, that the entities in question evolve in accordance with specified laws. This is a view I argued for decades ago (see Maxwell, 1968).

8. All the possible phenomena, predicted by any dynamical physical theory, T, may be represented by an imaginary "space", S, each point in S corresponding to a particular phenomenon, a particular kind of physical system evolving in time in the way predicted by T. In order to specify severely disunified rivals to T that fit all available evidence just as well as T does, all we need do is specify a region in S that consists of phenomena that have not been observed, and then replace the phenomena predicted by T with anything we care to think of. Given any T, there will always be infinitely many such disunified rivals to T. This point is inherent in Nelson Goodman's "new paradox of induction" (see Goodman, 1954), although the kind of empirically successful disunified rivals considered by Goodman in his discussion of "grue" and "bleen" are but one kind of a number of kinds of disunified theories, as we shall see in section 5.3. There is a vast philosophical literature on the underdetermination of theory by evidence. For an excellent recent discussion, and reference to further literature, see Howson (2000, especially ch. 1, pp. 30–4, 75–7, and ch. 5). See also Lipton (2004).

9. For a more detailed discussion of empirically successful *ad hoc* rivals to accepted theories, see Maxwell (1974; 1993a; 1998, pp. 51–4).

10. Vicente (2010) claims that this argument cannot establish that physics assumes the truth of some more or less unified physical theory, and the falsity of all seriously disunified rival theories. It is, however, especially in the context of practical applications that physics requires predictions of accepted theories to be true, and clashing predictions of rival theories to be false. In contexts such as bridge-building, for example, we do indeed need to assume that relevant accepted laws will yield true predictions, and better empirically established but disunified rivals, that predict the bridge will collapse, are false. (Human lives are at stake.) In making such an assumption, against the evidence, we implicitly assume the truth of a metaphysical thesis concerning the unity of phenomena. For a more detailed rebuttal of Vicente's claim, see Maxwell (2010b, pp. 673–4).

11. It may be objected that the universe might have been genuinely disunified, so that physics could consist only of a great number of physical laws. In this case, it may be argued, physics could not be construed as making a metaphysical assumption about underlying unity. But even in this counterfactual situation, endlessly many very much more disunified but empirically more successful rival laws could easily be formulated: these would have to be rejected on non-empirical grounds, or physics would drown in an ocean of rival laws. The persistent rejection of such much more disunified but empirically more successful rivals would involve the methods of physics making an implicit metaphysical assumption, to the effect that nature is unified to some extent at least (all grossly disunified laws being false). It is necessary to make some such assumption, however disunified the totality of accepted laws may be – even if the assumption made is rather weak in character, in that only gross disunity is denied.

12. See Maxwell (1984a, p. 224; 1998, p. 21).

13. See the previous chapter, section 4.1, note 3 and associated texts for various attempts to solve the problem, and criticisms of those attempts.

14. The account of theoretical unity given here and in the last chapter simplifies the account given in Maxwell (1998, chs. 3 and 4), where unity is explicated as "exemplifying physicalism", where physicalism is a metaphysical thesis asserting that the universe has some kind of unified dynamic structure. Explicating unity in that way invites the charge of circularity, a charge that is not actually valid (see Maxwell, 1998, pp. 118–23 and 168–72). The account given here forestalls this charge from the outset.

15. This point is of fundamental importance for the problem of induction. Traditionally, the problem is interpreted as the problem of justifying exclusion of empirically successful theories that are *ad hoc* in sense (1): how can evidence from the past provide grounds for any belief about the future? This makes the problem seem highly "philosophical", remote from any problem realistically encountered in scientific practice. But the moment it is appreciated that the problem of justifying exclusion of empirically successful theories that are *ad hoc* in sense (1) is just an extreme, special case of the more general problem of excluding empirically successful theories that are *ad hoc* in senses (1) to (8), it becomes clear that this latter problem is a scientific problem, a problem of theoretical physics itself. For the implications of this crucial insight, and for a proposal as to how the problem of induction is to be solved that exploits this insight, see Maxwell (1998, especially chs. 4 and 5). See also Chapter 7 of the present work, and Maxwell (2004a, appendix, section 6; 2007, ch. 14, section 6; 2017b; 2017a, especially ch. 9).

16. It may seem that there is rather a jump here, from T referring to any fundamental dynamical physical theory in (1) to (8), to T referring to a "theory of everything". However, the proper way to apply (1) to (8) is to the totality of fundamental physical theory (whether this consists of many or just one theory), and thus, in a sense, to candidate "theories of everything". (If a range of phenomena has no theory, then empirical laws governing these phenomena must be treated as theories.) If we do not do this, disunity could always be evaded, as far as (4) to (7) are concerned at least, by chopping a theory disunified to degree N into N distinct unified theories. When it comes to non-empirical considerations governing choice of theory, what matters is the way individual theories fit into the totality of fundamental physical theory – the degree of unity of the whole of fundamental physical theory.
17. Although, even in this case, one could imagine that there are different degrees of unification of space-time and matter.
18. Is physicalism(n,N) to be interpreted so as to be compatible with stronger versions of physicalism, such as physicalism(n+1, N-1) – assuming here that n < 8 and N > 1? If we want the different versions of physicalism to constitute metaphysical theses that are, as far as possible, mutually exclusive, then we should interpret physicalism(n,N) to be incompatible with stronger, more unified versions of physicalism. But if we want physicalism to play the role in physics of excluding more or less disunified theories, then it will be convenient to interpret physicalism(n,N) to be compatible with stronger, more unified versions of the doctrine. In this second case, physicalism(n,N) has the role of excluding theories more disunified than (n,N), but not theories more unified than this. In what follows, physicalism is to be interpreted in this latter way, as we will be considering the role physicalism has in excluding disunified theories from physics.
19. This explication of the "unity" and "explanatory power" of theories improves on proposals put forward by Friedman (1974), Kitcher (1981), Watkins (1984) and others referred to in section 5.1 and note 3 of the previous chapter. For a critical assessment of these and other proposals, see Salmon (1989) and Maxwell (1998, pp. 61–8). Maxwell (1998) also contains a detailed account of my positive theory of explanatory power (see especially chs. 3 and 4).
20. For further discussion, see Maxwell (1998, pp. 80–9, 131–40, 257–65, and additional works referred to therein).
21. Note 5 above indicates how my view differs from Popper's "metaphysical research programmes".
22. See Chapter 8 for further discussion of these issues.
23. Lagrangianism is discussed in Maxwell (1998, pp. 88–9).
24. See Maxwell (2004a, pp. 198–205) for a suggestion along these lines.
25. In practice, physical theory is persistently used to correct clashing experimental results, in that theory is used to reveal that experimental equipment is not working properly.
26. What does "increase the conflict" mean here? It means that the kind or degree of unity of the totality of fundamental physical theory gets worse with respect to the currently accepted metaphysical thesis at level 3. (This thesis asserts that there is a certain kind of unity in nature. The more the totality of physical theory departs from this kind of unity, the greater the "conflict" with the thesis at level 3.)
27. See Maxwell (1974, especially part II) and, more recently, Maxwell (1998, pp. 17–9). Others, too, have argued that the methods of science improve with improving knowledge, but have done so only within the framework of standard empiricism (see, for example, Boyd, 1980).
28. No attempt is being made in this chapter, I hope it is clear, to solve the problem of induction. I merely seek to rebut the objection that the problem cannot conceivably be solved, granted AOE. In Chapter 7, I will, however, argue that AOE solves the problem of induction – as I have argued elsewhere (see Maxwell, 1998, ch. 5; 2004a, appendix, section 6; 2007, ch. 14, pp. 400–30; 2017b; and 2017a, especially ch. 9).
29. I here echo the two reasons for accepting meta-knowability given in Chapter 2.
30. I assume that this is being read before 2050!
31. Consider a universe such that progress in theoretical physics requires infinitely many theoretical revolutions, each revolution leading to the degree of unity of the totality of physical theory going up by one. In such a universe, meta-knowability is true, since it possesses a general feature which, once discovered, would aid progress in physics; nevertheless, it is also the case that physicalism($N = \infty$) is true. Physicalism holds, meta-knowability holds, but the universe is infinitely disunified, infinitely incomprehensible.

Chapter 6

1. See also Maxwell (1972a; 1974; 1984a, or 2007, especially ch. 14; 1993a; 2001a, ch. 3 and appendix 3; 2004a; and especially 1998; 2017a; and 2017b. For summaries, see Maxwell 1999a; 2000b; 2009d; 2010a, ch. 5; 2011b; 2013; 2014a; 2015a; 2016a).
2. See Chapter 3. See also Maxwell (1998, pp. 219–23; 2017a, ch. 11; 2017b, ch. 5).
3. Elsewhere I have defended an objectivist, realist account of value (see Maxwell, 1984a, ch. 10; 1999b; and 2001a, ch. 2). This does not, however, affect the present argument.
4. Bassi et al. (2013) discuss a number of fundamentally probabilistic versions of quantum theory, and how they can be tested experimentally; Gao (2017) expounds a probabilistic version of quantum theory. Both works provide many references to earlier papers and books in the field.
5. There is, it is true, the additional requirement of simplicity. This, however, presupposes unity, and is not as methodologically significant as unity.
6. McAllister might, of course, reject SE and defend MAI in such a way that MAI acknowledges that science makes a substantial, permanent metaphysical assumption about the nature of the universe – namely that the universe is such that no *ad hoc* theory is true. But at once two major problems arise: What precisely is this assumption in view of the fact that there is no sharp distinction between the *ad hoc* and the non-*ad hoc*? What is the justification for making this assumption? In order to answer these questions satisfactorily, it is necessary to adopt AOE, which involves abandoning those parts of MAI which clash with AOE.

Chapter 7

1. See Chapter 1 of the present work for a discussion of Popper's attempt at solving the problem of induction. The problem was also discussed in Chapter 5. David Hume formulated and discussed the problem in Hume (1959, vol. 1, part III). For somewhat more recent discussions of the problem, and surveys of somewhat more recent literature on the subject, see Kyburg (1970), Swain (1970), Watkins (1984), Howson (2000) and Vickers (2016).
2. These include problems of unification and verisimilitude, the problem of rational scientific discovery, the problem of saying what it is that science has discovered about the ultimate nature of reality, problems concerning rationality and the nature of social inquiry, and, most important of all, the discovery that academic inquiry as it exists at present is profoundly defective when viewed from the standpoint of its capacity to help us learn how to become more civilized, there being an urgent need to bring about a revolution in the overall aims and methods of inquiry if we are to have what we so urgently need, a kind of inquiry rationally devoted to helping us acquire wisdom (see, for example, Maxwell, 1976a; 1980; 1984a or 2007; 1984b; 1985b; 1987; 1991; 1992a; 1994b; 1997b; 1998, ch. 3, 4 and 6; 2000a; 2000b; 2002b; 2003a; 2003b; 2003c; 2004a; 2008; 2010a; 2011b; 2012; 2013; 2014a; 2016a; 2016b; 2017a; 2017b).
3. Work of mine related to my proposed solution to the problem of induction has received some critical attention: see, for example, Kneller (1978, pp. 80–91), Longuet-Higgins (1984), Collingridge (1985), Easlea (1986), Midgley (1986), Smart (2000), Roush (2001), Hodgson (2002), Muller (2004), Iredale (2005), McHenry (2009b), Vicente (2010), Müürsepp (2014).
4. I sketched how to solve the circularity problem in Chapter 5, section 5.9. An earlier defence against the charge of circularity was brief and unsatisfactory (see Maxwell, 1998, pp. 166–8).
5. See Maxwell (2017a, especially ch. 9). See also Maxwell (2017b). For earlier attempts at solving, or contributing to the solution to, the problem of induction, see Maxwell (1968; 1972a; 1974; 1977a; 1979; 1984a, or 2007, especially ch. 14; 1993a; 1997a; 1998; 1999a; 2000b; 2001a, ch. 3 and appendix 3; 2002b; 2002e; 2004a; 2009d; 2010a, ch. 5; 2011b; 2013; 2015a; 2016a).
6. One of the great mistakes made by endlessly many attempts at solving the problem of induction is to assume unthinkingly that science is wholly in order just as it is, the task being to find some way of justifying existing valid methods of science. What the long-standing insolubility of the problem of induction is trying to tell us, in my view, is that the orthodox conception of science, taken for granted by scientists and non-scientists alike, is untenable, and needs

to be changed. Before the problem of induction can be solved, we need to change the currently accepted conception of science; indeed, we need to change, not just our conception of science, but science itself. Properly conceived, the problem of induction involves formulating a new conception of the aims and methods of science, more rigorous than current conceptions, which is such that the problem of induction no longer arises. The task is not to justify the status quo, but to change the status quo so that the problem of justification no longer arises. Popper's failed attempt at solving the problem stands head and shoulders above the rest just because it fits this prescription: it consists of a new view about the aims and methods of science, a new philosophy of science, namely falsificationism.

7. Aim-oriented empiricism was first expounded in Maxwell (1974). It was spelled out in greater detail in Maxwell (1984, 1998).
8. For the AOE account of what theoretical unification means, see Chapter 4.
9. In what follows "accept T" implies, not just "accept T as a working hypothesis for further research", but also "accept T for the purposes of technological and other practical applications, including contexts where human life may depend on the predictions of T being true".
10. All the possible phenomena, predicted by any dynamical physical theory, T, may be represented by an imaginary "space", S, each point in S corresponding to a particular phenomenon, a particular kind of physical system evolving in time in the way predicted by T. In order to specify severely disunified rivals to T that fit all available evidence just as well as T does, all we need do is specify a region in S that consists of phenomena that have not been observed, and then replace the phenomenon predicted by T with anything we care to think of. Given any T, there will always be infinitely many such disunified rivals to T.
11. For a more detailed discussion of empirically successful *ad hoc* rivals to accepted theories, see Maxwell (1974; 1993a; 1998, pp. 51–4; 2017a, ch. 4).
12. The two prescriptions for formulating empirically more successful rivals to accepted unifying theories, indicated in this and the previous paragraph, can of course be combined to create further more empirically successful disunified rival theories.
13. This needs to be amended in two ways. In the first place any number of approximate disunified theories would be implied by the true, unified "theory of everything" (supposing there is such a thing). If, for example, we take Newtonian theory as standing in for the true, unified theory of everything, then we can derive from this theory a combination of appropriately approximate versions of Kepler's laws of planetary motion and Galileo's laws of terrestrial motion, which, put together, constitute a disunified theory, true if Newtonian theory is true. "Science assumes that the universe is such that no disunified theory is true", must be amended to read, "Science assumes that the universe is such that no precise disunified theory is true". But even this amendment is not sufficient. Newtonian theory implies different, precise laws of planetary motion for the different planets in their orbits round the sun. Gather these precise laws together, and one has a precise but thoroughly disunified theory of planetary motion. These laws are, however, specific to precise specifications of initial conditions: initial relative positions and velocities, and masses. These laws cease to be true (granted that Newtonian theory is true), if the initial conditions are changed even slightly (in a way which does not obtain during the orbit of the planet in question). In other words, these laws cease to be consequences of Newtonian theory the moment they are interpreted to apply to a range of values of initial conditions (of the indicated kind), however minute this range of values might be. We must amend the statement in the text to read: "Science assumes that the universe is such that no precise disunified theory is true, where the components of the disunified theory are interpreted to apply to a range of values of initial conditions (of the appropriate kind), however minute that range might be." What does "of the appropriate kind" mean here? It means this. Variables that characterize initial conditions are "of the appropriate kind", if the component laws of the disunified theory apply only to different values of these variables. These variables are such, in other words, that in order to move from conditions for which one component law of the disunified theory applies, to conditions for which another component law applies, it is the value of these variables that we need to change. The range of values must not correspond to different states of the physical system that evolve from just one precise initial state of the system.
14. For a much more detailed exposition of this refutation of standard empiricism, see Maxwell (1998, ch. 2). See also Maxwell (2017a, ch. 4).
15. For discussion of the claim that Kuhn and Lakatos defend versions of SE, see Maxwell (1998, p. 40). Bayesianism might seem to reject SE, in acknowledging both prior and posterior probabilities. But Bayesianism tries to conform to the spirit of SE as much as possible, by regarding

prior probabilities as personal, subjective and non-rational, their role in theory choice being reduced as rapidly as possible by empirical testing (see Maxwell, 1998, p. 44).
16. See, for example, Maudlin (2007); Ladyman and Ross (2007); Chakravartty (2007); Lange (2009); Kuhlmann (2010); Achinstein (2010); Ladyman (2012); Kincaid, Ladyman and Ross (2013); Morganti (2013); Mumford and Tugby (2013); Dawid (2014); Robus (2015); Brown (2016); Andersen and Arenhart (2016); Scerri (2016); Schrenk (2016).
17. What does it mean to say that U_2 is an "improvement" over U_1, U_2 being "closer" to the true characterization of the unity of nature, T (supposing there is such a thing) than U_1 is? This problem can be solved by exploiting the solution to the problem of verisimilitude, which I have spelled out in some detail elsewhere (see Maxwell, 2007, pp. 393–400; 2017a, ch. 8). The problem of verisimilitude is this: What does it mean to say that physics makes progress towards the truth, if it advances from one false theory to another? What does it mean, in other words, to say that fundamental physical theory T_2 is closer to the truth than T_1, if both T_2 and T_1 are false? Let T, as before, represent the truth – the true "theory of everything". Then we can say that T_2 is closer to the truth than T_1 if T approximately implies T_2, and T_2 approximately implies T_1, but T_1 does not approximately imply T_2. The key notion of "approximately implies" can be illustrated by means of the way Newtonian theory approximately implies Kepler's law that planets move in ellipses. There are three steps. First, Newtonian theory is restricted to systems consisting of rather few bodies that move within a finite spatial region. Second, the masses of all the bodies but one tend to zero which, in the limit, leads to the bodies of zero mass tracing out ellipses round the remaining massive body. Third, the resulting laws are reinterpreted to apply to systems of bodies such that one body (the sun) is very much more massive than all the others (the planets), these others nevertheless having masses greater than zero. (It is this third step that introduces error.) More generally, T_2 approximately implies T_1 if a theory simulating the empirical predictions of T_1 can be extracted from T_2 by means of a finite number of steps of the kind just illustrated: T_2 is restricted in scope; non-zero physical variables tend to zero; laws applicable to precise values of variables are reinterpreted to apply to a range of values of physical variables. (For more details, see the works to which I have just referred.) We can now exploit this solution to the problem of verisimilitude to explicate what it means to say that the metaphysical thesis of unity, U_2, is closer to T than U_1.

Let T_2 and T_1 be theories that accord with, or exemplify, U_2 and U_1 respectively. U_2 can be said to be closer to T than U_1 if T_2 is closer to the truth, T, than T_1 is, and there are no theories, T_2* and T_1*, exemplifying U_2 and U_1 respectively such that T_1* is closer to the truth than T_2*.
18. For example, Gordon Fleming (personal communication).
19. How can level 3 assumptions, or assumptions higher up in the hierarchy, both influence, and be influenced by, level 2 theories? What makes this possible is the feature of the hierarchy about to be indicated in the text, namely that, as one goes up the hierarchy, assumptions are more and more firmly upheld. Level 2 theories that accord with the best available level 3 assumption tend to be favoured over rivals that do not so accord. Nevertheless, a level 2 theory that clashes with the current level 3 assumption, but (a) accords with the level 4 assumption, and (b) is more empirically successful than theories that are in accord with the best level 3 assumption, will be accepted, and will lead to the rejection, or modification, of the level 3 assumption with which it clashes. Consider, however, a theory that clashes, not just with level 3, but with level 4 as well, even though it is compatible with level 5, in such a way that no version of the idea that the universe is physically comprehensible, at level 4, can be rendered compatible with the theory. Such a theory would have to meet with far greater, sustained empirical success before it led to the overthrow of the current level 4 assumption. It would have to lead to empirical research programmes across a broad front of natural science even more successful than current science, based on the current level 4 assumption, before it would become acceptable. And this would be the case even more, given a theory that clashes with the current level 5 assumption. In short, an assumption at a given level may, for much of the time, determine choices lower down in the hierarchy; but every now and again, it may itself be revised, because the revised version accords better with the assumption above or is more empirically fruitful, or, more likely, both of these simultaneously.
20. Corresponding to each cosmological thesis, at level 3 to 7, there is a more or less problematic aim for theoretical physics: to specify that cosmological thesis as a true, precise, testable, experimentally confirmed "theory of everything". The aim corresponding to level 7 is relatively unproblematic: circumstances will never arise such that it would serve the interests of acquiring knowledge to revise this aim. As one descends the hierarchy of cosmological

assumptions, the corresponding aims become increasingly problematic, increasingly likely to be unrealizable, just because the corresponding assumption becomes increasingly likely to be false. Whereas upper-level aims and methods will not need revision, lower-level aims and methods, especially those corresponding to level 3, will need to be revised as science advances. Thus lower-level aims and methods evolve within the fixed framework of upper-level aims and methods.

21. Some features of AOE may seem reminiscent of Laudan's "normative naturalism" (see Laudan, 1984, 1987). There are, however, marked differences: "normative naturalism" is not committed to physicalism, and does not postulate the hierarchy of aims and methods of AOE, which makes the rational assessment of low-level aims and methods possible. I might add that Laudan's "normative naturalism" is derived from AOE, which was first expounded in a colloquium I gave at the University of Pittsburgh in 1972, chaired by Laudan, the text of which became my paper "The Rationality of Scientific Discovery" (Maxwell, 1974). Rescher has defended the view that science makes metaphysical presuppositions (see Rescher, 1973, 1977, 1987); his views also differ substantially from AOE. For an excellent survey of methodological views, including those of Laudan and Rescher (but excluding AOE) see Nola and Sankey (2000).

22. At this point I confess that in *The Comprehensibility of the Universe* (Maxwell, 1998, pp. 192–3) I give a third argument for accepting meta-knowability which does, perhaps, contain a whiff of circularity, in that it appeals to the apparent success of science. This suffices, I now think, to make this argument circular.

23. What justifies the claim that physicalism has been more fruitful for theoretical physics than any rival idea? This is justified by the point made in section 7.2. All new, revolutionary, fundamental physical theories have been accepted because they (a) have brought greater unity to physics, and (b) have been more empirically successful, than any rivals – (a) being just as important as (b). In other words, the persisting non-empirical requirement for acceptance of revolutionary theory has been enhanced exemplification of physicalism (as far as theoretical physics as a whole is concerned). What irony that scientific revolutions – just that which convinced Kuhn (1970a) that there are ruptures in science with nothing theoretical surviving each rupture – actually demonstrate just the opposite: the persistent and increasingly successful search for unity, the assumption of underlying unity being repeatedly reinforced by each successive revolution. It may be asked: But how can revolutionary theories reinforce physicalism when the totality of physical theory has always, up till now, clashed with physicalism? The answer: If physicalism is true, then all physical theories that only unify a restricted range of phenomena, must be false. Granted the truth of physicalism, and granted that theoretical physics advances by putting forward theories of limited but ever increasing empirical scope, then it follows that physics will advance from one false theory to another, all theories being false until a unified theory of everything is achieved (which just might be true). The successful pursuit of physicalism requires progressive increase in both empirical scope and unity of the totality of fundamental physical theory. It is just this which the history of physics, from Galileo to today, exemplifies – thus demonstrating the unique fruitfulness of physicalism.

24. For further details of the argument for AOE, see Maxwell (1998; 2001a, ch. 3 and app. 3; 2002d; 2003d; 2007, ch. 14; and especially 2017a and 2017b).

25. Laudan (1977, 1984), inspired by Maxwell (1974, and earlier personal communication), does argue for changing methods of science within an SE view. But because of the anti-realist and SE character of his view, its lack of the hierarchical "meta-methodological" character of AOE, Laudan cannot do justice to the idea that new aims-and-methods need to be appraised so that those selected improve on earlier aims-and-methods, there being positive feedback between improving knowledge and improving knowledge-about-how-to-improve-knowledge – a key feature of the rationality of science, according to AOE, and one which helps account for the explosive growth of modern science.

26. See note 2 for references to recent attempts to solve the problem within the framework of SE.

27. Galileo's laws of terrestrial motion and Kepler's laws of planetary motion are contradicted by Newtonian theory, in turn contradicted by special and general relativity. The whole of classical physics is contradicted by quantum theory, in turn contradicted by quantum field theory. Science advances from one false theory to another. Viewed from an SE perspective, this seems discouraging and is often called "the pessimistic induction". Viewed from an AOE perspective, as I have already mentioned, this mode of advance is wholly encouraging, since it is required by AOE. Granted physicalism, the only way a dynamical theory can be precisely true of any

restricted range of phenomena is to be true of all phenomena. All physical theories must be false until we obtain a theory of everything!
28. See especially Maxwell (2017b); see also Maxwell (2017a).

Chapter 8

1. Good introductory accounts of OQT, increasingly technical, are Squires (1986), Gillespie (1973), and Feynman et al. (1965). See also Maxwell (1998, appendix).
2. The argument that the failure of OQT to solve the wave/particle problem leads remorselessly to multiple, severe defects in the theory was developed by me in a series of papers (see Maxwell, 1972b, 1973b, 1976b, 1982, 1988, 1993c, 1994a, 1995). Some of the points were developed independently by John Bell (1987). Bell tended to restrict himself, however, to the point that, if quantum theory is about measurement, it is inherently imprecise. I discuss Bell's contribution in Maxwell (1992c).
3. Rival interpretations of quantum theory include Bohm's interpretation, according to which quantum systems are both particles and waves; Everett's many-worlds interpretation; decoherence; and consistent histories. None of these, in my view, provides us with a satisfactory version of quantum theory. For critical surveys and further literature, see Squires (1986), Rae (2002, ch. 13) and Bacciagaluppi (2003). For a fairly recent exposition and defence of Everett's interpretation, see Wallace (2012). For a review, see Bacciagaluppi and Ismae (2015).
4. Popper has suggested that probabilism is the key to understanding wave/particle duality, and has put forward a propensity interpretation of quantum theory (see Popper, 1957b, 1967, 1982b). His interpretation of quantum theory is, however, unsatisfactory and quite different from the one I advocate here. For my criticisms of Popper, see Maxwell (1976b, pp. 285–6; 1985, pp. 41–2).
5. Elsewhere (Maxwell, 1976b, pp. 283–6; 1988, pp. 44–8) I have indicated how the notion of probabilistic physical property, or propensity, that is being presupposed here, amounts to a probabilistic generalization of the notion of deterministic, necessitating property explicated in Maxwell (1968; see also Maxwell, 1998, pp. 141–55). I might add, no doubt controversially, that in my view, my 1968 paper gives the definitive account of how dispositional, necessitating properties in physics should be conceived. This viewpoint, in particular, makes no appeal to Kripke's (1981) fallacious considerations concerning identity and necessity; for a refutation of Kripke, see Maxwell (2001a, appendix 2). Much subsequent work on dispositional properties in science is vitiated by a failure to take my earlier work into account, and a reliance instead on Kripke.
6. For a survey of more recent proposals, and attempts to test them experimentally, see Bassi et al. (2013); see also Gao (2017). These authors, including Ghirardi et al. and Penrose, do not stress, however, as I do in this chapter, that probabilism provides a very natural solution, potentially, to the key quantum wave/particle problem.
7. See Maxwell (1972b; 1973a; 1973b; 1976b; 1982; 1988; 1993c; 1994a; 1995; 1998, ch. 7; 2004b; 2011a).
8. See also Maxwell (1974; 1984, ch. 9, or 2007, chs. 9 and 14; 1998; 2004b; 2017a; 2017b).

Chapter 9

1. As the argument of this chapter develops, it will emerge that the position I wish to defend ought really to be called specio-universalism, rather than just universalism. It is, in a sense, an admixture of specialism and universalism. Or rather, it ought really to be called critical specio-universalism, but that is too much of a mouthful to repeat throughout the chapter. I should add that, having selected "universalism" to stand for the view I wish to defend, I was somewhat dismayed to find that the *Oxford Concise Dictionary* defines "universalism" as the Christian view that everyone will be saved. This has the merit of being more humane than those views which consign many to eternal damnation. Nevertheless, in what follows, please ignore this Christian interpretation of the term. "Universalism", here, means what I say it means.

2. These fundamental problems may of course be formulated a little differently from this without affecting the overall argument. I shall argue, in fact, that these problems need to be understood, at the most fundamental level, as personal and interpersonal, or social, problems which we encounter in our lives. The exact form in which problem (3), for example, arises for any individual will depend upon the circumstances in which the individual finds himself. "How can I get enough to eat?", "How can I find worthwhile, productive work to engage in?", "How can I give and receive love?", "How can my life be of value if I am to grow old and die?", "How can I escape being killed?", "What am I to do with my life?", "How can I develop my present pursuits so that I achieve more successfully that which is of real value?" – these can all be regarded as possible variants of problem (3).
3. "… the one method of all rational discussion … is that of stating one's problem clearly and of examining its various proposed solutions critically" (Popper, 1959, p. 16).
4. We may regard a problem as having the form of an aim we seek to realize and some provisional idea for a route to the realization of our aim, which fails, however, to enable us to achieve the aim. As a result of representing problems in this fashion, we may well adopt the idea that rationality involves quite essentially seeking to improve our aims and methods as we act by imaginatively developing and critically scrutinizing possible and actual aims and methods. For an exposition of this somewhat more sophisticated "aim-oriented" conception of rationality – and its implications for intellectual inquiry – see Chapters 4, 5, 7 and 8, and Maxwell (1976a, esp. ch. 9; 1984a, chs. 3 and 4).
5. See Maxwell (2010a, ch. 2; 2014c, ch. 2; 2017b, ch. 3).
6. Rules (a) and (b) specify universalism, pure and simple. It is the addition of rules (c) and (d) that transforms universalism into what may be called specio-universalism (see note 1). Specialism results when rule (c) alone is implemented, rules (a), (b) and (d) being ignored, at least as far as fundamental problems are concerned.
7. See Maxwell (2017b, especially ch. 3) for an account of the history of modern philosophy along these lines.
8. I return below to this issue of the failure of philosophy to keep universalism alive.
9. I am grateful to L. Briskman for provoking me into discussing this objection explicitly.
10. Problem P_1 is more fundamental than P_2 if (a) the solution to P_1 solves P_2, but not vice versa; and (b) the solution to P_1 is unified or coherent in some significant, substantial sense of these terms, and not just a jumble of disconnected items. An example of a unified or coherent solution is a unified physical theory that solves a range of problems in physics (see Maxwell, 2014b, pp. 14–5). For what it means to say that a physical theory is unified, see Chapter 4, and also Maxwell (1998, chs. 3 and 4; 2017a, ch. 5).
11. I even put this forward as a psycho-neurological hypothesis: our wonderful unconscious problem-solving capacity, which we exhibit so effortlessly in life whenever we perceive, understand, speak and act, is due to the fact that a fundamentalist hierarchical structure is programmed, as it were, into the neurological structure of our brains. This has evolved as a result of natural selection (problem-solving ability – and above all the ability to solve relevant problems, procured by the fundamentalist hierarchical structure – having great survival value). Unfortunately, at present, nothing like so intelligent a structure is built into scientific, academic inquiry – or into much conscious thought – in that here, lamentably, specialism prevails. In particular, we have failed to build the hierarchical structure of universalism into our civilization. Not surprisingly, this civilization, or world order, at present exhibits a terrifying failure to recognize and resolve its fundamental problems – problems most relevant to the achievement of what is of most value-even to the extent that its very survival is now in doubt.
12. Some modern writers have done full justice to the great potential value of living and working in a small community or "tribe": see, for example, Turnbull (1976), Schumacher (1973), and "A Blueprint for Survival" (*The Ecologist*, 1972). Popper's failure to recognize this potential must be due partly to his being unacquainted with the anthropological evidence. He asserts that "the main element" of the tribal "magical attitude towards social custom" is "the lack of distinction between the customary or conventional regularities of social life and the regularities found in 'nature'", this often being associated with "the belief that both are enforced by a supernatural will". Social customs are rigidly maintained, there being a superstitious fear of change, magical "taboos rigidly regulating [and dominating] all aspects of life". Significantly, Popper adds that "comparatively infrequent changes have the character of religious conversions or convulsions, or of the introduction of new magical taboos" (Popper, 1966a, vol. I, p. 172). It is striking that Turnbull finds all these Popperian characteristics of tribal life dominating the

life of agricultural Bantu tribes in Central Africa. Turnbull describes just such a rigid, taboo-ridden, superstitious, compulsive, fearful, ritualistic way of life. Turnbull's really remarkable discovery, however, is that all this is entirely absent in the Pygmy hunting and gathering tribal way of life. The Pygmies' lives are imbued with a quite extraordinary spontaneity, grace and trust, there being a complete absence of superstition, compulsive ritual or fearful observance of taboo. Turnbull argues, in my view entirely convincingly, that it is the development of agriculture which is responsible for this dramatic difference in the whole way of life. Hunting and gathering tribes can afford to live spontaneously, from day to day, trusting in the forest to provide food for tomorrow. Agricultural tribes, on the contrary, live in a state of constant battle with the environment and must perform persistent, long-term agricultural work before food and reward are eventually forthcoming. (M. Harris, in his *Cannibals and Kings*, comes to the conclusion, from a consideration of archaeological evidence, that early hunting and gathering tribes "enjoyed relatively high standards of comfort and security" [Harris, 1978, p. 17], having more leisure than later agricultural tribes.) Thus, it is not closeness to Nature, but the exact opposite, departure from day-to-day dependency on Nature, the development of agricultural technology, which creates rigidity, taboo and ritual. In any case, the Pygmies decisively refute Popper's contention that tribal life is invariably rigid, ritualistic and irrational. In many ways, in fact, our modern "open" societies in the industrially advanced West are closer, at the institutional level, to the Bantu reliance on rigidly maintained ritual and taboo, than to the Pygmy reliance on spontaneous instinct and skill. And – of particular relevance to the theme of this chapter – this is perhaps especially true of modern specialized scientific, academic research. Rigidly maintained taboo and ritual, broken only by "comparatively infrequent changes" having "the character of religious conversions or convulsions" – this corresponds almost exactly to specialist scientific research as described and documented by Kuhn (1970a). Even the vocabulary is the same. Kuhn describes scientific revolutions as infrequent episodes of crisis, inducing intense anxiety while they last, the process of acquiring the new paradigm constituting a kind of irrational religious conversion.

At present one perhaps needs the serene self-assurance and lucidity of an Einstein (acquired as a result of sustained, instinctive fundamentalist thought) to recapture the spontaneity and trust of the Pygmy way of life in the modern scientific world. It is clear that in Einstein's case scientific curiosity arose spontaneously from the heart in response to a feeling of "rapturous amazement at the harmony of natural law". (In a letter to Gertrud Warschauer in 1952, Einstein wrote: "You have given me great joy with the little book about Faraday. This man loved mysterious Nature as a lover loves his distant beloved. In his day there did not yet exist the dull specialization that stares with self-conceit through hornrimmed glasses and destroys poetry" [In Dukas and Hoffmann, 1979, p. 42]). And Einstein found no difficulty in conceiving himself as a part of Nature. When asked to respond to the question "If, on your death bed, you looked back on your life, by what facts would you determine whether it was a success or failure?", Einstein replied: "Neither on my death bed nor before will I ask myself such a question. Nature is not an engineer or contractor, and I myself am a part of Nature" (In Dukas and Hoffmann, 1979, p. 92).

13. The destructive impact of industrially more advanced ways of life on primitive or so-called primitive ways of life has been, and is at present, all too often, blatant and brutal. But it can also be subtle and unintended. For a perceptive account of this in connection with the importation of Western economic ideas and practices, see Schumacher (1973).

14. Einstein was always aware of the instinctively fundamentalist character of childish thinking – as well as of the childish origins of mature fundamentalist thought – associated, for him, essentially with curiosity provoked by a sense of wonder, together with scepticism concerning the received dogmas of the adult world. In explanation of his own fundamentalist thinking concerning the structure of the physical universe, he once remarked that ordinarily only children take such problems seriously. He, however – a late developer – continued to pursue such elementary questions; and, as an adult, naturally, was better equipped to come up with improved answers. On another occasion he remarked: "There exists a passion for comprehension, just as there exists a passion for music. That passion is rather common in children, but gets lost in most people later on. Without this passion, there would be neither mathematics nor natural science" (Einstein, 1973, p. 342). And in connection with his own education, in a well-known passage, he remarks:

> In this field … [of physics] I soon learned to scent out that which was able to lead to fundamentals and to turn aside from everything else, from the multitude of things which

clutter up the mind and divert it from the essential. The hitch in this was, of course, the fact that one had to cram all this stuff into one's mind for the examinations, whether one liked it or not. This coercion had such a deterring effect [upon me] that, after I had passed the final examination, I found the consideration of any scientific problems distasteful to me for an entire year. In justice I must add, moreover, that in Switzerland we had to suffer far less under such coercion, which smothers every truly scientific impulse, than is the case in many another locality. There were altogether only two examinations; aside from these, one could just about do as one pleased. This was especially the case if one had a friend, as did I, who attended the lectures regularly and who worked over their content conscientiously. This gave one freedom in the choice of pursuits until a few months before the examination, a freedom which I enjoyed to a great extent having gladly taken into the bargain the bad conscience connected with it as by far the lesser evil. It is, in fact, nothing short of a miracle that the modern methods of instruction have not yet entirely strangled the holy curiosity of inquiry; for this delicate little plant, aside from stimulation, stands mainly in need of freedom; without this it goes to wrack and ruin without fail. It is a very grave mistake to think that the enjoyment of seeing and searching can be promoted by means of coercion and a sense of duty. (Einstein, "Autobiographical Notes", in Schilpp, 1969, p. 17).

15. See Maxwell (2014c, ch. 1) for a suggestion as to how critical fundamentalist education might be conducted for five- to ninety-five–year-olds.
16. Kuhn, for example, argues that the instigation of the specialized, autonomous puzzle solving of the specialist is essential for scientific progress (see Kuhn, 1970a, pp. 21, 24, 37, and 64–5).
17. As we shall see below, there is a further vital point of difference. Universalism asserts that inquiry can only be really intellectually rigorous if it is recognized that inquiry (thought, problem solving), at the most fundamental level, goes on in life as an integral part of our personal and social lives, actively helping us to discover and achieve what is of most value in life, potentially and actually, as we live.
18. A further clarification, to be elaborated below. Universalism conceives of intellectual progress, fundamentally, in personal and social terms, in terms of progress in our achievement of what is of value in life, in terms of the progress in our personal and social thinking actively associated with and guiding our endeavours to achieve what is of value, on a personal and worldwide basis.
19. A remark about the first and last of these "fundamentalists". Einstein once said: "I want to know how God created this world. I'm not interested in this-or-that phenomenon, the spectrum of this-or-that element. I want to know His thoughts, the rest are details" (see Salaman, 1979, p. 22). In *The Seventh Enemy: The Human Factor in the Global Crisis*, R. Higgins (1978) outlines with devastating clarity and force six basic threats to the future of civilization – six fundamental worldwide problems which we must somehow resolve on a worldwide basis if mankind is to survive. His "seventh enemy" is our human incapacity to acknowledge and respond to these fundamental problems, on both individual and social, political or institutional, levels. Thus, on a worldwide basis, life on earth is at present almost lunatically irrational in the most elementary fashion (since it fails to put into practice the two most elementary rules of rational problem solving).
20. It must be emphasized that this modern meaning of the phrase, introduced by Price, is a typical specialist perversion of the original fundamentalist meaning intended, for example, by Robert Boyle in the seventeenth century when he writes:

> The "Invisible College" consists of persons that endeavour to put narrow-mindedness out of countenance by the practice of so extensive a charity that it reaches unto every thing called man, and nothing less than an universal good-will can content it. And indeed they are so apprehensive of the want of good employment that they take the whole body of mankind for their care. But … there is not enough of them. (Quoted in Werskey, 1978, p. 13)

21. It is noteworthy, for example, that Higgins (1978) is obliged to break all conventional academic boundaries in order to articulate our basic global problems. It is also noteworthy that these problems discussed by Higgins and others do not receive sustained, influential discussion as an integral part of the orthodox scientific, academic enterprise. In recent years this may, however, have changed a bit; the growing menace of climate change has led some experts to speak out about the problem in non-specialist terms. There is, in recent times, more discussion of global

problems in universities – as I shall indicate below. Some determined academics do manage to struggle against the stifling constraints of specialism.

22. I refer here, of course, to the dominant schools of philosophy in Britain and the USA since the war, ordinary-language philosophy, conceptual analysis, logical empiricism and descriptive metaphysics, as practised by, for example, Gilbert Ryle, J. L. Austin, A. J. Ayer, Elizabeth Anscombe, Geoffrey Warnock, R. M. Hare, Anthony Kenny, P. F. Strawson, Rudolf Carnap, Carl Hempel, W.V.O. Quine, Donald Davidson, and many others. Somewhat more recently there have been indications of some improvement in this tradition. J.J.C. Smart (1963), Mary Midgley (1979), Thomas Nagel (1986), Daniel Dennett (1993) and David Chalmers (1996) have all produced work that can be regarded as making valuable contributions to the fundamental problem "How do we fit into the world and how have we come to be?" In contrast to this tradition, there is so-called Continental philosophy: idealism, phenomenology, existentialism, Marxism, the Frankfurt School, structuralism, post-structuralism, postmodernism, and the work of G.W.F. Hegel, Arthur Schopenhauer, Søren Kierkegaard, Friedrich Nietzsche, Edmund Husserl, Martin Heidegger, Jean-Paul Sartre, Maurice Merleau-Ponty, Michel Foucault, Jacques Derrida, and many others. Some of this work can be held to put problems of living at the heart of thought, but as a body of work it is deeply flawed by anti-scientific and anti-rationalist attitudes, an allegiance to idealism, and a readiness to aspire to profundity by means of bombastic obscurity. These prevalent flaws disqualify this body of work from constituting serious discussion of fundamental problems within the context of universalism. See, however, Bakewell (2016) for a delightful account of the lives and work of the existentialists. Karl Popper, as I have already indicated, has been highly critical of both philosophical traditions. He has argued that real philosophical problems have their roots outside philosophy, in science, politics, art, social life, and in his work he tackles fundamental problems with exemplary clarity and intellectual integrity (especially in Popper, 1959, 1961, 1963 and 1966a). Popper has also vehemently criticized specialization in such remarks as, "Specialization may be a great temptation for the scientist. For the philosopher it is the mortal sin" (Popper, 1963, p. 136); and "If the many, the specialists, gain the day, it will be the end of science as we know it – of great science. It will be a spiritual catastrophe comparable in its consequences to nuclear armament" (Popper, 1994, p. 72). So vehement was Popper's condemnation of specialization that he failed to see its value, indeed its necessity – and failed to see how the damaging irrationality of specialism can be overcome by implementing rule (d) of the four rules of rational problem solving I formulated in section 9.2. For a more detailed criticism of academic philosophy, on the grounds that it fails to keep alive rational tackling of fundamental problems – and fails to solve two key philosophical problems as a result – see Maxwell (2017b, chs. 3 and 4).
23. Something like this account is presupposed, or propounded, by G. Ryle (1967), Ayer (1969), ch. 1, pp. 1–18) and Whiteley (1955, pp. 5–6).
24. An amusing indication of this is the way in which philosophers tend to acknowledge, apologetically or critically, that philosophy still concerns itself with the problems discussed by, for example, Plato, whereas other disciplines successfully solve initial problems and move on to new problems, thus making progress. The failure of philosophy to progress in this way is only problematic if philosophy is conceived in specialist terms. From the standpoint of the fundamentalist or Enlightenment conception of philosophy, it is of course precisely the basic task of philosophy to keep alive, throughout the whole of intellectual inquiry, and throughout our culture and social life, a sustained concern with our four fundamental problems.
25. Since the first version of this chapter was published, in 1980, some academics have become concerned that members of the public should contribute to discussion concerning science policy options. Thus, in the UK, the Royal Society produced a report on the future of nanotechnology, the result of a collaboration of scientists and non-scientists. The Economic and Social Research Council has funded a research programme, Science in Society, which has explored issues having to do with public engagement with science. There is a rather general recognition that communication between science and the public should go in both directions. It is not good enough for scientists merely to inform the public about science. These developments do not, however, amount to academia as a whole engaging with the social world in a two-way discussion about how our problems of living are to be tackled in increasingly cooperatively rational ways.
26. It is this feature of universalism which led me to suggest the view should really be called "specio-universalism" (see note 1).
27. Failure to put universalism into practice has, as a consequence, that sustained discussion of our fundamental global problems does not take place in a way that can influence the priorities

of scientific research. As a result, research devoted to such matters as halting global warming, or alleviating global poverty, and the diseases of the global poor, is neglected, in favour of research funded by, and devoted to the interests of, the military, government, industry and commerce.
28. For a more detailed and sophisticated advocacy of this critical fundamentalist conception of science, see Chapters 2 and 5, and Maxwell (1974; 1976a; 1977a; 1979; 1984a; 1998; 2004a; and 2007, especially ch. 14). For a critical assessment, see Kneller (1978, pp. 36–8, 80–7, 90–9); Muller (2008) and my reply (Maxwell, 2009b); McHenry (2009a); and Vicente (2010), Pandit (2010) and my reply (Maxwell, 2010b). For more recent expositions see Maxwell (2013, 2017a, 2017b).
29. Contrast Russell's uncritical or inflexible "postulational" approach with the critical, flexible postulationism of aim-oriented empiricism, which stresses that science, in order to be rational, must continuously articulate, develop and criticize metaphysical blueprints for science as an integral part of scientific inquiry, and in the light of ostensible scientific progress, thus enabling us to improve our aims and methods as our scientific knowledge and understanding of the world improves.
30. See also Popper (1983, section 23; 1982b, sections 20–8).
31. See note 28.
32. This important point can be established quite simply as follows. Science is centrally concerned to solve the problem "What kind of world is this?" If science is to tackle this problem rationally, priority needs to be given to proposing and critically assessing possible solutions – thus developing a tradition of rational cosmology like that represented, for example, by Popper in his "Back to the Presocratics" (1963, ch. 5). This leads, however, to the development of a number of rival imprecise possible solutions – rival cosmologies – with no indication as to how we are to make these vague ideas precise and choose between them. In order to proceed, we need to put into practice the third and fourth rules of rational problem solving: each vague solution needs to generate preliminary, subordinate, specialized problem solving. If one such approach begins to achieve apparent spectacular specialized success, then this entitles us to take this general approach especially seriously. Thus the spectacular specialized successes of Kepler and Galileo entitle us to take especially seriously their common vague cosmological presupposition: "the book of Nature is written in the language of [simple] mathematics." If science is to proceed rationally, however, it is essential that there continues to be an interplay between our best ideas as to how the overall problem is to be solved, and our best solutions to subordinate problems. In particular, our assessment of possible solutions to subordinate problems – testable scientific laws and theories – must not be dissociated from our assessment of untestable, metaphysical ideas as to how the overall problem is to be solved. Popper, however, violates this elementary, general requirement for rationality, in insisting that assessment of scientific laws and theories is dissociated from assessment of metaphysical ideas. Furthermore, it is precisely this irrational insistence which creates, for Popper, the insoluble problem of induction. The impossibility of assessing scientific laws and theories solely with respect to empirical success is a special case of the general irrationality of attempting to assess possible solutions to subordinate problems independently of vague ideas about how to solve the overall problem. The problem of induction, in short, is a product of specialism, the insolubility of the problem, as traditionally conceived, an indication of the irrationality of specialism as far as science is concerned.
33. Ironically enough, Popper does come close to acknowledging the Russellian point that the methods of science make implicit metaphysical presuppositions about the nature of the world (see Popper, 1959. pp. 252–4), despite explicit disavowals elsewhere (see Popper, 1963, p. 54). He fails, however, to emphasize that critical rationalism requires that we explicitly articulate these metaphysical presuppositions, so that they may be criticized, and thus, we may hope, improved, as an integral part of science, so that the methods of science may be improved with our improving knowledge. Just this way of doing science was instigated by Einstein in developing the special and general theories of relativity, as we saw in Chapter 3. The invariance and symmetry principles of modern physics – which can be interpreted as either methodological or metaphysical principles – are a development of Einstein's profound innovation. However, modern physics, and modern science quite generally, fail to put into practice, explicitly and fully, Einstein's way of doing science, in that they fail to articulate and criticize actual and possible aims and methods – or philosophies of science – as an integral part of science itself. The institutional reorganization that this requires – namely philosophy of science pursued as an integral part of science itself – has not been carried out. This is of course in part due to the

fact that the scientific community accepts Popper's falsificationist demarcation criterion for dividing off science from non-science. Views about what ought to be the aims and methods of science – philosophies of science – not being themselves testable theories in any straightforward sense, have no place in science itself according to traditional, and Popperian, empiricism. Thus scientific integrity at present demands that discussion of aims and methods be excluded from science, instead of demanding that this discussion constitutes an integral part of science (as required by aim-oriented empiricism or universalism). At present, by and large, science departments and departments of history and philosophy of science do not speak to each other (although in recent years a bit of dialogue has been instigated). To this extent Popper, rather than Einstein, is institutionalized. This institutionalization of Popperian methodology prevents us from developing a genuinely rational, fundamentalist science.

34. For Einstein's advocacy of universalism see, for example, Einstein (1973, part v). See also Chapter 3 of the present book.
35. See note 16.
36. It should be noted that the basic objection to Kuhn's prescription for science applies with almost equal force to Lakatos's prescription as outlined in his (1970). Lakatos's problem is to reconcile the dogmatism of Kuhn's normal science, on the one hand, with the anti-dogmatic, critical falsificationism of Popper, on the other hand, taking into account especially Feyerabend's important point that in order to test a given theory severely, we need to possess, and even develop, alternative theories (see Feyerabend, 1965). Lakatos's solution is to prescribe for science simultaneous competing fragments of Kuhnian normal science – competing research programmes – thus doing justice simultaneously to Kuhnian dogmatism and Feyerabendian pluralism. Lakatos makes it abundantly clear, however, that ultimately only relative empirical success ought to decide the fate of research programmes within science. There is thus no essential role, within Lakatos's conception of science, for sustained critical development of our best metaphysical answer to the problem "What kind of world is this?", so that the hard cores of research programmes could be assessed in part in terms of this answer. Lakatos advocates a kind of competitive specialism. In terms of our obstacle-course analogy, Lakatos sees science as a number of competing individuals, with different routes in mind, stumbling blindly from A to B.
37. For a powerful criticism of the idea that the social sciences should be value-neutral, see Easlea (1973, pp. 167–78). Essentially the same point is made by Schumacher (1973), when he argues that economic thinking must reflect or presuppose some philosophy of life, some view as to what is of value in life. For the point that explicit articulation and criticism of value assumptions implicit in the aims of research is actually essential for the whole of science if it is to be objective and rational, see Maxwell (1976a, chs, 5 and 7; 1977b; 1984a or 2007; 2004a; 2014b; 2016b). Values are even implicit, it should be noted, in the aims of a science as apparently remote from ordinary life as pure theoretical physics. The question "What kind of world is this?", may be interpreted in such a fashion that merely developing theories, like quantum theory, which predict more and more phenomena more and more accurately constitutes satisfactory progress towards answering the question. Einstein asked for much more from theoretical physics: he sought to capture, in a "wildly speculative way" the "thoughts of God". He did not know that the universe has a coherent, unified structure; rather, the mere possibility of discovering such a structure seemed to him to be of such supreme value that to abandon the search for it seemed to be a profound betrayal of the noblest aspirations of theoretical physics. Thus Einstein's judgement that quantum theory is unsatisfactory, in that it abandons micro-realism, was in part based on a value judgement. (For an endorsement of Einstein's judgement on this point see Maxwell, 1976b; 1982; 1988; 1994a; 1998, ch. 7; see also Chapter 8 of the present book.)
38. A number of writers – for example, Koyré (1973), Burtt (1932) and Buchdahl (1969) – have advocated a view which might be called "metaphysical presuppositionism", according to which the natural sciences do make substantial metaphysical presuppositions about the world. These writers fail, however, to emphasize the crucial point that scientific rationality demands sustained, explicit, critical development of such metaphysical presuppositions as an integral pan of science itself – the essential tenet of universalism, and aim-oriented empiricism.
39. See Mannheim (1952); Merton (1970); Mathias (1972); Teich and Young (1973); Hagstrom (1965); Bloor (1976); Mulkay (1979); Barnes, Bloor and Henry (1996).
40. On the one hand there are those who pursue sociology of science and "externalist" history of science merely in order to add to specialist knowledge within sociology and history. These

writers tend to decry the significance of epistemology and the study of scientific method. (A notable example of this is to be found in Bloor, 1976.) From the standpoint of the fundamentalist viewpoint defended in section 9.7, this approach entirely misses the point. For, according to the view advocated below, the basic task of the social sciences is to help us develop more rational institutions and ways of life, a more rational world. A central task of the social sciences, in other words, is to propose and critically assess possible institutional and social changes designed to help people all the better to discover and achieve what is of value in life – that is, to help people solve rationally the problems of living which they encounter in seeking to achieve that which is of value in life. The social sciences, on this view, ought thus fundamentally to be institutional or social epistemology or methodology. What is being attempted in this chapter in connection with one institution – the scientific, academic enterprise – should be attempted quite generally in connection with institutions associated with politics, the law, the media, commerce, industry and international relations. Far from the sociology of science taking over from the philosophy of science, on the contrary, sociology – and the social sciences quite generally – need to become the philosophy and methodology of institutional, social pursuits and enterprises. Granted that our concern is to develop better solutions to problems (3) and (4), a central task of the social sciences and humanities ought to be to help us develop fundamentalist, or aim-oriented rationalistic, institutions quite generally – including aim-oriented rationalistic academic institutions (see Maxwell, 1976a, chs. 8 and 9; 1984a or 2007, chs. 5–8; 2004a, chs. 3 and 4; 2014b).

On the other hand, there are those Marxist-inclined writers who wish to commit science to socialist or Marxist objectives and who seek to "radicalize" science (see, for example, Easlea, 1973; Rose and Rose, 1976). These writers see social and cultural reality in terms of competing class interests: the dominant class ensuring that even culture and science serve its own class interests, this situation being maintained, in part, by means of the institutionalized lie that science is an objective, value-neutral search for truth, serving no special class interests. There is clearly some truth in this allegation. The moment we view scientific and technological research on a worldwide basis, it becomes clear that very little such research is devoted to serving the interests of the millions upon millions of desperately poor people in the developing world. In so far as such research does serve social interests, it is the interests of those who live in industrially advanced, relatively wealthy countries which are served – even to the point of increasing the misery of the underprivileged, as in the case, perhaps, of the tin miners of Bolivia. The fundamental defect of this Marxist conception of intellectual inquiry, however, is that it commits intellectual inquiry to socialist or Marxist social theory and objectives, and thus prevents intellectual inquiry from itself scrutinizing these social, political and evaluative presuppositions, even to the point of improving on them.

We might view the matter as follows. (1) Standard empiricists, like Hempel and Popper, reject the existence of permanent metaphysical presuppositions inherent in science. (Even Kuhn and Lakatos only allow for temporary metaphysical presuppositions to be assessed ultimately in terms of the empirical success of the specialist research they support; thus Kuhn and Lakatos ultimately also advocate standard empiricism.) This is dishonest, as the insolubility of the problem of induction indicates. (2) Metaphysical presuppositionists, like Russell, Koyré, Burtt and Buchdahl, do acknowledge the existence of long-term, comprehensive metaphysical presuppositions implicit in science. This is more honest. These writers fail, however, to emphasize the crucial importance of articulating and critically developing such presuppositions as an integral part of science. In addition, these writers fail to acknowledge the existence of value-presuppositions implicit in science. This is dishonest. (3) Easlea, Schumacher and others do acknowledge the existence of such value-presuppositions implicit in science. This is more honest still. These writers fail, however, to emphasize the crucial importance of articulating and critically developing such presuppositions as an integral part of intellectual inquiry – thus failing to advocate a rational, critical fundamentalist version of the philosophy of wisdom. In addition, merely to acknowledge that value-presuppositions are implicit in intellectual inquiry is to fail to acknowledge that intellectual inquiry is itself a part of personal, social life, a kind of personal, social action, pursued in order to realize personal, social goals. This is dishonest. (4) Radical Marxists go further, in that they do conceive of, and pursue, intellectual inquiry as an aspect of personal, social action, designed to help achieve personal, social objectives. According to these writers, in capitalist societies intellectual inquiry is devoted primarily to helping to attain the objectives of capitalism: in their intellectual work these writers seek to act in such a way as to help overthrow capitalism, thus creating a socialist society and a socialist intellectual inquiry devoted to helping to realize socialist goals. In so far as these writers see and pursue intellectual inquiry as an aspect of life, social reality, social action, their vision and practice is even more honest still. These writers fail, however, to acknowledge the crucial importance of articulating and critically developing basic socialist presuppositions and objectives. They fail to confront obvious major problems inherent in the idea of a socialist society – such as the problem of centralized, bureaucratic power. This is dangerously dishonest. In particular, as a result of this failing, these writers fail to emphasize the fundamental importance of seeking to develop ways of life, institutions, societies, which progressively develop the aims and methods of personal, institutional and social life – thus enhancing our capacity to achieve that which is of value in life. These writers presuppose answers to problems (3) and (4), instead of seeking to develop a fundamentalist, rational society which enables us to discover improved answers to these problems, as we live.

In short, despite their diversity, the four positions just outlined have one crucial failing in common: they all fail to emphasize that rational action involves quite essentially seeking to improve our aims and methods as we act – the key tenet of aim-oriented rationality (see Maxwell, 1976a, 1984a or 2007, 2004a, 2014b).

41. For a fascinating account of such a hunting and gathering tribal life, see Turnbull (1976).
42. Higgins writes (1978, pp. 21–45) especially clearly and convincingly on this point, in part from personal experience.
43. A humane, cooperative, mutually understanding, pluralistic society presupposes and is, in a sense, presupposed by, universalism. If two people, two societies or two cultures, giving different answers to our four fundamental problems, are to act humanely and cooperatively together, there must be mutual understanding; this requires that each is able to imagine, at least as a possibility, that the other's answers are correct (or at least an improvement, in certain respects, over his own). This in turn requires that each recognizes the genuineness of the four fundamental problems. If each is to learn from the other, then each must acknowledge the genuineness of the four fundamental problems. On the other hand, to recognize that these problems are genuine is to imagine at least the possibility of answers different from one's own being given – which is to imagine a pluralistic society, at least as a possibility.

Only universalism can do justice to the Socratic and Kantian idea that Reason forms a basis for the unity of mankind.

44. "Reason, like language, can be said to be a product of social life" (Popper, 1966a, vol. 2, p. 225). See also the discussion of the claim that science is necessarily social in character, "Robinson Crusoe science", however successful, being necessarily only "revealed science" in that it must lack objectivity, pluralistic criticism (ibid., pp. 216–20). Unfortunately, Popper in his later work fails lamentably to develop these anticipations of the point stressed in this chapter (see, for example, note 40), that reason, epistemology, thought, intellectual inquiry, all need to be conceived of, and developed, as personal and social in character, in the world, a part of life.

If we adopt the view advocated in section 9.7 that the aim of intellectual inquiry is to help us achieve wisdom, life of value, then the fundamental aim of intellectual inquiry becomes a personal, social aim, and the problems of intellectual inquiry become, fundamentally, personal, social problems of living. Our central task, in pursuing intellectual inquiry, becomes to help develop more rational, wiser ways of living, institutions, social orders. The split between personal, social aims and intellectual aims – the split between personal, social action and thought – disappears. Popper, however, holds that the basic aim of intellectual inquiry is to develop impersonal, objective knowledge. This leads him to develop his "world 3" theory of the intellectual domain. As a result, and quite disastrously, the fundamental personal and social problems of intellectual inquiry – problems we encounter in helping to develop life of value, a wiser world – are transformed into the pseudo metaphysical-neurological problems of how "world 3" can interact with the mind and the brain (see Popper, 1972; Popper and Eccles, 1977).

Universalism, in sharp disagreement with Popper, recognizes just one world. Within universalism, Popper's conceptually incoherent psycho-neurological thesis that world 3 interacts with world 1 via world 2, can be replaced by the kind of conceptually coherent psycho-neurological postulate indicated in note 11, or by a version of this postulate which asserts that aim-oriented rationalism is programmed into the neurological structure of our brains – or at least needs to be so "programmed" if we are to be able to achieve what is of value in life. (For an exposition of aim-oriented rationalism, see the Prologue and Chapter 10. See also Maxwell, 1976a; 1984a or 2007; 2000a; 2001a, ch. 9; 2004a, chs. 3 and 4; 2010a, chs. 6 and 9; 2014b;

2016b; 2017a, ch. 14; 2017b, ch. 8.) There is only one world; it is in this world that universalism and aim-oriented rationalism need to flourish. In order to help achieve this, it is essential that we see critical fundamentalist intellectual standards, aim-oriented rationality, as something embedded, actually and potentially, in this world. It is essentially this insight that we need to see implicit in much of Popper's *The Open Society and Its Enemies* (1966a) in order to appreciate the real value of that work, and in order to make sense of Popper's wonderful suggestion that we should see intellectual evolution as a development of biological evolution. In the circumstances it is somewhat tragic that Popper should have gone on, with the development of his three-world view, to reject explicitly the insight that reason needs to be seen as materially and socially embodied in this one world.

45. According to Popper, we must learn to live with the intense emotional strain of civilization, as the price that must be paid for reason, for the open society, for civilization. Any attempt to introduce social and cultural changes which alleviate this strain must be fiercely resisted, as such changes must inevitably lead to totalitarianism. However, as indicated in section 9.2, this is because Popper fails to conceive of the possibility of universalism, and is led as a result to defend a seriously irrational and undesirable conception of rational inquiry and civilization. See Maxwell (1984a, pp. 189–99, or 2007, pp. 213–22) for a related criticism of Popper's social and political rationalist philosophy.

For purposes of clarification, I should indicate four further main differences between the viewpoint being advocated in this chapter, and views advocated or presupposed in Popper's writings. The chief difference, unquestionably, is simply this. I advocate that the basic aim of rational inquiry is to enhance wisdom. Here I part company not only with Popper, but with the whole Western tradition, in that this tradition gives to rational inquiry the basic aim of enhancing knowledge (human welfare, enlightenment and progress being only secondary and uncertain by-products of the basic and prior achievement of knowledge). As a result of giving priority to wisdom – to our living, actual capacity to discover and achieve what is of value in life – I am led to locate rational inquiry, at the most fundamental level, within and amidst our lives, personal and interpersonal or social. This leads me to stress the fundamental importance of aim-oriented rationalism designed to help us improve our aims, and thus our lives, as we live (aim-oriented empiricism being simply a special case of aim-oriented rationalism, applicable to science). This is in marked contrast to Popper's conception of reason, which he has called critical rationalism (falsificationism being simply, for Popper, a special case of critical rationalism, applicable to science). Since for me rational inquiry has, as its basic task, to help us achieve what is of value in life, I hold that all intellectual values need ultimately to be founded in human value – especially in the supreme value of each individual person, and the good things that can go on between people once this is recognized – rigorous, objective inquiry being, as though by definition, a universal tool perfectly designed to help us all maximally to achieve, or grow, life of value. I thus disagree absolutely with Popper's thesis – in effect a standard component of the philosophy of knowledge – that purely scientific values should be distinguished as sharply as possible from human or extra-scientific values. (This is Popper's sixteenth thesis in his "Reason or Revolution" [1976b, pp. 96–8].) The purpose of the present essay is to argue that the rationality, the intellectual rigour, the objectivity, of inquiry is essentially bound up with the capacity of inquiry to help us resolve those problems of living we need to resolve in order to achieve what is of value in life. That which is of value in life is primary: intellectual value is a reflection of – or is subservient to – primary value in life. Ideas – including the idea of this chapter – are spectacles intended to help us to see clearly what is of value in existence, actually and potentially; they are forks and spades designed to help us to cultivate what is of value in our lives, in reality. Like spectacles, ideas are to be assessed in terms of whether they serve to clarify or blur our vision; like forks and spades, ideas are to be assessed in terms of their use, their success in practice. The idea that intellectual value is dissociated from value in life, in the world, quite fundamentally misconceives the proper value and use of ideas: pushed to the extreme, this becomes Plato's doctrine of the forms. (The Popperian, Western doctrine of the autonomy of intellectual value devolves, in fact, from Plato's doctrine.) The idea that intellectual value needs to be conceived of as dissociated from value in the world – and not as integral to and contributing to value in the world – receives support no doubt from the desire of many intellectuals to find in intellectual work some kind of escape from the world, a quiet and transparent refuge. Given Popper's defence of the orthodox doctrine concerning the autonomy of purely scientific or intellectual value, it is not at all surprising that he should call Hume's thesis that "Reason is, and ought only to be, the slave of the passions" a "horrifying doctrine" (Popper, 1977, p. 132). An upholder of aim-oriented rationalism and the philosophy of wisdom would wish to make only minor adjustments to Hume's thesis. Either reason should be held to be the slave of that most profound passion of ours of all to participate in life of value; or reason ought perhaps to be called the "good servant" or "enlightened tutor" of the passions – reason itself the outcome of our cooperative, balancing, or resolving, of our passion for a whole, authentic life of value. As a result of conceiving of inquiry and reason as being an essential, active component of human life, more or less realized in practice in our personal and social actions, I am led to avoid the conceptual incoherence of Popper's three-world view, as indicated in notes 11 and 44, and in this note above. In general, the viewpoint that I wish to advocate is much closer to Einstein's than to Popper's, taking into account especially the emphasis that Einstein came to place in his later life on the fundamental importance of developing a living ethical culture, and a kind of education designed to help us acquire and participate in such a culture:

> It is not enough to teach a man a speciality. Through it he may become a kind of useful machine but not a harmoniously developed personality. It is essential that the student acquire an understanding of and a lively feeling for values. He must acquire a vivid sense of the beautiful and of the morally good. Otherwise he – with his specialized knowledge – more closely resembles a well-trained dog than a harmoniously developed person. He must learn to understand the motives of human beings, their illusions, and their sufferings in order to acquire a proper relationship to individual fellow-men and to the community. (Einstein, "Education for Independent Thought", in Einstein, 1973, p. 66)

46. See Isaac Bashevis Singer (1975a, 1975b, 1975c, 1977a, 1977b, 1977c).
47. It is doubtless commitment to this kind of oracular conception of reason which leads both Kuhn and Feyerabend, in their rather different ways, to characterize science as irrational. The same mistake is implicit in almost all forms of relativism.
48. In this chapter I am of course arguing that even though the goal is difficult to attain, nevertheless it is a matter of supreme importance that we seek to build universalism, rather than specialism, into the institutional structure of the scientific, academic enterprise, and education. Indeed what I am proposing goes much further than this. We need to build universalism, and aim-oriented rationalism, into our whole way of life, into society as a whole, into the human world. The basic aim of intellectual inquiry ought to be to devote reason to the enhancement of wisdom. This programme can scarcely begin to be put into practice, however, as long as specialism rather than universalism is built into the institutional structure of the scientific, academic enterprise.
49. See Gray (1972) for the suggestion that the aim of education should be the achievement of wisdom in life. Gray fails to point out, however, that intellectual inquiry is at present profoundly irrational and defective when judged from the standpoint of having as its basic aim to help us achieve wisdom.
50. For a detailed exposition and critical assessment of the two contrasting philosophies of inquiry of the philosophy of knowledge and the philosophy of wisdom, see Maxwell (1984a or 2007). I have subsequently come to call these two contrasting philosophies of scientific and academic inquiry "knowledge-inquiry" and "wisdom-inquiry". In Maxwell (1984a or 2007) I expound in detail the two views, and argue that knowledge-inquiry prevails in academia, despite its damaging irrationality, and despite the clearly articulated, more rigorous and humanly valuable alternative of wisdom-inquiry. This situation has not changed much since 1984, or 2007. Both knowledge-inquiry and wisdom-inquiry can be taken to hold that the basic aim of inquiry is to help promote human welfare by means of research and education. Knowledge-inquiry holds that this is to be done by, in the first instance, the acquisition of knowledge. First, knowledge is to be acquired; once acquired, it can be applied to help solve social problems. The pursuit of knowledge must be decisively shielded from potentially corrupting influences of the social world, otherwise authentic, objective factual knowledge will degenerate into mere propaganda and dogma, and will cease to be of human value. Wisdom-inquiry holds that the four elementary rules of rational problem solving of section 9.2 need to be put into practice. Priority needs to be given to the tasks of (a) articulating, and improving the articulation of, problems of living, and (b) proposing and critically assessing possible solutions, possible actions, policies, political programmes, ways of living, philosophies of life. Problems of knowledge are subordinate to problems of living. Wisdom-inquiry puts specio-universalism into practice, whereas knowledge-inquiry is a version of specialism (in that it dissociates problems of knowledge from more fundamental problems of living).

51. For a much more detailed depiction of the damaging irrationality of the philosophy of knowledge – or knowledge-inquiry as I have come to call it – see Maxwell (1984a, or 2007, chs. 2 and 3).
52. Only if intellectual priority is given to the task of proposing and criticizing possible and actual personal, social actions, policies, aims and methods, institutional enterprises, ideologies – problems of knowledge and technology being tackled as subordinate to our fundamental personal, social problems of living – can intellectual inquiry overcome these defects. For an elaboration of this point, see Maxwell (1984a or 2007; 2004a; 2014b; 2016b).
53. For a development of this point see my *What's Wrong With Science?* (Maxwell, 1976a), the subtitle of which reads, "Towards a People's Rational Science of Delight and Compassion". See too Maxwell (1984a or 2007).
54. Provocations of or contributions to the "science wars" include Barnes (1974); Barnes, Bloor and Henry (1996); Bloor (1976); Brown (2001); Feyerabend (1978 and 1987); Gross and Levitt (1994); Gross, Levitt and Lewis (1996); Harding (1986); Koertge (1998); Latour (1987); Pickering (1984); Segerstrale (2000); Shapin (1994); Sokal (2008); Sokal and Bricmont (1998).
55. It is highly significant that "philosophy of science" exists as a discipline, but "philosophy of inquiry" does not. What kind of inquiry can best help us create a good world? – to quote the title of one of my papers (Maxwell, 1992a) – is a question of fundamental significance for the future of humanity that is ignored by philosophers, and ignored by academics more generally. Only bureaucrats, business corporations and politicians think about the overall aims of academic inquiry; we should not be surprised that the level of this thought and decision-making does not amount to much when judged from the standpoint of its intellectual and humanitarian value – its value for the long-term interests of humanity.
56. For more on this theme, see Maxwell (2015b). For Popper's views on specialization, see Maxwell (2016a).
57. See Maxwell (1976a, 1984a, 1998, 2004a, 2010a, 2014b, 2014c, 2016b). See, too, the flood of my articles on this issue, available online at http://discovery.ucl.ac.uk/view/people/ANMAX22.date.html and https://philpapers.org/profile/17092
58. See www.ucl.ac.uk/research/wisdom-agenda
59. For further details, see Maxwell (2014b, ch. 4).
60. See note 25.

Chapter 10

1. See also Popper (1963, pp. 193–200; 1972, pp. 119 and 243; 1976a, pp. 115–6).
2. This seems to me to be a reasonable brief characterization of the basic idea of the French Enlightenment. It seems to me to be implicit in much that the *philosophes* did and said. I have not, however, found any philosophe explicitly asserting this to be the basic idea of the Enlightenment. And it was, I admit, only after I had developed my own version of the basic Enlightenment idea, partly as a result of learning from and criticizing Popper, that I came to the conclusion – in hunting for predecessors – that the basic idea could be traced back to the Enlightenment.
3. Three magnificent works on the Enlightenment are Peter Gay, *The Enlightenment: An Interpretation* (1973); Jonathan Israel, *Democratic Enlightenment* (2011); and P. N. Furbank, *Diderot: A Critical Biography* (1993).
4. It may be objected that it is not science that is the cause of our global problems but rather the things that we do, made possible by science and technology. This is obviously correct. But it is also correct to say that scientific and technological progress is the cause. The meaning of "cause" is ambiguous. By "the cause" of event E we may mean something like "the most obvious observable events preceding E that figure in the common-sense explanation for the occurrence of E". In this sense, human actions (made possible by science) are the cause of such things as people being killed in war or destruction of tropical rainforests. On the other hand, by the "cause" of E we may mean "that prior change in the environment of E which led to the occurrence of E, and without which E would not have occurred". If we put the twentieth century into the context of human history, then it is entirely correct to say that, in this sense, scientific and technological progress is the cause of distinctively twentieth-century disasters: what

has changed, what is new, is scientific knowledge, not human nature. Yet again, from the standpoint of theoretical physics, "the cause" of E might be interpreted to mean something like "the physical state of affairs prior to E, throughout a sufficiently large spatial region surrounding the place where E occurs". In this third sense, the sun continuing to shine is as much a part of the cause of war and pollution as human action or human science and technology.

5. See Hayek (1979) and Fargaus (1993, introduction).
6. For accounts of Romantic opposition to the Enlightenment, see Berlin (1980, 1999) and Gascardi (1999).
7. Elsewhere I have argued for this thesis in much greater detail (see Maxwell, 1984a, or 2007; see also Maxwell, 2004a, 2014b, 2016b).
8. Significant, in this connection, is Popper's detection of an authoritarian streak in even apparently anti-authoritarian traditional conceptions of empiricism and reason upheld by, for example, Bacon and Descartes: see Popper (1963, introduction).
9. For a novelist's fantasy of such a "rational" society, see Zamyatin (1972).
10. This point is especially emphasized and further developed by Feyerabend (1965). Popper, too, emphasizes that, in order to make sense of the idea of severe testing, we need to appeal to crucial experiments (see, for example, Popper, 1963, p. 112).
11. See, for example, Popper (1959, preface to the English edition; 1963, chs. 8 and 10).
12. Elsewhere I have spelled out in more detail than I am able to do here why this revolution is needed, and what it would amount to: see Maxwell (1984a or 2007). See also Maxwell (1976a, ch. 3; 1991; 1992a; 1997b; 2000a; 2001, ch. 9; 2004a; 2014b; 2016b).
13. For a somewhat more detailed discussion of rational problem solving see Maxwell (1984a, pp. 67–75, or 2007, pp. 80–8).
14. Can problem-solving rationality be regarded as a generalization of Popper's falsificationist conception of scientific method? In Popper (1959, pp. 276–8) there is a discussion of the "'inductive' direction" of the growth of science: science proceeds by putting forward laws and theories of ever greater empirical content. This could be regarded as a case of putting rule 3 into practice: science proceeds by, initially, tackling highly specialized, restricted problems and moves towards tackling problems that are increasingly general and fundamental in character. But rule 2 cannot be put into practice in science, in the context of "justification", since this would involve considering metaphysical theses as a part of scientific knowledge (as Popper himself in effect notes), which goes against Popper's demarcation criterion. This point comes up again in section 10.5.
15. These rules are not sufficient for rationality, in part because of a lack of specific detail about how to improve aims and methods when aims are problematic, and in part because the list of rules is by no means complete (see Maxwell, 1984a, pp. 69–75, or 2007, pp. 82–8). A very important additional rule is: 5. In seeking to solve a problem, P*, search for an analogous, already solved problem, P; if such a problem is found, modify the solution, S, appropriately, taking the similarities and differences between P and P* into account, so that S becomes S*, and consider this as a candidate solution to P*.
16. As I am using the term, a conflict is only resolved "cooperatively" if it is resolved "justly".
17. Social inquiry needs, of course, to tackle problems of knowledge of the social world subordinate to the tackling of social problems of living, in accordance with rule 3.
18. "Realize" is intentionally ambiguous in that it here means both "to apprehend" and "to make real" – both aspects of inquiry being included, inquiry pursued for its own sake, and for the sake of other, practical ends.
19. For a more detailed discussion of the nature, significance and intellectually fundamental character of "personalistic" understanding, and its role in "wisdom-inquiry", see Maxwell (1984a, pp. 172–89 and 264–75, or 2007, pp. 196–-213 and 285–96; 2001a, chs. 5–7).
20. Popper holds that the methods of the natural and social sciences "are fundamentally the same ... The methods always consist in offering deductive causal explanations, and testing them (by way of predictions)" (Popper, 1961, p. 131). Popper does not seem to have appreciated that this is starkly at odds with his passionate advocacy of the open society, the society of diversity, tolerance, reason, democracy and the rule of law. Even if we restrict social inquiry to the pursuit of knowledge and understanding of social phenomena, still there are a number of reasons for holding that social inquiry must differ fundamentally from natural science. First, in pursuing social inquiry we study ourselves; this is not the case when we pursue natural science. Results of social inquiry may well directly influence what is studied, namely ourselves; this does not happen as far as natural science is concerned (in that our scientific theories do

not directly influence nature). Again, causal explanations of natural phenomena enable us to manipulate nature so that it comes to serve our ends; analogous knowledge of social phenomena would provide the means to manipulate people. Some manipulation may be necessary, even desirable, in certain contexts, but this should not be the primary way we interact with one another. Yet again, and reinforcing this point, the kind of knowledge and understanding we should seek to acquire of people differs profoundly from that acquired by natural science: within social inquiry, primacy should be given to what I have called "personalistic understanding" (and others have called empathic understanding or "theory of mind"): see note 19 and associated texts. Our capacity to acquire personalistic understanding of each other is absolutely essential for our humanity, for cooperative action, for friendship, and love; the open society is inconceivable without it. And yet nothing like personalistic understanding of nature arises within natural science – apart from that element of it that is relevant for the understanding of sentient animals. There are no legal or moral constraints on "torturing nature to reveal her secrets" in experiments (as seventeenth-century natural philosophers put it); there are such constraints when it comes to the human world. Finally, we may well hold, with Popper, that the social sciences and the humanities, even when restricted to improving knowledge and understanding of the social world, should nevertheless be pursued in such a way as to aid the promotion of the open society, for example, in seeking to acquire knowledge of human suffering, or in seeking to assess the success and failure of policies and political programmes. But the moment social science is given this role of acquiring knowledge relevant to the promotion of the open society, it becomes distinct from natural science. Nature has nothing comparable to our struggle to make progress towards a more open or enlightened world, and that suffices to distinguish natural and social science. For all these reasons, social inquiry, even when restricted to the pursuit of knowledge, must differ radically from natural science. In this section (section 10.4), I have taken this argument much further. The primary task of social inquiry and the humanities, I have argued, is to help promote the open society – or, as I have put it, the cooperatively rational society. Social inquiry and the humanities should indeed seek to improve our knowledge and understanding of aspects of the social world; but their primary task is to promote cooperatively rational tackling of problems of living. Social inquiry is primarily social methodology or social philosophy, and only secondarily devoted to the pursuit of knowledge.

21. Given any accepted scientific theory, whether Newtonian theory, general relativity, quantum theory or the standard model, endlessly many rival theories can be concocted in each case that are even more falsifiable, better corroborated but, if anything, even more seriously *ad hoc* (i.e. lacking in simplicity) than the accepted theory, in the following way. Taking Newtonian theory (NT) as an example of an accepted theory, here are two examples of grossly *ad hoc* rival theories. NT*: "Everything occurs as NT asserts, until the first second of the year 2100, when an inverse cube law of gravitation will abruptly hold." NT**: "Everything occurs as NT asserts, except for systems consisting of gold spheres, each having a mass of 1,000 tons, interacting with each other gravitationally in outer space, in a vacuum, within a spherical region of 10 miles: for these systems, Newton's law of gravitation is repulsive, not attractive." It is easy to see that there are infinitely many such rivals to NT, all just as empirically successful (at the moment) as NT. The predictions of NT may be represented as points in a multidimensional space, each point corresponding to some specific kind of system (there being infinitely many points). NT has only been corroborated for a minute region of this space. In order to concoct a (grossly *ad hoc*) rival to NT, just as well corroborated as NT, all we need do is identify some region in this space that includes no prediction of NT that has been verified, and then modify the laws of NT arbitrarily, for just that identified region. Rival theories, of the above type, can easily be concocted that satisfy falsificationist requirements for being more acceptable than NT. NT, like most accepted physical theories, yields predictions that clash with observation or experiment, and thus are ostensibly falsified. We can always concoct new theories, in the way just indicated, doctored to yield the "correct" predictions. We can add on independently testable auxiliary postulates, thus ensuring that the new theory has greater empirical content than the old one. And no doubt this excess content will be corroborated. For further examples and discussion, see previous chapters, and Maxwell (1998, pp. 47–54; 2017a, ch. 5).

22. But does implementing Popper's methodological "principle of simplicity" really commit science to the metaphysical thesis that the universe is simple? Suppose, instead of adopting Popper's principle, science adopted the principle "In order to be acceptable, a new physical theory must postulate that the universe is made up of atoms". This methodological principle is

upheld in such a way that even though theories are available which postulate fields rather than atoms, and which are much more empirically successful than any atomic theory, nevertheless these rival field theories are all excluded from science. Would it not be clear that science, in adopting and implementing the methodological principle of atomicity in this way, is making the assumption that the universe is made up of atoms, whether this is acknowledged or not? How can this be denied? Just the same holds if science adopts and implements Popper's methodological principle of simplicity.

23. Followers of Popper have proved strangely impervious to recognition of this crucial Popperian criticism of Popper. No hint of it is to be found, for example, in Miller (2006), or Musgrave (2004). Even sympathetic critics of Popper, such as Rowbottom (2011), are oblivious to this crucial point. (Darrell Rowbottom denies it explicitly [personal communication].)

24. See also Maxwell (2004a, chs. 1 and 2; 2007, ch. 14; 2013; 2017a, chs. 6–9).

25. It is not enough just to be critical: we need to be critically critical. Criticism needs to be directed at those points in a possible solution to a problem where it is most likely to be fruitful.

26. See Maxwell (1998, pp. 78–89, 159–63, 217–23; and especially 2017b, ch. 5).

27. In holding that metaphysical theses and philosophies of science are an integral part of science itself, AOE implies that Popper's principle of demarcation (Popper, 1963, ch. 11) is to be rejected. Popper's demarcation proposal, apart from being untenable, is in any case too simplistic, in that it reduces to one a number of distinct demarcation issues. Popper rolls into one the distinct tasks of demarcating (a) good from bad science, (b) science from non-science, (c) science from pseudoscience, (d) rational from irrational inquiry, (e) knowledge from mere speculation, (f) knowledge from dogma (or superstition, or prejudice, or popular belief), (g) the empirical from the metaphysical, and (h) factual truth from non-factual (analytic) truth. (a) to (d) involve demarcating between disciplines, whereas (e) to (h) involve demarcating between propositions.

28. And as a bonus (a consequence of its intellectual rigour) AOE turns out to be both necessary and sufficient to solve the problem of induction, and other major associated problems of scientific progress, such as the problem of verisimilitude, and the problem of what it means to say of a scientific theory that it is explanatory – as we have seen in previous chapters. See also Maxwell (1998; 2004a; 2007, ch. 14; 2013; 2017a; 2017b). AOE has emerged from criticism of falsificationism over many years (beginning with Maxwell, 1972a and 1974).

29. Figure 2.1 represents science as making a hierarchy of problematic metaphysical assumptions. We may equally, however, interpret AOE as attributing to science a hierarchy of problematic aims. At each level in the diagram, the aim is to transform the metaphysical thesis at that level into a precise, testable physical "theory of everything".

30. See Maxwell (1976a; 1984a, or 2007, ch. 5; 2001a, ch. 9; 2014b; 2016b).

31. Science specifically, and academic inquiry more generally, misrepresent basic aims in just this way (see Maxwell, 1984a; 2002a; 2004a).

32. See Chapter 2 and Maxwell (1976a, 1977a, 1984b, 1987, 1991, 1992a, 1992b, 1997b, 1998, 2000a, 2000b, 2001a, 2001b, 2002a, 2002b, and especially 1984a or 2007). See also Maxwell (2014b, 2016b).

33. Fundamentally, this is due to the profound difficulty of discovering what is achievable (by increasingly civilized means), and of value. People hold conflicting views about what is achievable and of value, all too often in a highly dogmatic way, ignoring the profoundly problematic character of the whole idea of civilization. Many well-known views that have been proposed as to what constitutes Utopia, an ideally civilized society, have been unrealizable, horrifically undesirable, or both, attempts to realize such ideals, when taken up in practice, leading to various kinds of hell on earth (as in Hitler's Germany, Stalin's Russia or Mao's China). Furthermore, it is not just that people have conflicting interests, values and ideals; even our very best ideas as to what constitutes civilization embody (and need to embody) conflicting ideals. Thus freedom and equality, even though interrelated, may nevertheless clash. It would be an odd notion of individual freedom which held that freedom was for some, and not for others; and yet if equality is pursued too single-mindedly this will undermine individual freedom, and will even undermine equality, in that a privileged class will be required to enforce equality on the rest, as in the Soviet Union. A basic aim of legislation for civilization, we may well hold, ought to be to increase freedom by restricting it: this brings out the inherently problematic character of the aim of achieving civilization. One thinker who has stressed the inherently contradictory character of the idea of civilization is Isaiah Berlin (see, for example, Berlin, 1980, pp. 74–9). Berlin thought the problem could not be solved, but this was because he was

ignorant of aim-oriented rationality. In depicting ideals of civilization at a hierarchy of levels, aim-oriented rationality provides the means for progressively improving resolutions to inherently conflicting ideals, such as freedom and equality.

34. This is the objection that most academics will wish to raise against the conception of inquiry implied by the "Improved Popperian Enlightenment" and the "New Enlightenment". It will be made by all those who hold that academic inquiry quite properly seeks to make a contribution to human welfare by, first, acquiring knowledge and then, secondarily, applying it to help solve human problems.
35. For a development of this point, see Maxwell (1984, pp. 174–81). In some respects it accords with Popper's views on the biological and evolutionary origins of human thought (see, for example, Popper, 1972, ch. 7).
36. For further discussion see Maxwell (1984a, pp. 189–98).
37. See Maxwell (1984a, pp. 63–4, 85–91 and 117–8), for further discussion of this issue. See also Maxwell (1976a, especially chs. 1 and 8–10).
38. This objection has been made by N. Rescher (personal communication) and Durant (1997).
39. This is a modified version of the list to be found in Maxwell (2004a, pp. 119–21).

References

Most of the author's published papers are available at: http://discovery.ucl.ac.uk/view/people/ANMAX22.date.html or https://philpapers.org/profile/17092

"A Blueprint for Survival". 1972. *The Ecologist* 2(1), January.
Abott, P. and Wallace, C. 1990. *An Introduction to Sociology: Feminist Perspectives*. London: Routledge.
Achinstein, P. 2010. *Evidence, Explanation and Realism*. Oxford: Oxford University Press.
Agassi, J. 2014. *Popper and his Popular Critics: Thomas Kuhn, Paul Feyerabend and Imre Lakatos*. Berlin: Springer.
Aitchison, I. and Hey, A. 1982. *Gauge Theories in Particle Physics*. Bristol: Adam Hilger.
Andersen, F. and Arenhart, J. 2016. "Metaphysics Within Science: Against Radical Naturalism". *Metaphilosophy* 47: 159–80.
Aspect, A., Grangier, P. and Roger, G. 1982. "Experimental Realization of Einstein-Podolsky-Rosen-Bohm *Gedankenexperiment*: A New Violation of Bell's Inequalities". *Physical Review Letters* 49: 91–4.
Atkins, P. W. 1983. *Molecular Quantum Mechanics*. Oxford: Oxford University Press.
Ayer, A. J. 1965. *The Problem of Knowledge*. Middlesex: Penguin Books.
Ayer, A. J. 1969. *Metaphysics and Common Sense*. London: Macmillan.
Ayer, A. J. et al. 1967. *The Revolution in Philosophy*. London: Macmillan.
Bacciagaluppi, G. 2003. "The Role of Decoherence in Quantum Mechanics". In *The Stanford Encyclopedia of Philosophy (Winter 2003 Edition)*, ed. Edward N. Zalta. http://plato.stanford.edu/archives/win2003/entries/qm-decoherence/
Bacciagaluppi, G. and Ismae, J. 2015. "Review of *The Emergent Multiverse* by David Wallace". *Philosophy of Science* 82(1): 129–48.
Bakewell, S. 2016. *At the Existentialist Café: Freedom, Being and Apricot Cocktails*. London: Chatto and Windus.
Barker, S. F. 1957. *Induction and Hypothesis: A Study of the Logic of Confirmation*. Ithaca, NY: Cornell University Press.
Barnes, B. 1974. *Scientific Knowledge and Sociological Theory*. London: Routledge & Kegan Paul.
Barnes, B., Bloor, D. and Henry, J. 1996. *Scientific Knowledge: A Sociological Analysis*. Chicago, IL: University of Chicago Press.
Bartelborth, T. 2002. "Explanatory Unification". *Synthese* 130: 91–107.
Bassi, A., Lochan, K., Satin, S., Singh, T. P. and Ulbricht, H. 2013. "Models of Wave-function Collapse, Underlying Theories, and Experimental Tests. *Reviews of Modern Physics* 85: 471–527.
Bell, J. S. 1964. "On the Einstein Podolsky Rosen Paradox". *Physics* 1: 195–200 (reprinted in Bell, 1987).
Bell, J. S. 1987. *Speakable and Unspeakable in Quantum Mechanics*. Cambridge: Cambridge University Press.
Berlin, I. 1980. *Against the Current: Essays in the History of Ideas*. London: Hogarth Press.
Berlin, I. 1999. *The Roots of Romanticism*. London: Chatto and Windus.
Bird, A. 2000. *Thomas Kuhn*. Chesham: Acumen.
Bloor, D. 1976. *Knowledge and Social Imagery*. London: Routledge & Kegan Paul.
Bohm, D. 1952 "A Suggested Interpretation of the Quantum Theory in Terms of 'Hidden' Variables". *Physical Review* 85: 166–79 and 180–93.
Bohm, D. and Aharonov, Y. 1957. "Discussion of Experimental Proof for the Paradox of Einstein, Rosen and Podolsky". *Physical Review* 108: 1070–6.
Born, M. 1971. *The Born–Einstein Letters*. London: Macmillan.

Boyd, R. 1980. "Scientific Realism and Naturalistic Epistemology". In *Proceedings of the Biennial Meeting of the Philosophy of Science Association*, vol. 2, 613–62. Chicago, IL: University of Chicago Press.
Braithwaite, R. 1953. *Scientific Explanation*. Cambridge: Cambridge University Press.
Brown, J. R. 2001. *Who Rules in Science?* Cambridge, MA: Harvard University Press.
Brown, M. 2016. "The Abundant World: Paul Feyerabend's Metaphysics of Science". *Studies in History and Philosophy of Science* 57: 142–54.
Brown, R. B., Deletic, A. and Wong, T. 2015. "Interdisciplinarity: How to Catalyse Collaboration". *Nature* 525: 315–7.
Buchdahl, G. 1969. *Metaphysics and the Philosophy of Science: The Classical Origins*. Oxford: Blackwell.
Burtt, E. A. 1932. *The Metaphysical Foundations of Modern Science*. London: Routledge & Kegan Paul (first published 1924).
Cartwright, N. 1999. *The Dappled World*. Cambridge: Cambridge University Press.
Catton, P. and Macdonald, G., eds. 2004. *Karl Popper: Critical Appraisals*. London: Routledge.
Chakravartty, A. 1999. "Review of N. Maxwell, *The Comprehensibility of the Universe*". *Times Higher Educational Supplement*, 24 September, 24.
Chakravartty, A. 2007. *A Metaphysics for Scientific Realism: Knowing the Unobservable*. Cambridge: Cambridge University Press.
Chalmers, D. 1996. *The Conscious Mind*. Oxford: Oxford University Press.
Chart, D. 2000. "MAXWELL, N. – The Comprehensibility of the Universe". *Philosophical Books* 41(4): 283–5.
Chui, H. and Hoffmann, W. F., eds. 1964. *Gravitation and Relativity*. New York: Benjamin.
Cohen, M. 2015. "On Schupbach and Sprenger's Measures of Explanatory Power". *Philosophy of Science* 82: 97–109.
Cohen, R. S., Feyerabend, P. and Wartofsky, M. W., eds. 1976. *Essays in Memory of Imre Lakatos*. Dordrecht: Reidel.
Collingridge, D. 1985. "Reforming Science". *Social Studies of Science* 15: 763–9.
Craver, C. and Darden, L. 2013. *In Search of Mechanisms: Discoveries Across the Life Sciences*. Chicago, IL: University of Chicago Press.
D'Agostino, F. 2010. *Naturalizing Epistemology: Thomas Kuhn and the "Essential Tension"*. Basingstoke: Palgrave Macmillan.
Dasgupta, S. 2016. "Symmetry as an Epistemic Notion (Twice Over)". *British Journal for the Philosophy of Science* 67: 837–78.
Dawid, R. 2014. *String Theory and the Scientific Method*. Cambridge: Cambridge University Press.
Dennett, D. 1993. *Consciousness Explained*. London: Penguin Books.
Deutscher, M. 1968. "Popper's Problem of an Empirical Base". *Australasian Journal of Philosophy* 46: 277–88.
Dilworth, C. 2007. *The Metaphysics of Science*. Dordrecht: Springer.
Duhem, P. 1962. *The Aim and Structure of Physical Theory*. New York: Atheneum.
Dukas, H. and Hoffmann, B., eds. 1979. *Albert Einstein: The Human Side*. Princeton, NJ: Princeton University Press.
Dupré, J. 1995. *The Disorder of Things*. Cambridge, MA: Harvard University Press.
Durant, J. 1997. "Beyond the Scope of Science". *Science and Public Affairs*, Spring, 56–7.
Easlea, B. 1973. *Liberation and the Aims of Science: An Essay on Obstacles to the Building of a Beautiful World*. London: Chatto & Windus.
Easlea, B. 1986. "Review of From Knowledge to Wisdom". *Journal of Applied Philosophy* 3: 139–40.
Einstein, A. 1905. "Zur Elektrodynamik bewegter Körper". *Annalen der Physik* 17: 891–921.
Einstein, A. 1969. "Autobiographical Notes" and "Replies to Criticisms". In *Albert Einstein: Philosopher-Scientist*, edited by P. A. Schilpp, 1–94. LaSalle, IL: Open Court.
Einstein, A. 1973. *Ideas and Opinions*. London: Souvenir Press.
Einstein, A. 1982. "Autobiographical Notes". In *Albert Einstein: Philosopher-Scientist*, edited by P. A. Schilpp, 1–94. La Salle, IL: Open Court (first published 1949).
Einstein, A., Lorentz, H. A., Minkowski, H. and Weyl, H. 1952. *The Principle of Relativity*. New York: Dover.
Einstein, A., Podolsky, B. and Rosen, N. 1935. "Can Quantum-mechanical Description of Reality be Considered Complete?" *Physical Review* 47: 777–80.
Everett, H. 1957. "'Relative State' Formulation of Quantum Mechanics". *Reviews of Modern Physics* 29(3): 454–62.

Fargaus, J., ed. 1993. *Readings in Social Theory*. New York: McGraw-Hill.
Ferreira, P. G. 2014. *The Perfect Theory: A Century of Geniuses and the Battle over General Relativity*. Boston, MA: Houghton Mifflin Harcourt.
Feyerabend, P. 1965. "Problems of Empiricism". In *Beyond the Edge of Certainty*, edited by R. Colodny. Vol. 2 of *University of Pittsburgh Series in the Philosophy of Science*, 145–260. Englewood Cliffs, NJ: Prentice-Hall.
Feyerabend, P. 1968–9. "On a Recent Critique of Complementarity". *Philosophy of Science* 35: 309–31 and 36: 82–105.
Feyerabend, P. 1970. "Problems of Empiricism II". In *The Nature and Function of Scientific Theories*, edited by R. Colodny. Vol. 4 of *University of Pittsburgh Series in the Philosophy of Science*, 275–354. Pittsburgh, PA: University of Pittsburgh Press.
Feyerabend, P. 1978. *Against Method*. London: Verso.
Feyerabend, P. 1987. *Farewell to Reason*. London: Verso.
Feynman, R., Leighton, R. and Sands, M. 1965. *The Feynman Lectures on Physics*, vol. 2. Reading, MA: Addison-Wesley.
Fine, A. 1986. *The Shaky Game*. Chicago, IL: University of Chicago Press.
Franklin, A. 1986. *The Neglect of Experiment*. Cambridge: Cambridge University Press.
Franklin, A. 1990. *Experiment: Right or Wrong*. Cambridge: Cambridge University Press.
French, A. P., ed. 1979. *Einstein: A Centenary Volume*. London: Heinemann.
Friedman, M. 1974. "Explanation and Scientific Understanding". *Journal of Philosophy* 71: 5–19.
Friedman, M. 1983. *Foundations of Space-Time Theories*. Princeton, NJ: Princeton University Press.
Fromm, E. 1960. *The Fear of Freedom*. London: Routledge & Kegan Paul.
Fromm, E. 1963. *The Sane Society*. London: Routledge & Kegan Paul.
Fuller, S. 2003. *Kuhn vs. Popper*. Cambridge: Icon.
Furbank, P. N. 1993. *Diderot: A Critical Biography*. London: Minerva.
Gao, S. 2017. *The Meaning of the Wave Function: In Search of the Ontology of Quantum Mechanics*. Cambridge: Cambridge University Press. https://arxiv.org/abs/1611.02738
Gascardi, A. 1999. *Consequences of Enlightenment*. Cambridge: Cambridge University Press.
Gattei, S. 2008. *Thomas Kuhn's "Linguistic Turn" and the Legacy of Logical Empiricism: Incommensurability, Rationality and the Search for Truth*. Aldershot: Ashgate.
Gattei, S. 2009. *Karl Popper's Philosophy of Science: Rationality without Foundations*. London: Routledge.
Gay, P. 1973. *The Enlightenment: An Interpretation*. London: Wildwood House.
Ghirardi, G. C. and Rimini, A. 1990. "Old and New Ideas in the Theory of Quantum Measurement". In *Sixty-Two Years of Uncertainty*, edited by A. Miller, 167–91. New York: Plenum Press.
Ghirardi, G. C., Rimini, A. and Weber, T. 1986. "Unified Dynamics for Microscopic and Macroscopic Systems". *Physical Review* D34: 470–91.
Gillespie, T. D. 1973. *A Quantum Mechanical Primer*. Aylesbury: International Textbook.
Glymour, C. 1980. *Theory and Evidence*. Princeton, NJ: Princeton University Press.
Godfrey-Smith, P. 2016. "Popper's Philosophy of Science: Looking Ahead". In *The Cambridge Companion to Popper*, edited by J. Shearmur and G. Stokes, 104–24. Cambridge: Cambridge University Press.
Goodman, N. 1954. *Fact, Fiction and Forecast*. London: Athlone Press.
Gray, J. G. 1972. *The Promise of Wisdom: An Introduction to Philosophy of Education*. New York: Harper & Row.
Griffiths, D. 1987. *Introduction to Elementary Particles*. New York: Wiley.
Gross, P. and Levitt, N. 1994. *Higher Superstition: The Academic Left and Its Quarrels with Science*. Baltimore, MD: Johns Hopkins University Press.
Gross, P., Levitt, N. and Lewis, M., eds. 1996. *The Flight from Science and Reason*. Baltimore, MD: Johns Hopkins University Press.
Grünbaum, A. 1963. *Philosophical Problems of Space and Time*. London: Routledge and Kegan Paul.
Haberman, F. W. 1972. *Nobel Lectures: Peace, 1951–1970*, vol. 3. Amsterdam: Elsevier.
Hacohen, M. H. 2001. *Karl Popper – The Formative Years, 1902–1945: Politics and Philosophy in Interwar Vienna*. Cambridge: Cambridge University Press.
Hagstrom, W. O. 1965. *The Scientific Community*. New York: Basic Books.
Harding, S. 1986. *The Science Question in Feminism*. Milton Keynes: Open University Press.
Harré, R. 1986. *Varieties of Realism*. Oxford: Blackwell.
Harris, M. 1978. *Cannibals and Kings: The Origins of Cultures*. London: Fontana.

Hayek, F. 1979. *The Counter-revolution of Science*. Indianapolis, IN: Liberty Press.
Hempel, C. G. 1965. *Aspects of Scientific Explanation: And other Essays in the Philosophy of Science*. New York: Free Press.
Higgins, R. 1978. *The Seventh Enemy: The Human Factor in the Global Crisis*. London: Hodder & Stoughton.
Hodgson, D. 2002. "Review of *The Human World in the Physical Universe*". *Journal of Consciousness Studies* 9: 93–4.
Hoffmann, B. 1973. *Albert Einstein: Creator and Rebel*. London: Hart-Davis and MacGibbon.
Holton, G. 1973. *Thematic Origins of Modern Science*. Cambridge, MA: Harvard University Press.
Hooker, G. A. 1987. *A Realistic Theory of Science*. Albany, NY: State University of New York Press.
Horwich, P., ed. 1993. *World Changes: Thomas Kuhn and the Nature of Science*. Cambridge, MA: MIT Press.
Howson, C. 2000. *Hume's Problem*. Oxford: Clarendon Press.
Howson, C. and Urbach, P. 1993. *Scientific Reasoning*. La Salle, IL: Open Court.
Hoyningen-Huene, P. 1993. *Reconstructing Scientific Revolutions: Thomas S. Kuhn's Philosophy of Science*. Chicago, IL: University of Chicago Press.
Hull, D. L. 1988. *Science as a Process*. Chicago, IL: University of Chicago Press.
Hume, D. 1959. *A Treatise of Human Nature*. London: J. M. Dent and Sons.
Huxley, A. 1938. *Ends and Means: An Enquiry into the Nature of Ideals and into the Methods Employed for their Realization*. London: Chatto & Windus.
Iredale, M. 2005. "Our Neurotic Friend". *The Philosopher's Magazine* 31: 86–7.
Isham, C. 1989. *Lectures on Groups and Vector Spaces for Physicists*. London: World Scientific.
Israel, J. 2011. *Democratic Enlightenment*. Oxford: Oxford University Press.
Jarvie, I. 2001. *The Republic of Science: The Emergence of Popper's Social View of Science 1935–1945*. Amsterdam: Rodopi.
Jarvie, I., Milford, K. and Miller, D., eds. 2006. *Karl Popper: A Centenary Assesssment*, vols. I, II and III, Aldershot: Ashgate.
Jarvie, I. and Pralong, S., eds. 1999. *Popper's Open Society after Fifty Years*. London: Routledge.
Jeffreys, H. 1957. *Scientific Inference*. Cambridge: Cambridge University Press.
Jeffreys, H. and Wrinch, D. 1921. "On Certain Fundamental Principles of Scientific Enquiry". *Philosophical Magazine* 42: 269–98.
Jones, H. 1990. *Groups, Representations and Physics*. Bristol: Adam Hilger.
Juhl, C. F. 2000. "Review of *The Comprehensibility of the Universe: A New Conception of Science* by Nicholas Maxwell", *International Philosophical Quarterly* 40(4): 517–8.
Kadvany, J. D. 2001. *Imre Lakatos and the Guises of Reason*. Durham, NC: Duke University Press.
Kampis, G., Kvasz, L. and Stöltzner, M., eds. 2002. *Appraising Lakatos: Mathematics, Methodology, and the Man*. London: Kluwer Academic.
Kant, I. 1961. *Critique of Pure Reason*, translated by N. K. Smith. London: Macmillan and Co.
Keuth, H. 2005. *The Philosophy of Karl Popper*. Cambridge: Cambridge University Press.
Kincaid, H., Ladyman, J. and Ross, D., eds. 2013. *Scientific Metaphysics*. Oxford: Oxford University Press.
Kitcher, P. 1981. "Explanatory Unification". *Philosophy of Science* 48: 507–31.
Kitcher, P. 1993. *The Advancement of Science*. New York: Oxford University Press.
Kneller, G. F. 1978. *Science as a Human Endeavor*. New York: Columbia University Press.
Koertge, N., ed. 1998. *A House Built on Sand: Exposing Postmodernist Myths about Science*. Oxford: Oxford University Press.
Koyré, A. 1965. *Newtonian Studies*. London: Chapman and Hall.
Koyré, A. 1973. *Metaphysics and Measurement*. London: Methuen.
Kripke, S. 1981. *Naming and Necessity*. Oxford: Blackwell.
Kuhlmann, M. 2010. *The Ultimate Constituents of the Material World: In Search of an Ontology for Fundamental Physics*. Frankfurt: Ontos-Verlag.
Kuhn, T. S. 1970a. *The Structure of Scientific Revolutions*. Chicago, IL: University of Chicago Press (first published 1962).
Kuhn, T. S. 1970b. "Logic of Discovery or Psychology of Research?" and "Reflections on my Critics". In *Criticism and the Growth of Knowledge*, edited by I. Lakatos and A. Musgrave, 1–23 and 231–78. Cambridge: Cambridge University Press.
Kuhn, T. S. 1977. *The Essential Tension*. Chicago, IL: University of Chicago Press.
Kumar, M. 2008. *Quantum: Einstein, Bohr and the Great Debate about the Nature of Reality*. Cambridge: Icon Books.

Kyburg, H. 1970. *Probability and Inductive Logic*. London: Collier-Macmillan.
Ladyman, J. 2012. "Science, Metaphysics and Method". *Philosophical Studies* 160: 31–51.
Ladyman, J. and Ross, D., with D. Spurrett and J. Collier. 2007. *Every Thing Must Go: Metaphysics Naturalized*. Oxford: Oxford University Press.
Lakatos, I. 1968a. "The Changing Problem of Inductive Logic". In *The Problem of Inductive Logic*, edited by I. Lakatos, 315–417. Amsterdam: North Holland.
Lakatos, I. 1968b. "Criticism and the Methodology of Scientific Research Programmes". *Proceedings of the Aristotelian Society* 69: 149–86.
Lakatos, I. 1970. "Falsification and the Methodology of Scientific Research Programmes". In *Criticism and the Growth of Knowledge*, edited by I. Lakatos and A. Musgrave, 91–195. Cambridge University Press, Cambridge.
Lakatos, I. 1971. "History and its Rational Reconstructions", in *PSA 1970 in Memory of Rudolf Carnap: Proceedings of the 1970 Biennial Meeting Philosophy of Science Association, Boston Studies*, vol. 8, edited by Roger C. Buck and Robert S. Cohen, pp. 91–136. Dordrecht: D. Reidel Publishing Co.
Lakatos, I. 1978. *The Methodology of Scientific Research Programmes*, edited by J. Worrall and G. Currie. Cambridge: Cambridge University Press.
Lakatos, I. and Feyerabend, P. 1999. *For and Against Method*, edited by M. Motterlini. Chicago, IL: University of Chicago Press.
Lakatos, I. and Musgrave, A., eds. 1970. *Criticism and the Growth of Knowledge*. Cambridge: Cambridge University Press.
Lanczos, C. 1974. *The Einstein Decade (1905–1915)*. London: Paul Elek (Scientific Books) Ltd.
Lange, M. 2009. *Laws and Lawmakers: Science, Metaphysics and the Laws of Nature*. Oxford: Oxford University Press.
Larvor, B. 1998. *Lakatos: An Introduction*. New York: Routledge.
Latour, B. 1987. *Science in Action*. Milton Keynes: Open University Press.
Laudan, L. 1977. *Progress and Its Problems*. Berkeley, CA: University of California Press.
Laudan, L. 1980. "A Confutation of Convergent Realism". *Philosophy of Science* 48: 19–48.
Laudan, L. 1984. *Science and Values*. Berkeley: CA: University of California Press.
Laudan, L. 1987. "Progress or Rationality? The Prospects for Normative Naturalism". *American Philosophical Quarterly* 24: 19–31.
Lenski, G. et al. 1995. *Human Societies: An Introduction to Macrosociology*. New York: McGraw-Hill.
Levinson, P. 1982. *In Pursuit of Truth: Essays on the Philosophy of Karl Popper on the Occasion of His 80th Birthday*. Atlantic Highlands, NJ: Humanities Press.
Lipton, P. 2004. *Inference to the Best Explanation*, 2nd ed. London: Routledge.
Longuet-Higgins, L. 1984 "For goodness sake". *Nature* 312: 204.
Mach, E. 1970. "The Guiding Principles of My Scientific Theory of Knowledge and Its Reception by My Contemporaries". In *Physical Reality*, edited by S. Toulmin, 28–43. New York: Harper Torchbooks.
MacIntyre, A. 2009. "The Very Idea of a University: Aristotle, Newman, and Us". *British Journal of Educational Studies* 57(4): 358.
Macionis, J. and Plummer, K. 1997 *Sociology: A Global Introduction*. New York: Prentice Hall.
Mandl, F. and Shaw, G. 1984. *Quantum Field Theory*. New York: Wiley.
Mannheim, K., 1952. *Essays in the Sociology of Knowledge*, Routledge & Kegan Paul, London.
Marcum, J. 2015. *Thomas Kuhn's Revolutions: A Historical and an Evolutionary Philosophy of Science*. London: Bloomsbury.
Mathias, P., ed. 1972. *Science and Society 1600–1900*. London: Cambridge University Press.
Maudlin, T. 1996. "On the Unification of Physics". *Journal of Philosophy* 93: 129–44.
Maudlin, T. 2007. *The Metaphysics within Physics*. Oxford: Oxford University Press.
Maxwell, N. 1968. "Can there be Necessary Connections between Successive Events?" *British Journal for the Philosophy of Science* 19: 1–25.
Maxwell, N. 1972a. "A Critique of Popper's Views on Scientific Method". *Philosophy of Science* 39: 131–52.
Maxwell, N. 1972b. "A New Look at the Quantum Mechanical Problem of Measurement". *American Journal of Physics* 40: 1431–5.
Maxwell, N. 1973a. "Alpha Particle Emission and the Orthodox Interpretation of Quantum Mechanics". *Physics Letters* 43A: 29–30.
Maxwell, N. 1973b. "The Problem of Measurement – Real or Imaginary?" *American Journal of Physics* 41: 1022–5.

Maxwell, N. 1974 "The Rationality of Scientific Discovery, Parts 1 and II". *Philosophy of Science* 41: 123–53 and 247–95.
Maxwell, N. 1975. "Does the Minimal Statistical Interpretation of Quantum Mechanics Resolve the Measurement Problem?" *Methodology and Science* 8: 84–101.
Maxwell, N. 1976a. *What's Wrong With Science? Towards a People's Rational Science of Delight and Compassion*. Hayes: Bran's Head Books, Hayes.
Maxwell, N. 1976b. "Towards a Micro Realistic Version of Quantum Mechanics". *Foundations of Physics* 6: 275–92 and 661–76.
Maxwell, N. 1977a. "Articulating the Aims of Science". *Nature* 265: 2.
Maxwell, N. 1977b. "Science and Values". *Times Higher Educational Supplement*, 4 November, 27.
Maxwell, N. 1979. "Induction, Simplicity and Scientific Progress". *Scientia* 114: 629–53 (Italian trans. pp. 655–74).
Maxwell, N. 1980. "Science, Reason, Knowledge and Wisdom: A Critique of Specialism". *Inquiry* 23: 19–81.
Maxwell, N. 1982. "Instead of Particles and Fields: A Micro Realistic Quantum 'Smearon' Theory". *Foundations of Physics* 12: 607–31.
Maxwell, N. 1984a. *From Knowledge to Wisdom: A Revolution in the Aims and Methods of Science*. Oxford: Blackwell.
Maxwell, N. 1984b. "Guiding Choices in Scientific Research". *Bulletin of Science, Technology and Society* 4: 316–34.
Maxwell, N. 1985a. "Are Probabilism and Special Relativity Incompatible?" *Philosophy of Science* 52: 23–43.
Maxwell, N. 1985b. "From Knowledge to Wisdom: The Need for an Intellectual Revolution". *Science, Technology and Society Newsletter* 21: 55–63.
Maxwell, N. 1986. "The Fate of the Enlightenment: Reply to Kekes". *Inquiry* 29: 79–82.
Maxwell, N. 1987. "Wanted: A New Way of Thinking". *New Scientist* 14: 63.
Maxwell, N. 1988. "Quantum Propensiton Theory: A Testable Resolution of the Wave/Particle Dilemma". *The British Journal for the Philosophy of Science* 39: 1–50.
Maxwell, N. 1991. "How Can We Build a Better World?" In *Einheit der Wissenschaften*, edited by J. Mittelstrass, 388–427. Berlin: Walter de Gruyter.
Maxwell, N. 1992a. "What Kind of Inquiry Can Best Help Us Create a Good World?" *Science, Technology and Human Values* 17: 205–27.
Maxwell, N. 1992b. "What the Task of Creating Civilization has to Learn from the Success of Modern Science: Towards a New Enlightenment". *Reflections on Higher Education* 4: 47–69.
Maxwell, N. 1993a. "Induction and Scientific Realism: Einstein versus van Fraassen". *British Journal for the Philosophy of Science* 44: 61–79, 81–101 and 275–305.
Maxwell, N. 1993b. "Does Quantum Mechanics Undermine, or Support, Scientific Realism?" *The Philosophical Quarterly* 43: 139–57.
Maxwell, N. 1993c. "Beyond Fapp: Three Approaches to Improving Orthodox Quantum Theory and An Experimental Test". In *Bell's Theorem and the Foundations of Modern Physics*, edited by A. van der Merwe, F. Selleri and G. Tarozzi, 362–70. Singapore: World Scientific.
Maxwell, N. 1994a. "Particle Creation as the Quantum Condition for Probabilistic Events to Occur". *Physics Letters* A 187: 351–355.
Maxwell, N. 1994b. "Towards a New Enlightenment: What the Task of Creating Civilization has to Learn from the Success of Modern Science". In *Academic Community: Discourse or Discord?* edited by R. Barnett, 86–105. London: Jessica Kingsley.
Maxwell, N. 1995. "A Philosopher Struggles to Understand Quantum Theory: Particle Creation and Wavepacket Reduction". In *Fundamental Problems in Quantum Physics*, edited by M. Ferrero and A. Van der Merwe, 205–14. London: Kluwer Academic.
Maxwell, N. 1997a. "Must Science Make Cosmological Assumptions if it is to be Rational?" In *The Philosophy of Science: Proceedings of the Irish Philosophical Society Spring Conference*, edited by T. Kelly, 98–146. Maynooth: Irish Philosophical Society.
Maxwell, N. 1997b. "Science and the Environment: A New Enlightenment". *Science and Public Affairs*, Spring, 50–6.
Maxwell, N. 1998. *The Comprehensibility of the Universe: A New Conception of Science*. Oxford: Oxford University Press.
Maxwell, N. 1999a. "Has Science Established that the Universe is Comprehensible?" *Cogito* 13: 139–45.
Maxwell, N. 1999b. "Are There Objective Values?" *The Dalhousie Review* 79: 301–17.

Maxwell, N. 2000a. "Can Humanity Learn to Become Civilized? The Crisis of Science without Civilization". *Journal of Applied Philosophy* 17: 29–44.

Maxwell, N. 2000b. "A new conception of science". *Physics World* 13(8): 17–8.

Maxwell, N. 2000c. "Observation, Meaning and Theory: Review of *For and Against Method* by Imre Lakatos and Paul Feyerabend". *Times Higher Education Supplement* 1,427: 30.

Maxwell, N. 2001a. *The Human World in the Physical Universe*. Lanham, MD: Rowman and Littlefield.

Maxwell, N. 2001b. "Can Humanity Learn to Create a Better World? The Crisis of Science without Wisdom". In *The Moral Universe (Demos Collection 16)*, edited by T. Bentley and D. Stedman Jones, 149–56. London: Demos.

Maxwell, N. 2001c. "Weinert's Review of *The Comprehensibility of the Universe*". *Philosophy* 76: 297–303.

Maxwell, N. 2002a. "Is Science Neurotic?" *Metaphilosophy* 33: 259–99.

Maxwell, N. 2002b. "The Need for a Revolution in the Philosophy of Science". *Journal for General Philosophy of Science* 33: 381–408.

Maxwell, N. 2002c. "Cutting God in Half". *Philosophy Now* 35, March/April, 22–5.

Maxwell, N. 2002d. "The Need for a Revolution in the Philosophy of Science". *Journal for General Philosophy of Science* 33(2): 381–408.

Maxwell, N., 2002e. "Karl Raimund Popper". In *British Philosophers, 1800–2000*, edited by P. Dematteis, P. Fosl and L. McHenry. Columbia, NY: Bruccoli Clark Layman.

Maxwell, N. 2003a. "Two Great Problems of Learning". *Teaching in Higher Education* 8: 129–34.

Maxwell, N. 2003b. "Science, Knowledge, Wisdom and the Public Good". *Scientists for Global Responsibility Newsletter* 26: 7–9.

Maxwell, N. 2003c. "Do Philosophers' Love Wisdom?" *The Philsophers' Magazine* 22, 2nd quarter, 22–4.

Maxwell, N. 2004a. *Is Science Neurotic?* London: Imperial College Press.

Maxwell, N. 2004b. "Does Probabilism Solve the Great Quantum Mystery?" *Theoria* 19/3(51): 321–36.

Maxwell, N. 2006. "Aim-Oriented Empiricism: David Miller's Critique". *PhilSci Archive*. http://philsci-archive.pitt.edu/3092/

Maxwell, N. 2007. *From Knowledge to Wisdom: A Revolution for Science and the Humanities* (2nd ed., revised and extended, of Maxwell, 1984). London: Pentire Press.

Maxwell, N. 2008. "Do We Need a Scientific Revolution? *Journal for Biological Physics and Chemistry* 8(3): 95–105.

Maxwell, N. 2009a. "Are Universities Undergoing an Intellectual Revolution?" *Oxford Magazine* 290, June, 13–6. https://www.ucl.ac.uk/from-knowledge-to-wisdom/essays/areuniversities

Maxwell, N. 2009b. "Muller's Critique of the Argument for Aim-Oriented Empiricism". *Journal for General Philosophy of Science* 40: 103–14.

Maxwell, N. 2009c. "Replies and Reflections". In *Science and the Pursuit of Wisdom: Studies in the Philosophy of Nicholas Maxwell*, edited by L. McHenry, 249–313. Frankfurt: Ontos Verlag.

Maxwell, N. 2009d. "The Metaphysics of Science: An Account of Modern Science in Terms of Principles, Laws and Theories" (review of a book by Craig Dilworth). *International Studies in the Philosophy of Science* 23(2): 228–32.

Maxwell, N. 2010a. *Cutting God in Half – And Putting the Pieces Together Again: A New Approach to Philosophy*. London: Pentire Press.

Maxwell, N. 2010b. "Reply to Comments on *Science and the Pursuit of Wisdom*". *Philosophia* 38: 667–90.

Maxwell, N. 2010c. "Wisdom Mathematics". *Friends of Wisdom Newsletter* 6: 1–6. www.knowledge-towisdom.org/Newsletter%206.pdf

Maxwell, N. 2011a. "Is the Quantum World Composed of Propensitons?" In *Probabilities, Causes and Propensities in Physics*, edited by Mauricio Suárez, 219–41. Dordrecht: Synthese Library, Springer.

Maxwell, N. 2011b. "A Priori Conjectural Knowledge in Physics". In *What Place for the A Priori?*, edited by Michael Shaffer and Michael Veber, 211–40. Chicago, IL: Open Court

Maxwell, N. 2012. "Arguing for Wisdom in the University: An Intellectual Autobiography". *Philosophia* 40(4): 663–704.

Maxwell, N. 2013. "Has Science Established that the Cosmos is Physically Comprehensible?" In *Recent Advances in Cosmology*, edited by A. Travena and B. Soen, 1–56. New York: Nova Publishers Inc. http://discovery.ucl.ac.uk/view/people/ANMAX22.date.html

Maxwell, N. 2014a. "Unification and Revolution: A Paradigm for Paradigms". *Journal for General Philosophy of Science* 45(1): 133–49.
Maxwell, N. 2014b. *How Universities Can Help Create a Wiser World: The Urgent Need for an Academic Revolution*. Exeter: Imprint Academic.
Maxwell, N. 2014c. *Global Philosophy*. Exeter: Imprint Academic.
Maxwell, N. 2015a. "What's Wrong With Aim-Oriented Empiricism?" *Acta Baltica Historiae et Philosophiae Scientiarum* 3: 5–31.
Maxwell, N. 2015b. "What's Wrong with Science and Technological Studies? What Needs to be Done to Put It Right?" In *Physics, Astronomy and Engineering: A Bridge between Conceptual Frameworks*, edited by R. Pisano and D. Capecchi, vii–xxxvii. Netherlands: Springer.
Maxwell, N. 2016a. "Popper's Paradoxical Pursuit of Natural Philosophy". In *Cambridge Companion to Popper*, edited by J. Shearmur and G. Stokes, 170–207. Cambridge: Cambridge University Press.
Maxwell, N. 2016b. *Two Great Problems of Learning: Science and Civilization*. Rounded Globe. https://roundedglobe.com
Maxwell, N. 2017a. *Understanding Scientific Progress*. Saint Paul, MN: Paragon House.
Maxwell, N. 2017b. *In Praise of Natural Philosophy: A Revolution for Thought and Life*. Montreal: McGill-Queen's University Press.
McAllister, J. 1989. "Truth and Beauty in Scientific Reason". *Synthese* 78: 25–51.
McAllister, J. 1990. "Dirac and the Aesthetic Evaluation of Theories". *Methodology and Science* 23: 87–102.
McAllister, J. 1991. "The Simplicity of Theories: Its Degree and Form". *Journal for General Philosophy of Science* 22: 1–14.
McAllister, J. W. 1996. *Beauty and Revolution in Science*. Ithaca, NY: Cornell University Press.
McHenry, L. 2000. "Review: The Comprehensibility of the Universe: A New Conception of Science by Nicholas Maxwell". *Mind* 109: 162–6.
McHenry, L. 2009a. "Popper and Maxwell on Scientific Progress". In *Science and the Pursuit of Wisdom: Studies in the Philosophy of Nicholas Maxwell*, edited by L. McHenry, 233–47. Franfurt: Ontos Verlag.
McHenry, L., ed. 2009b. *Science and the Pursuit of Wisdom: Studies in the Philosophy of Nicholas Maxwell*. Frankfurt: Ontos Verlag.
Mead, M. 1943. *Coming of Age in Samoa: A Study of Adolescence and Sex in Primitive Societies*. Harmondsworth: Penguin Books.
Mellor, D. H. 1991. 'The Warrant of Induction". In *Matters of Metaphysics*, 254–68. Cambridge: Cambridge University Press.
Merton, R. K. 1970. *Science, Technology and Society in Seventeenth-Century England*. New York: Harper & Row.
Meyerson, E. 1930. *Identity and Reality*. London: George Allen and Unwin.
Midgley, M. 1979. *Beast and Man: The Roots of Human Nature*. Sussex: Harvester Press.
Midgley, M. 1986. "Is Wisdom Forgotten?" *University Quarterly: Culture, Education and Society* 40: 425–7.
Miller, D. 1974. "Popper's Qualitative Theory of Verisimilitude". *British Journal for the Philosophy of Science* 25: 166–77.
Miller, D. 1994. *Critical Rationalism*. Chicago, IL: Open Court.
Miller, D. 1997. "Sir Karl Raimund Popper, C. H., F. B. A." *Biographical Memoirs of Fellows of the Royal Society* 43: 367–409.
Miller, D. 2006. *Out of Error*. London: Ashgate.
Misner, C. W., Thorne, K. and Wheeler, J. A. 1973. *Gravitation*. San Francisco, CA: W. H. Freeman.
Morganti, M. 2013. *Combining Science and Metaphysics: Contemporary Physics, Conceptual Revision, and Common Sense*. London: Palgrave Macmillan.
Moriyasu, K. 1983. *An Elementary Primer for Gauge Theory*. Singapore: World Scientific.
Morrison, M. 2000. *Unifying Scientific Theories*. Cambridge: Cambridge University Press.
Mulkay, M. 1979. *Science and the Sociology of Knowledge*. London: George Allen & Unwin.
Muller, F. A. 2004. "Maxwell's Lonely War". *Studies in History and Philosophy of Modern Physics* 35: 109–10 and p. 117.
Muller, F. A. 2008. "In Defence of Constructivism: Maxwell's Master Argument and Aberrant Theories". *Journal of General Philosophy of Science* 39: 131–56.
Mumford, S. and Tugby, M. 2013. *Metaphysics and Science*. Oxford: Oxford University Press.
Musgrave, A. 1993. *Common Sense, Science and Scepticism*. Cambridge: Cambridge University Press.

Musgrave, A. 2004. "How Popper (Might Have) Solved the Problem of Induction". In *Karl Popper: Critical Appraisals*, edited by P. Catton and G. Macdonald, 16–27. London: Routledge.
Müürsepp, P. 2014. "Review". *Dialogue and Universalism* 2: 247.
Nagel, T. 1986. *The View from Nowhere*. Oxford: Oxford University Press.
Newton, I., 1962. *Principia, vol. II*, translated by A. Motte and F. Cajori. Berkeley, CA: California University Press (first published 1687).
Newton-Smith, W. 1981. *The Rationality of Science*. London: Routledge and Kegan Paul.
Nickles, T., ed. 2003. *Thomas Kuhn*. Cambridge: Cambridge University Press.
Niiniluoto, I. 2016. "Unification and Confirmation". *Theoria* 31: 107–23.
Nola, R. and Sankey, H. 2000. "A Selective Survey of Theories of Scientific Method". In *After Popper, Kuhn and Feyerabend*, edited by R. Nola and H. Sankey, 1–65. Dordrecht: Kluwer.
North, J. 1965. *The Measure of the Universe*. Oxford: Clarendon Press.
Norton, J. 1984. "How Einstein Found His Field Equations: 1912–1915". *Historical Studies in the Physical Sciences* 14: 253–316.
O'Hear, A., ed. 1995. *Karl Popper: Philosophy and Problems*. Cambridge: Cambridge University Press.
O'Hear, A., ed. 2004. *Karl Popper: Critical Assessments of Leading Philosophers*, vols. I, II and III, London: Routledge.
Pais, A., 1982. *Subtle is the Lord*. Oxford: Oxford University Press.
Pandit, G. L. 2010. "How Simple is it for Science to Acquire Wisdom According to its Chicest Aims?" *Philosophia* 38: 649–66.
Penrose, R. 1986. "Gravity and State Vector Reduction". In *Quantum Concepts in Space and Time*, edited by C. Isham and R. Penrose, 129–46. Oxford: Oxford University Press.
Penrose, R. 2004. *The Road to Reality*. London: Jonathan Cape.
Penrose, R. and MacCallum, M. 1973. "Twistor Theory: An Approach to the Quantization of Fields and Space-time". *Physics Reports* 6(4): 241–315.
Pickering, A. 1984. *Constructing Quarks*. Chicago, IL: University of Chicago Press.
Pitts, J. B. 2016. "Einstein's Physical Strategy, Energy Conservation, Symmetries, and Stability: 'But Grossmann and I believed that the conservation laws were not satisfied'". *Studies in History and Philosophy of Science Part B: Studies in History and Philosophy of Modern Physics* 54: 52–72.
Popper, K. R. 1957a. "The Aim of Science". *Ratio* 1: 24–35.
Popper, K. R. 1957b. "The Propensity Interpretation of the Calculus of Probability and the Quantum Theory". In *Observation and Interpretation*, edited by S. Körner, 65–70. London: Butterworth.
Popper, K. R. 1959. *The Logic of Scientific Discovery*. London: Hutchinson (first published 1934).
Popper, K. R. 1961. *The Poverty of Historicism*. London: Routledge and Kegan Paul (first published 1957).
Popper, K. R. 1963. *Conjectures and Refutations*. London: Routledge and Kegan Paul.
Popper, K. R. 1966a. *The Open Society and Its Enemies*. London: Routledge and Kegan Paul (first published 1945).
Popper, K. R. 1966b. "A Theorem on Truth Content". In *Mind, Matter, and Method*, edited by P. Feyerabend and G. Maxwell, 343–53. Minneapolis, MN: University of Minnesota Press.
Popper, K. R. 1967. "Quantum Mechanics without 'The Observer'". In *Quantum Theory and Reality*, edited by M. Bunge, 7–44. Berlin: Springer.
Popper, K. R. 1970. "Normal Science and Its Dangers". In *Criticism and the Growth of Knowledge*, edited by I. Lakatos and A. Musgrave, 51–8. Cambridge: Cambridge University Press.
Popper, K. R. 1972. *Objective Knowledge*. Oxford: Oxford University Press.
Popper, K. R. 1974. "Replies to Critics". In *The Philosophy of Karl Popper*, edited by P. A. Schilpp, vol. 2, 961–1197. LaSalle, IL: Open Court.
Popper, K. R. 1976a. *Unended Quest*. London: Fontana/Collins.
Popper, K. R. 1976b. "Reason or Revolution?" In *The Positivist Dispute in German Sociology*, 288–300. London: Heinemann Educational.
Popper, K. R. 1977. "How I see Philosophy". In *Philosophers on their Own Work*, edited by A. Mercier, M. Svilar and P. Lang, vol. 3, Las Vegas, NV: Peter Lang.
Popper, K. R. 1982a. *The Open Universe: An Argument for Indeterminism*, vol. 2 of *The Postscript to The Logic of Scientific Discovery*. London: Hutchinson.
Popper, K. R. 1982b. *Quantum Theory and the Schism in Physics*, vol. 3 of *The Postscript*. London: Hutchinson.
Popper, K. R. 1983. *Realism and the Aim of Science*, vol. 1 of *The Postscript*. London: Hutchinson.

Popper, K. R. 1994. *The Myth of the Framework*. London: Routledge.
Popper, K. R. 2009. *The Two Fundamental Problems of the Theory of Knowledge*, edited by T. E. Hansen and translated by A. Pickel. London: Routledge (first published in German in 1979).
Popper, K. R. and Eccles, J. 1977. *The Self and Its Brain*. London: Springer-Verlag.
Price, D. de S. 1961. *Science Since Babylon*. New Haven, CT: Yale University Press.
Przibram, K., ed. 1967. *Letters on Wave Mechanics*. London: Vision Press.
Rae, A. I. M. 2002. *Quantum Mechanics*. Bristol: Institute of Physics Publishing.
Reichenbach, H. 1958. *The Rise of Scientific Philosophy*. Berkeley, CA: University of California Press.
Reichenbach, H. 1961. *Experience and Prediction*. Chicago, IL: University of Chicago Press (first published 1938).
Renn, J. ed., 2007. *The Genesis of General Relativity*, vols. 1–4. Dordrecht: Springer.
Rescher, N. 1973. *The Primacy of Practice*. Oxford: Blackwell.
Rescher, N. 1977. *Methodological Pragmatism*. Oxford: Blackwell.
Rescher, N. 1987. *Scientific Realism*. Dordrecht: Reidel.
Rescher, N. 2001. *Nature and Understanding: The Metaphysics and Methods of Science*. Oxford: Clarendon Press.
Robus, O. 2015. "Does Science License Metaphysics?" *Philosophy of Science* 82: 845–55.
Rogers, K. 2009. "Metaphysics and Methodology: Aim-Oriented Empiricism". In *Science and the Pursuit of Wisdom: Studies in the Philosophy of Nicholas Maxwell*, edited by L. McHenry, 217–32. Franfurt: Ontos Verlag.
Rose, H. and Rose, S., eds. 1976. *The Radicalization of Science*. London: Macmillan.
Roush, S. 2001. *The Philosophical Review* 110: 85–7.
Rowbottom, D. P. 2011. *Popper's Critical Rationalism*. London: Routledge.
Rudner. R. S. 1961. "An Introduction to Simplicity". *Philosophy of Science* 28: 109–19.
Russell, B. 1948. *Human Knowledge: Its Scope and Limits*. London: Allen and Unwin.
Rylance, R. 2015. "Global Funders to Focus on Interdisciplinarity". *Nature* 525: 313–5.
Ryle, G. 1949. *The Concept of Mind*. London: Hutchinson.
Ryle, G. 1967. "Introduction". In *The Revolution in Philosophy*, edited by A. J. Ayer, 1–11. London: Macmillan.
Salaman, E. 1979. "Memories of Einstein". *Encounter*, April, 18–23.
Salmon, W. 1968. "The Justification of Inductive Rules of Inference". In *The Problem of Inductive Logic*, edited by I. Lakatos, 24–43. Amsterdam: North Holland.
Salmon, W. 1989. *Four Decades of Scientific Explanation*. Minneapolis, MN: University of Minnesota Press.
Scerri, E. 2016. *A Tale of Seven Scientists and a New Philosophy of Science*. Oxford: Oxford University Press.
Scheffler, I. 1963. *The Anatomy of Inquiry*. New York: Alfred A. Knopf.
Schiff, M. 2005. "Is Science Neutoric?" *Choice*, 42, (11/12) 2005–6.
Schilpp, P., ed. 1969. *Albert Einstein: Philosopher-Scientist*. LaSalle, IL: Open Court.
Schilpp, P. A., ed. 1974. *The Philosophy of Karl Popper*. LaSalle, IL: Open Court.
Schrenk, M. 2016. *Metaphysics of Science*. Abingdon: Routledge.
Schumacher, E. F. 1973. *Small is Beautiful: A Study of Economics as if People Mattered*. London: Blond & Briggs.
Schurz, G. 1999. "Explanation as Unification". *Synthese* 120: 95–114.
Schurz, G. 2014. "Unification and Explanation: Explanation as a Prototype Concept. A Reply to Weber and van Dyck, Gijsbers, and de Regt". *Theoria* 29: 57–70.
Schurz, G. 2015. "Causality and Unification: How Causality Unifies Statistical Regularities". *Theoria* 30: 73–95.
Schutz, B. F. 1988. *A First Course in General Relativity*. Cambridge: Cambridge University Press.
Segerstrale, U., ed. 2000. *Beyond the Science Wars*. Albany, NY: State University of New York Press.
Shankland, R. 1963. "Conversations with Albert Einstein". *American Journal of Physics* 31: 37–47.
Shanks, N. 2000. *Metascience* 9: 294–8.
Shapin, S. 1994. *A Social History of Truth*. Chicago, IL: University of Chicago Press.
Shearmur, J. and Stokes, G., eds. 2016. *The Cambridge Companion to Popper*. Cambridge: Cambridge University Press.
Singer, I. B. 1975a. *The Manor*. Harmondsworth: Penguin Books.
Singer, I. B. 1975b. *The Estate*. Harmondsworth: Penguin Books.
Singer, I. B. 1975c. *A Friend of Kafka*. Harmondsworth: Penguin Books.

Singer, I. B. 1977a. *The Séance and Other Stories*. Harmondsworth: Penguin Books.
Singer, I. B. 1977b. *A Crown of Feathers and Other Stories*. Harmondsworth: Penguin Books.
Singer, I. B. 1977c. *Enemies: A Love Story*. Harmondsworth: Penguin Books.
Smart, J. J. C. 1963. *Philosophy and Scientific Realism*. London: Routledge and Kegan Paul.
Smart, J. J. C. 2000. "The Comprehensibility of the Universe". *British Journal for the Philosophy of Science* 51: 907–11.
Sober, E. 2015. *Ockham's Razors: A User's Manual*. Cambridge: Cambridge University Press.
Sokal, A. 2008. *Beyond the Hoax*. Oxford: Oxford University Press.
Sokal, A. and Bricmont, J. 1998. *Intellectual Impostures*. London: Profile Books.
Squires, E. 1986. *The Mystery of the Quantum World*. Bristol: Adam Hilger.
Stachel, J. 1980. "Einstein and the Rigidly Rotating Disk". In *General Relativity and Gravitation*, edited by A. Held, vol. 1, 3–15. New York: Plenum Press.
Stachel, J., ed. 1998. *Einstein's Miraculous Year: Five Papers That Changed the Face of Physics*. Princeton, NJ: Princeton University Press.
Stachel, J. 2007. "The First Two Acts". In *The Genesis of General Relativity*, edited by J. Renn, vols. 1–4, 81–111. Dordrecht: Springer.
Sterkenburg, T. 2016. "Solomonoff Prediction and Occam's Razor". *Philosophy of Science* 83: 459–79.
Stove, D. 1982. *Popper and After: Four Modern Irrationalists*. Oxford: Pergamon Press.
Swain, M. ed. 1970. *Induction, Acceptance and Rational Belief*. Dordrecht: Reidel.
Swinburne, R. G., ed. 1974. *The Justification of Induction*. London: Oxford University Press.
Teich, M. and Young, R. M., eds. 1973. *Changing Perspectives in the History of Science*. London: Heinemann, London.
Tichy, P. 1974. "On Popper's Definition of Verisimilitude". *British Journal for the Philosophy of Science* 25: 155–60.
Tischler, H. 1996. *Introduction to Sociology*. Orlando, FL: Harcourt Brace.
Tumulka, R. 2006. "A Relativistic Version of the Ghirardi–Rimini–Weber Model." *Journal of Statistical Physics* 125: 821–40.
Turnbull, C. 1976. *The Forest People*. London: Picador.
Van Fraassen, B. 1980. *The Scientific Image*. Oxford: Clarendon Press.
Van Fraassen, B. 1985. "Empiricism in the Philosophy of Science". In *Images of Science*, edited by P. M. Churchland and A. Hooker, 245–308. Chicago, IL: University of Chicago Press.
Van Noorden, R. 2015. "Interdisciplinary Research by the Numbers". *Nature* 525: 306–7.
Vicente, A. 2010. "An Enlightened Revolt: on the Philosophy of Nicholas Maxwell". *Philosophia* 38: 631–48.
Vickers, J. 2016. "The Problem of Induction". *The Stanford Encyclopedia of Philosophy Archive*, edited by E. N. Zalta. https://plato.stanford.edu/archives/spr2016/entries/induction-problem/
Votsis, I. 2015. "Unification: Not Just a Thing of Beauty". *Theoria* 30: 97–114.
Wallace, D. 2012. *The Emergent Multiverse*. Oxford: Oxford University Press.
Watkins, J. 1958. "Confirmable and Influential Metaphysics". *Mind* 67: 344–65.
Watkins, J. 1984. *Science and Scepticism*. Princeton, NJ: Princeton University Press.
Weber, E. 1999. "Unification". *Synthese* 118: 479–99.
Weber, E. and Lefevere, M. 2014. "The Role of Unification in Micro-Explanations of Physical Laws". *Theoria* 29: 41–56.
Weinert, F. 2000. *Philosophy* 75: 296–309.
Werskey, G. 1978. *The Visible College*. London: Allen Lane.
Weyl, H. 1963. *Philosophy of Mathematics and Natural Science*. New York: Athenium (first published in German in 1926).
Wheeler, J. A. 1968. "Superspace and the Nature of Quantum Geometrodynamics". In *Battelle Rencontres*, edited by C. M. DeWitt and J. A. Wheeler, 242–307. London: Benjamin.
Wheeler, J. A. 1972. *Geometrodynamics*. London: Academic Press.
Whiteley, C, H. 1955. *An Introduction to Metaphysics*. London: Methuen.
Whitrow, G. 1973. *Einstein: The Man and His Achievement*. New York: Dover Press.
Wigner, E. P. 1970. *Symmetries and Reflections*. London: MIT Press.
Worrall, J. 1989. "Why Both Popper and Watkins Fail to Solve the Problem of Induction". In *Freedom and Rationality: Essays in Honour of John Watkins*, edited by F. D'Agostino, I. C. Jarvie and J. Watkins, 257–96. Dordrecht: Kluwer.
Wray, K. B. 2011. *Kuhn's Evolutionary Social Epistemology*. Cambridge: Cambridge University Press.

Wu, C. S., Ambler, E., Hayward, R. W., Hoppes, D. D. and Hudson, R. P. 1957. *Physical Review* 105: 1413.
Zahar. E. 1973. "Why Did Einstein's Programme Supersede Lorentz's?" *British Journal for the Philosophy of Science* 24: 95–123.
Zamyatin, Y. 1972. *We*. Harmondsworth: Penguin.
Ziman, J. 1968. *Public Knowledge: An Essay Concerning the Social Dimension of Science*. Cambridge: Cambridge University Press.

Index

a priori 57, 69, 96, 119, 120, 239, 259, 263, 273, 332n19
abstractness 179
academic inquiry 2, 6, 169, 233–4, 236–7, 240, 243, 248–9, 250–2, 254–7, 265, 272, 274, 275, 279–80, 284–5, 289, 294, 298, 300, 301–3, 305–6, 313, 316, 320–5
ad hoc theory 14, 16, 52, 53, 55, 56, 63, 95, 113, 146, 183, 184, 185, 202, 217, 307, 332n21
Adler, Alfred 11, 12
aesthetics *see* model of aesthetic induction
Agassi, Joseph 26
aim-oriented empiricism (AOE) xi, 1, 4–5, 6, 327n1, 337n7
 and aesthetic criteria 176
 alternative versions 163–5
 and blueprint 62, 66–7, 69–70, 72–5, 77–8, 81–5, 175–7, 183, 309–10
 comparison with MAI 173–83
 and comprehensibility 1, 30, 44, 59–61, 78, 104, 118–9, 141, 176, 181, 211
 consistent 66
 criteria as heuristically powerful 175
 critical assessment 66
 discovery of new fundamental theories 72–3
 diversity of scientific method 73–5
 and Einstein 5, 90–124, 331n2
 evolving aims and methods 68–70
 and explanatory power 180–3
 fundamental line of thought 189
 and general relativity 100–2
 generalized 174, 186, 198, 211
 hierarchy 74–5
 as improvement over falsificationism 66–75
 improving scientific practice 82
 and invocation of a model 178–9
 justification of hierarchical view 42–7, 63–4, 188–214, 329n17
 and Kuhn 76–82, 262–3
 and Lakatos 83–9, 156–7
 and metaphysical allegiance 177–8
 move from falsificationism to 56–66, 306–12
 nine-level version 159–63
 and non-empirical criteria of theory choice 174–6
 and problem of induction 188–98
 and relationship between science and philosophy of science 183
 rigour 66–7
 and science as having hierarchy of aims 174
 scientific method 68
 seven-level version 43, 44–7, 63, 91
 and simplicity 67–8, 178
 solution to circularity problem 64, 165–8, 201–2
 solving problem of induction 198–214
 and special relativity 98
 and symmetry 177
 as synthesis of Popper, Kuhn, Lakatos's views 42, 47–9
 and unity of theory 42–4, 177
 verisimilitude 70–2
 and visualization/abstractness 179–80
 see also empiricism; meta-knowability; partial knowability; physicalism; rational scientific discovery
aim-oriented rationalism (AOR) xi–xiv, 1, 312–15
 see also critical rationalism; rationalism
anti-naturalist doctrines 25, 26
AOE *see* aim-oriented empiricism
AOR *see* aim-oriented rationalism
approximate derivation 71–2, 139–40
Aristotle, Aristotelianism 18, 20, 25, 46, 270
assumptions 337n15, 338n19
 cosmological 195–6, 207
 hierarchy of 42–7, 159–61, 196–7, 199
 metaphysical 154, 166–7, 211, 258–9, 354n29
 scientific 192–6, 197, 199, 337n13
 and theories 42, 140–2, 191–2
atomic theory 62, 74, 77
Ayer, A.J. 11, 33, 344n22

Bacon, Francis 253, 259, 261, 285, 293
Bakewell, Sarah 344n22
Bartlborth, T. 128
Bartley, W.W. III 26
Bayesianism 145, 337n15
beauty 170–1, 176
Bell, John 36, 115
Berlin, Isaiah 354n33
biology 74, 77, 78, 81, 128
blueprint problem 103–4, 331n9
Bode's law 135
Bohm, David 115

369

Bohr, Niels 36, 72, 78, 81, 110, 219
Born, Max 109, 110, 215–16
Boscovich, Roger J. 84, 230
Braithwaite, Richard 182, 184, 185, 186, 199, 200
Broglie, Louis de 99, 110
bucket theory of the mind 31
Bühler, Karl 10, 11

Cambridge Environmental Initiative (CEI) 288–9
capitalism 21–4
Carnap, Rudolf 11, 259, 344n22
Cartesianism 213–14, 259
CEI *see* Cambridge Environmental Initiative
CEM *see* classical Maxwellian electrodynamics
Chalmers, David 344n22
chemistry 74, 77, 330n28
childish thinking 246, 342n14
circularity problem 165–8, 202
civilized world 38, 294, 300–1, 312, 316–20, 354n33
classical Maxwellian electrodynamics (CEM) 61, 92, 94, 96, 103, 154–5, 189, 190, 231–2
Comte, Auguste 294
cooperative action 234, 239–40, 241–6, 254–5, 256, 268, 269, 278, 290, 298, 300–1, 303
cosmological assumptions, conjectures or theses 42–7, 58–65, 146–9, 174, 151–69, 195–7, 206–7, 209–14, 338n20
  see also aim-oriented empiricism; assumptions; metaphysical; metaphysical thesis; physicalism; presuppositionism
Coulomb's law 52
critical rationalism ix, 202–3, 241–4, 345n33
  critical 203
  dogmatic 203
  see also aim-oriented rationalism; rationalism

Darwin, Charles 78, 80, 250
Davisson, Clinton 110
Dedekind, Richard 3
demarcation problem 3, 9, 13, 15, 16–17, 34, 38, 54–5, 65, 66, 144, 204, 307
Dennett, Daniel 344n22
Descartes, René 27, 237, 253, 259, 285
determinism 34–5
Diderot, Denis 254
Dilworth, Craig 328n8
Dirac, Paul 72, 81, 99, 170
Duhem, Pierre 259

Eccles, John 9, 32
Economic and Social Research Council (ESRC) 289
Eddington, Arthur 251
education 246, 350n49
Ehrenfest, Paul 78
Einstein, Albert xiii, 5, 6, 12, 45, 78–9, 80, 81, 128, 149, 179, 180, 197, 209, 231, 244, 251, 282, 332n21, 342n14, 345n33, 350n45

  as advocate for AOE 91–2, 115–24
  as constructive empiricist 90–1
  employment of AOE 91–2, 102–7, 331n2
  general relativity 12, 45, 61–2, 78, 100–2, 158–9, 180, 189, 250, 263
  and light quanta 110, 331n11
  new method of discovery 90–3
  quantum theory 107–15
  special relativity 45, 61, 94–100, 154, 189, 231, 250, 263
  'Autobiographical Notes' 92, 95, 98, 104, 107–8, 122
electrodynamics 99, 105–7
  Faraday interpretation 96
  Maxwellian 61, 92, 94, 96, 103, 154–5, 189, 230
  quantum 62, 72, 178, 189
electromagnetism, Faraday interpretation 95, 230
empirical laws 45, 111, 162, 191, 335n16
empiricism 4–5, 85, 93, 259, 260–1, 303, 330n36
  standard 91, 99–100, 122, 145, 148, 188–9, 192–3, 204–5, 209, 210–12, 261, 339n25, 347n40
  see also aim-oriented empiricism
Engineering and Physical Sciences Research Council (EPSRC) 289
Enlightenment 2, 6–7, 271, 292–5, 298–9, 351n2
Enlightenment Programme
  background 292–5
  implications for academic inquiry 320–1
  improved Popperian Enlightenment 297–306
  Popper's contribution to 295–7
  see also New Enlightenment
essentialism 17, 27, 179, 333n7
evolution, theory of 9, 32, 78, 140, 250
experts 31, 247–8, 274, 276
explanation 146, 193, 335n19
explanatory power 180–3

falsificationism ix, 1–2, 8, 13–15, 30, 47–8, 49–51, 204, 208, 296
  AOE as improvement over 66–75
  bare 50
  dressed 50
  move to AOE 56–66, 306–12
  refutation of bare 51–3
  refutation of dressed 54–6
  role of metaphysics in science 65
  theories 15–17, 45, 48, 50–1, 53, 62
Faraday, Michael 96, 103
Feigl, Herbert 11, 12
Feyerabend, Paul 31, 36, 83, 90
field theory 45, 106
Fine, Arthur 90, 109, 119, 330n36
FitzGerald, George F. 95
Frank, Philipp 11
Franklin, Allan 86
Freud, Sigmund 12, 251
Friedman, Michael 128
Fromm, Erich 251, 272
  The Fear of Freedom 271
Furtwängler, Philipp 10

Galileo Galilei 45, 46, 61, 99, 154, 157, 212, 237, 253, 272, 285, 329n15, 337n13, 339n27
Gay, Peter, *The Enlightenment: An Interpretation* 271
Gell-Mann, Murray 81
geology 74, 329n27
Germer, Lester 110
Ghirardi, Giancarlo 223
God 163, 164, 195, 263, 343n19, 346n37
Gödel, Kurt 11
Gomperz, Heinrich 11
Goodall, Jane 251
Goodman's paradox, 137–8
gravitation, theory of 100–2, 130, 135, 137, 194, 229, 331n5
Grünbaum, Adolf 33, 95

Hahn, Hans 10, 11
Hayek, Friedrich 251
Hegel, G.W.F. 18, 20, 25, 48, 270, 344n22
Heisenberg, Werner 35, 72, 87, 110, 136, 215, 219
Hempel, Carl 11, 259, 261, 344n22
Heraclitus 270
Higgins, Ronald 251, 343n21
Hitler, Adolf 13, 17, 20
Hume, David 13, 206, 207, 254, 349n45
Huxley, Aldous 265–6
Huygens, Christiaan 229

Illich, Ivan 251
indigenous culture 245–6, 342n13
induction, problem of 6, 13, 47, 68, 91, 148, 166, 330n1, 334n15, 335n28, 336n6
  and AOE 188–98, 354n28
  AOE solution to 198–214
  cosmological conjectures need acknowledgement and improvement 210–14
  methodological 203
  practical 203–10, 204
  solving with metaphysical presuppositions 187–214
  theoretical 203–4
  two versions of critical rationalism 202–3
instrumentalism 27–8, 35, 113, 179, 229
intellectual
  progress 249–50, 343n18
  rigour 145–9, 343n17
  standards 249–50
intellectual inquiry
  four fundamental problems 234–6, 238, 243, 251, 257, 258, 299, 341n2, 348n43
  fundamental aim 279, 302
  historical development 253–4
  intrinsic or cultural value of 282
  as irrational 285–6
  Marxist 347n40
  and problem-solving rationality 299–306
  pure and pragmatic 279, 282
  pursuance of knowledge not wisdom 284–5, 302
  and rational discussions 255–6
  as rigorous 249, 343n17
  and universalism 239, 269, 276–84, 348n43

value in life 304
invisible college 251, 269, 343n20

Jeffreys, Harold 128
Jung, Carl Gustav 251
justificational problem 207–10

Kant, Immanuel 20, 119, 120, 254, 259
Kepler, Johann 45, 61, 71, 154, 189, 253, 282, 337n13, 339n27
King, Martin Luther 276–7, 286
Kitcher, Philip 128
Klein-Gordon equation 99
knowledge 2, 27–8, 33, 166, 213, 280–4, 290, 300, 302, 305, 313–15
  see also philosophy of knowledge
knowledge-inquiry 306, 323, 350n50, 351n51
  see also philosophy of knowledge
Koestler, Arthur 251
Kraft, Julius 11
Kraft, Victor 11
Kuhn, Thomas 4, 5, 31, 33, 42, 47–8, 76–82, 145, 156, 173, 247, 260, 264, 330n36, 343n16, 346n36
  The Structure of Scientific Revolutions 47, 262–3

Lagrangianism 161, 162
Lakatos, Imre 4, 5, 26–7, 31, 33, 42, 47–8, 83–9, 145, 156, 192, 346n36
Lanczos, Cornelius 117
Laudan, Larry 157, 339n21, 339n25
Leibniz, Gottfried Wilhelm 229, 237, 253, 254, 259
Levy, Maurice 52, 53
London School of Economics (LSE) 26–7
Lorentz, Hendrik 92, 95, 105
Lorentz invariance 69, 99, 100
Lynas, Mark 251

McAllister, James 6, 128, 171, 173, 182, 184–5
Mach, Ernst 11, 69
MAI *see* model of aesthetic induction
Margenau, Henry 33
Margulis, Lynn 251
Marx, Karl 12, 18, 19–24, 270, 294
Marxism 10, 17, 22, 347n40
matrix mechanics 215
Maudlin, Tim 128
Maxwell, James Clerk 61, 72, 81, 92, 230
Maxwellian electrodynamics (ME) 92, 94, 96, 103, 107, 154
Mead, Margaret 251
  Coming of Age in Samoa 270
Medawar, Peter 33
Mellor, David Hugh 182, 184, 185, 186, 199, 200
Menger, Karl 11
meta-knowability 60–1, 64, 120, 164, 166–8, 200–2, 335n31, 339n22
metaphysical
  allegiance 177–8
  assumptions 166–7, 211, 258–9, 354n29
  minimalism 158–9
  presuppositionism 147–9, 157, 159, 192–4, 346n38, 347n40
  research programmes 49–50, 260

metaphysical thesis 56–66, 175, 213, 329n22, 335n26, 354n27
  acceptance 57–64
  assertions 56–7
  conflicting desiderata for acceptability of 151–2
  four considerations 58–64
  hierarchical view 159–63, 169
metaphysics of science 145, 333n7
Michelson, Albert 95
Midgley, Mary 344n22
Mill, John Stuart 259, 261, 294
Miller, David 70
mind-body problem 9
Minkowski space-time 61, 99, 101–2, 189
model of aesthetic induction (MAI) 171–2, 336n6
  assessment 184–6
  comparison with AOE 173–83
  conservative criteria 175
  and correcting scientific practice 183
  explanatory power 180–3
  and importance of aesthetic criteria 176
  and invocation of a model 178–9
  and metaphysical allegiance 177–8
  and non-empirical criteria of theory choice 173, 174–6
  and rationalist image of science 173
  reasons given in defence of 183
  and relationship between science and philosophy of science 183
  and simplicity 178
  and symmetry 177
  and unity 177
  as version of standard empiricism 174
  and visualizability/abstractness 179–80
Morley, Edward 95
mug's game (van Fraassen's) 188, 189, 194, 198, 202, 214
Musgrave, Alan 27, 31

Naess, Arne 11
Nagel, Ernest 11, 259
Nagel, Thomas 251, 344n22
Natural Environment Research Council (NERC) 289
natural science 4, 9, 74, 77–8, 81, 84, 128, 352n20
Nature 289
Neurath, Otto 11
New Enlightenment 355n34
  from critical to aim-oriented rationalism 312–13
  from falsificationism to AOE 306–12
  from knowledge to wisdom 313–15
  objections 315–20, 355n34
  putting into practice 321–5
Newton, Isaac 45, 78, 81, 92, 130, 137, 142, 229, 230, 237, 253, 285
  Principia Mathematica 68
Newtonian mechanics (NM) 94, 96, 99, 103, 107
Newtonian theory (NT) 51–3, 61, 71, 92, 100, 135–6, 145–6, 154, 157, 189, 190, 209, 231, 334n8, 337n10 & n13, 338n17, 353n21

non-empirical requirements 125–6
  further issues 127, 138–42
  objections to solution 134–8
  proposed solution to terminological problem 128–34
  seven aspects 126–7
  terminological problem 126, 127–8
normative naturalism 339n21

open society 8, 10, 17–25, 241–4, 260, 270–1, 291–2, 297, 349n44 & n45
oracular conceptions of reason 272–4, 350n47
orthodox quantum theory (OQT) 106, 107–15, 159, 179–80, 332n17, 340n2
  as best and worst of theories 215–19
  further questions concerning 227–8
  probabilism as cure for 219–27
  see also quantum theory
Orwell, George 251

Pais, Abraham 113
paradigms 76–83, 145, 262
parity conservation 86, 330n38 & n39
partial knowability 120, 164, 332n19, n20
particle theory 115
Pauli, Wolfgang 109
Penrose, Roger 223, 251
philosophes 2, 284, 285, 292, 293, 294–5
philosophy 28–9, 214, 236–8, 284
  academic development 252–3, 254, 288, 344n22 & n24
  analytic 188
  Continental 29, 237, 344n22
  reform of 237–8
  retreat into specialized triviality 238
  as specialized discipline 253–4, 344n22
  see also science of philosophy
philosophy of knowledge 280–4, 315, 349n45, 350n50, 351n51
  see also knowledge-inquiry
philosophy of science 4, 38, 49, 70, 126, 145, 287–8, 327n2, 328n8, 330n1, 333n4, 337n6, 345n33, 347n40, 351n55
  and AOE 65, 89, 311, 354n27
  and Einstein 121–3
  and Popper 1–3, 31, 37, 83
  relationship with science 183
philosophy of wisdom 279–80, 281, 282, 284, 290, 332n18, 347n40, 350n45 & n50
  see also wisdom-inquiry
physicalism 56, 70, 81, 83, 85, 144, 147, 152, 169, 179, 186, 212–13, 308, 311, 329n15, 339n27
  aberrant versions 168
  fruitfulness of 201, 339n23
  hierarchical view 60–4, 155–8, 159–63
  incompatibility 201
  and theoretical unity 175
  truth of 209–10
physics 42–7, 128–9, 170, 205, 332n21, 345n33
  aims and methods 69–70
  classical 111–13
  criteria 146
  metaphysical presuppositions 153–4
  and PIR 152–3

principles 68–9
   ultraviolet catastrophe 78
   unified theories 42–4, 57, 67–8, 149–51, 328n13, 334n14
PIR *see* principle of intellectual rigour
Planck, Max 78, 80, 92, 104, 105, 107, 110–11
Plato 18–19, 25, 32, 254, 270, 349n45
Podolsky, Boris 115
Poincaré, Henri 251, 259
Popper, Jenny Schiff 9
Popper, Josephine Henninger 'Hennie' 11
Popper, Karl ix–xi, 1–2, 4, 6, 42, 77, 82–3, 119, 124, 128, 144, 145, 156–7, 188, 192, 203, 204, 213, 251, 272, 282, 330n36, 344n22, 345n33, 352n20
   at the LSE 26–7
   basic argument in early work 29–30
   conjectures and refutations 27–9
   correcting/improving his Enlightenment 297–306
   criticism 14, 15–17, 41, 47–9, 328n10
   demarcation criteria 3, 9, 13, 15, 16–17, 34, 38, 54–5, 65, 66, 144, 204, 307
   early work 12–13
   and Einstein 90
   and the Enlightenment Programme 295–7, 320
   and falsificationism 1–2, 8, 13–15, 30, 47–8, 51–75, 204, 208, 296
   final years 37–8
   later work 30–5
   life 9–12
   most significant contribution 291–2
   philosophy of science 3, 37, 65
   principle of demarcation 54
   principle of simplicity 50, 54–6, 307, 353n22
   publications 8–9
   quantum theory 35–7
   radical revision of his Enlightenment Programme 306–15
   and rationality 270, 349n45
   reputation 37–8
   and 'silly' rival theories 53, 328n11
   three-world view 31–3, 348n44
   and verisimilitude 70
   Conjectures and Refutations 9, 16, 50, 261, 291, 292
   Logic of Scientific Discovery 8, 13–17, 50, 261, 291, 292, 306
   Objective Knowledge 31–3, 187, 283
   The Open Society and Its Enemies 8, 10, 17–25, 241–4, 260, 270, 291, 292, 297, 349n44 & n45
   The Postscript to The Logic of Scientific Discovery 34
   The Poverty of Historicism 8–9, 25–6, 291, 292
Popper, Simon Carl Siegmund 9, 10
PQT *see* propensitonn quantum theory
presuppositionism 147–9, 150, 152, 159, 161, 192–4, 204, 346n38
   solving problem of induction 187–214
Price, Derek de Solla 343n20
principle of demarcation 54

principle of equivalence 101, 120
principle of induction 88–9
principle of intellectual rigour (PIR) 147–9, 152–3, 166
principle of simplicity 50, 54–6, 67–8, 79, 87
principles 68–9, 120
pro-naturalist doctrines 25–6
probabilism 6, 219–27, 340n4 & n5 & n6
problem-solving 187, 263, 264, 336n2, 346n38
   basic rules 234–5, 237
   and four fundamental problems 234–6, 238, 243, 251, 257, 258, 299–300, 341n2, 348n43
   living vs knowledge 316–20
   rational 234–5, 255–6, 258, 262, 299–306, 341n4, 345n32, 352n14 & n15
   and specialism 257
   two-way flow 277
   unconscious capacity for 341n11
propensiton quantum theory (PQT) 226–7
propensitons 37, 221–2, 226, 231
pseudoscience 12
Putnam, Hilary 33

quantum chromodynamics 62, 72, 99, 189
quantum electrodynamics 62, 72, 178, 189
quantum electroweak theory 62, 72, 189
quantum field theory 45, 157, 178, 218
quantum theory 45, 62, 72, 78, 87, 105, 106, 155, 189, 250, 346n37
   confusions as part of historical pattern 229–32
   and Einstein 6, 106, 107–15
   and Popper 35–7
   relativistic 45
   see also orthodox quantum theory (OQT)
quantum wave/particle problem 110, 179, 220
Quine, W.V.O. 11, 33, 344n22

Ramsey, Frank 11
rational
   inquiry 241–4
   problem solving 234–5, 255–6, 258, 262, 299–306, 341n4, 345n32, 352n14 & n15
   society 29–30, 244, 291–2, 295–6
rational scientific discovery 5, 48, 65, 72–3, 81, 92–4, 100–7, 175, 186
rationalism 30, 255–6, 259, 270, 271, 291–2, 294, 296, 348n44, 34945
   oracular conceptions of 272–4, 350n47
   see also aim-oriented rationalism; critical rationalism
realism 28, 31, 106–7, 109, 113
   deterministic 34–5
   metaphysical 90
   scientific 90, 106, 109, 117
Reichenbach, Hans 11, 199, 200
relativism 245, 350n47
relativity
   general 12, 45, 61–2, 78, 100–2, 158–9, 180, 189, 250, 263
   special 45, 61, 94–100, 154, 189, 231, 250, 263

research programmes 83–9, 156, 289, 330n34 & n35
  Lakatos and AOE conceptions compared 83–8, 156–7
Riemann, Bernhard 52, 102
Rimini, Alberto 223
Romantic movement 294
Rosen, Nathan 115
Roszak, Theodore 251
Royal Society 344n25
Russell, Bertrand 11, 188, 251, 259–60, 345n29, 345n33

Sagan, Carl 251
Saint-Simon, Comte de 293
Salam, Abdus 81
Salaman, Esther 116
Schiff, Walter 13
Schilpp, Paul A. 33
Schlick, Moritz 11
Schopenhauer, Arthur 20, 344n22
Schrödinger, Erwin 72, 81, 87, 108, 109, 110, 136, 179, 215, 216, 221, 251
Schrödinger's cat 109, 332n15
Schumacher, Ernst Friedrich 251, 346n37
Schurz, Gerhard 128
science 12, 148, 156–7, 169, 211, 258, 283, 284
  answer to 'What kind of world is this?' 261–2, 345n32, 346n36, 346n37
  assumptions of 192–6, 197, 199, 337n13
  changing conception of 337n6
  critical fundamentalist conception of 258–9, 345n28
  engagement with the public 344n25
  externalist historians of 264–5, 346n40
  function of criticism 64–5
  and global problems 293, 351n4
  and Kuhn 47, 76–83
  and Lakatos 83–9
  meta-discipline approach 2–3
  and metaphysical conjectures 309–10
  metaphysics of 50, 65, 328n8
  and Popper 47–8
  presuppositions 74, 77
  puzzle solving 31, 78, 80, 257, 260, 262
  relationship with philosophy of science 183
  selection of theories 261–2
  severe testing/crucial experiments 291, 296, 352n10
  socially responsible 286–7
Science in Society 344n25
scientific metaphysics 143–4, 333n4 & n5
  alternative versions of AOE 163–5
  circularity problem solved 165–8
  concluding remarks 168–9
  conflicting desiderata for acceptability of theses 151–2
  empirically fruitful metaphysics 152–8
  hierarchical view 159–63
  intellectual rigour requires that presuppositions are explicit 145–9
  metaphysical minimalism 158–9
  unity of physical theory 149–51
scientific method 2, 12, 48, 49, 73–5, 259–60, 328n8

  see also aim-oriented empiricism; rational scientific discovery
scientific theory 5–6, 13–15
Selz, Otto 10
Shankland, Robert 109
simplicity 30, 56, 67, 68, 87, 126–7, 128, 129, 138–40, 145, 146, 178, 193, 259, 263, 307, 353n22
Singer, Isaac Bashevis 271, 272
  The Estate 271
  The Manor 271
Smart, J.J.C. 344n22
social engineering 24, 26
social inquiry 2, 4, 295, 297–8, 301, 352n17
social sciences 8–9, 17, 21, 25, 73, 169, 247, 250, 280, 289, 293, 294–5, 305, 346n37, 347n40, 352n20
Socrates 272
space-time 61, 61–2, 99, 101–2, 133, 150, 189, 194
specialism 6, 233, 246–9, 340n1, 341n6
  building a house analogy 277–8
  condemnation of 255, 260, 265, 344n22
  dominance and untenability 250–65
  experts 246–8
  and experts 274–5
  failure to criticize fundamentalist decisions 272
  failure to see fundamental problems 266–72
  harmfulness of 257–8
  and institutionalized viewpoints 275
  intellectual consequences 252, 344n22
  and intellectual standards 249–50
  and Kuhn 262–3
  non-rational influences 264
  and oracular conceptions of reason 272–4
  organization/institutional structure 251–2, 343n21
  and personal responsibility 266–72
  and problems of universalism 248–9
  reasons for prevailing 265–76
  support for 275–6
  see also specio-universalism
specio-universalism 255, 288–90, 340n1, 344n26
  integrating specialized and fundamental problem solving 236
  and philosophy 236–8
  and thinking in personal and social life 240–1
  see also specialism; universalism
Spinoza, Baruch 253, 254, 259, 272, 282
Stalin, Josef 17, 20
standard model 45
string theory (or M-theory) 62, 189
superposition 109–15, 159, 331n15
symmetry 137–8, 139, 142, 175, 177
Szasz, Thomas 251

Tarski, Alfred 11
theories
  aberrant 58, 63, 79, 130, 131, 137, 146, 147, 192
  ad hoc 14, 16, 52, 53, 54, 55, 56, 63, 95, 113, 146, 185

374 INDEX

assumptions 140–2
choosing 172, 173, 174–6
content 134–5, 138
degrees of unity 129–34, 332n4
empirical considerations 5, 46, 91, 104, 113, 145–6
exclusions 1, 14, 16, 30
explanatory 61
falsifiable 15–17, 45, 48, 50–1, 53, 62
new 72–3, 75, 77, 80–1, 87–8, 92–3, 100, 165–6
non-empirical requirements 125–42
and Popper 9, 12–17, 27–8, 29–30, 31–2, 34, 35, 36, 49, 51–6, 76, 90
probabilistic 35
properties of 172–3
pseudo and genuine 12
refutation 83
requirements 68–9
rival 145–6, 353n21
simplicity 15, 119, 129, 138–40, 142
symmetry 137–8, 139, 142
truth of 197, 199, 201, 206
unified 5, 42–4, 57, 59, 62, 67–8, 69–70, 106, 113–14, 119, 128, 135–6, 143, 146–7, 189–92, 204, 205, 328n13, 329n25, 334n10 & n11 & 14, 337n12
verifiable 8, 9, 12, 13, 14, 15, 29, 49, 50
theory of everything 34–5, 44–5, 56, 60, 62, 69–70, 71–2, 81, 127, 150, 330n32, 334n15, 338n20
Thirring, Hans 10
three-world view 31–3, 348n44
Tichy, Pavel 28, 70
Toulmin, Stephen 31
trends 26
tribal life 241–4, 268, 269, 341n12
Tyndall Centre for Climate Change Research 289

UCL *see* University College London
unity of method 73–5
unity of nature 4, 5, 30, 44, 81, 194
 presupposition of 146–9, 152–9, 192–6, 261–3
unity of society 241–3, 246
unity of theory 4, 6, 56, 57, 61–2, 128–40, 154, 204–5, 335n19
 and AOE 30, 44, 46, 67–70, 81, 144, 159–65, 196–202
 attempted solutions to problem of 128, 306–8
 degrees of 130–4
 demand for 175–6, 177, 180
 and Einstein 104, 116, 121, 122–3, 128, 149
 facets of 130–4, 150–1
 and Feynman 127
 and functions 136–7
 lawful 57, 58, 140
 metaphysical assumption 154
 nature of 126, 128–40, 149–51
 and physicalism 150–1
 problem of 87, 126–7, 129, 175
 and quantum theory 114
 and refutation of falsificationism 15–6, 30, 306–8

requirements 134–5, 138
and scientific realism 113
solution to problem of 128–40, 149–51, 335n19
and standard empiricism (SE) 145
terminological 126–8, 129, 130, 139
views of MAI and AOE compared 170–86
see also aim-oriented empiricism; physicalism; presuppositionism; symmetry
universalism 233, 340n1, 348n44
 adoption of 245–6
 advocacy of 234–46
 building a house analogy 277
 defence of 254–5
 development of intellectual inquiry 254
 failure to put into practice 256, 266–72, 344n27
 four fundamental questions or problems 234–5, 238–9, 239–41, 243, 341n2 & n10 & n11
 including in academic practice 288–90
 influence 251
 and inquiry as intellectually rigorous 249, 343n17
 and intellectual inquiry 239, 276–84
 and intellectual standards 249–50
 non-rational influences 264
 and personal responsibility 266–72
 questions cherished beliefs and values 272
 and reasons why specialism prevails 265–76
 and science 258
 and that which is of value 244–5
 thinking in personal and social life 240, 241–5
 see also specio-universalism
universe 52–3, 55, 135–6, 185–6, 328n13, 335n31
 comprehensibility of 163–4, 181
 hierarchy of assumptions 42–7
 as unified 194–5
University College London (UCL) 289
University of Oxford 289

value-presuppositions 347n40
Van Fraassen, Bas 90, 188, 198, 330n36
verisimilitude, problem of 28, 70–2, 329n25
Vienna Circle 11, 12
visualizability 179
Voltaire, François-Marie Arouet 254
Von Mises, Richard 11

Waismann, Friedrich 11
Watkins, John 26, 31, 33, 128
wave mechanics 110, 215–16
Weber, Erik 128
Weber, Tullio 223
Weber, Wilhelm Eduard 52
Wegener, Alfred 80
Weinberg, Steven 81
Weirstrass, Karl 3
Weyl, Hermann 128
Whitehead, Alfred North 251
Wirtinger, Wilhelm 10

wisdom 2, 233–4, 255, 302, 313–15, 348n44, 349n45, 350n48, 350n50
  see also philosophy of wisdom
'The Wisdom Agenda' (2011) 289
Wisdom, J.O. 26
wisdom-inquiry 306, 323–4, 350n50, 352n19
  see also philosophy of wisdom

Wittgenstein, Ludwig 11, 12, 27, 327n1
Worrall, John 208, 210
Wrinch, D. 128
Wu, Chien-Shiung 86

Zahar, Elie 95
Ziman, John 283